User's Manual for NFPA 921

IAAI Development Committee for the *User's Manual for NFPA 921*

Committee for the Second Edition

Rodney Pevytoe, Division of Criminal Investigations, WI, *Co-Chair*
David Crosbie, Burnsville Fire Department, MN, *Co-Chair*
John Comery, Bureau of Alcohol, Tobacco, Firearms and Explosives, OR
Dan Choudek, Onsite Engineering, MN
Robert Corry, sceneinvestigator.com
Guy E. "Sandy" Burnette Jr., PA, Tallahassee, FL
Hal C. Lyson, Robins, Kaplan, Miller & Ciresi L.L.P., MN

Committee for the First Edition

Carolyn M. Blocher, Editor, U.S. Tennessee Valley Authority Police, TN
Randy Bills, SEA, OH
Guy E. Burnette, Jr., Butler, Burnette, Pappas L.L.P., FL
Ed Comeau, Editor, Writer-Tech, MA
John Comery, Bureau of Alcohol, Tobacco, Firearms and Explosives, OR
Robert Corry, American Re, CT; Liaison, "interFIRE"
David Crosbie, Burnsville Fire Department, MN
Bill Grom, Bureau of Alcohol, Tobacco, Firearms and Explosives, PA
David Icove, Editor, U.S. Tennessee Valley Authority Police, TN
Pat Kennedy, Kennedy & Associates, FL
David Kircher, Essex County Prosecutor's Office, NJ
Jeff Long, Salt Lake City Fire Department, UT
Hal C. Lyson, Project Director (IAAI), Robins, Kaplan, Miller & Ciresi L.L.P., MN
John McDermott, Morris County Prosecutor's Office, NJ
Gregory Meisser, Bureau of Alcohol, Tobacco, Firearms and Explosives, MI
Russell Melton, Foley & Mansfield, MN
Gerard Naylis, FM Global, NJ; Liaison, IAAI and NFPA
Ted Nixon, Ward & Whitemore, FL
Jamie Novak, St. Paul Fire Department, MN
Don Perkins, Fire Cause Analysis, CA
Rodney Pevytoe, Division of Criminal Investigations, WI
Jack Sanderson, Fire Findings, MI
Dennis Smith, Kodiak, Inc., IN
Joe Toscano, American Re, CT; Liaison, "interFIRE"
Joe Zickerman, Fire Cause Analysis

National Fire Protection Association

International Association of Arson Investigators

User's Manual for NFPA 921

Guide for Fire and Explosion Investigations

Second Edition

JONES AND BARTLETT PUBLISHERS

Sudbury, Massachusetts

BOSTON TORONTO LONDON SINGAPORE

Jones and Bartlett Publishers
40 Tall Pine Drive
Sudbury, MA 01776
978-443-5000
www.jbpub.com

National Fire Protection Association
1 Batterymarch Park
Quincy, MA 02169
www.NFPA.org

Jones and Bartlett Publishers Canada
6339 Ormindale Way
Mississauga, Ontario L5V 1J2
Canada

Jones and Bartlett Publishers International
Barb House, Barb Mews
London W6 7PA
United Kingdom

Jones and Bartlett's books and products are available through most bookstores and online booksellers. To contact Jones and Bartlett Publishers directly, call 800-832-0034, fax 978-443-8000, or visit our website www.jbpub.com.

Substantial discounts on bulk quantities of Jones and Bartlett's publications are available to corporations, professional associations, and other qualified organizations. For details and specific discount information, contact the special sales department at Jones and Bartlett via the above contact information or send an email to specialsales@jbpub.com.

ISBN 13: 978-0-7637-4402-1
ISBN-10: 0-7637-4402-6

6048

Production Credits
Chief Executive Officer: Clayton E. Jones
Chief Operating Officer: Donald W. Jones, Jr.
President, Higher Education and Professional Publishing: Robert W. Holland, Jr.
V.P., Sales and Marketing: William J. Kane
V.P., Production and Design: Anne Spencer
V.P., Manufacturing and Inventory Control: Therese Connell
Publisher, Public Safety Group: Kimberly Brophy
Acquisition Editor: William Larkin
Editor: Jennifer Reed
Manufacturing and Inventory Coordinator: Amy Bacus
Reprint Editor: Jennifer Feltri
Director of Marketing: Alisha Weisman
Interior Design: Anne Spencer
Cover Design: Kristin E. Ohlin
Text Printing and Binding: Courier Company
Cover Printing: Courier Company
Photo Research Manager/Photographer: Kimberly Potvin

Printed in the United States of America
10 09 08 07 06   10 9 8 7 6 5 4 3 2 1

Contents

CHAPTER 14 Physical Evidence 183

CHAPTER 15 Origin Determination 203

CHAPTER 23 Motor Vehicle Fires 313

CHAPTER 24 Wildfire Investigations 335

Preface

Those of you who are reading this preface are to be commended. According to some studies, you are among only 15 percent of readers who take the time to obtain all of the information in a book. I congratulate you on being aware that the field of fire and explosion investigation requires continuous study.

Our job as fire and explosion investigators is no different from any discipline that grows through constant research. NFPA 921, *Guide for Fire and Explosion Investigations*, provides a resource to investigators to assist them in conducting examinations using a systematic procedure. Although it is not a standard or code and cannot be codified into law, NFPA 921 provides valuable information.

Through its leadership, the National Fire Protection Association (NFPA) has given us this valuable tool. Through the talents of its professional instructors, the International Association of Arson Investigators (IAAI) has cooperatively written this second edition of the *User's Manual* to further enhance your familiarity with the 2004 edition of NFPA 921. This *User's Manual* provides plain language explanations, suggestions for NFPA 921/interFIRE VR training, important questions at the end of each chapter, and hands-on activities throughout the text.

We hope that this manual will contribute to your investigative career. We encourage you to continue your review of the literature concerning all aspects of our profession. Research continues in the field of fire and explosion investigation, so we must further our education to ensure that we make proper determinations.

On behalf of the officers, directors, staff, and membership of the International Association of Arson Investigators, I applaud your pursuit of knowledge.

Michael A. Schlatman, IAAI-CFI
President, 2004–2005
International Association of Arson Investigators

Administration

▸▸▸ OBJECTIVES

Upon completion of Chapter 1, the user will be able to

- Understand the purpose, nature, and philosophy of NFPA 921
- Appreciate the importance of NFPA's process for developing documents
- Begin to acquire the vocabulary of fire and explosion investigations

NFPA 921, *Guide for Fire and Explosion Investigations,* 2004 edition, developed by the National Fire Protection Association (NFPA), can be used by anyone who is charged with the responsibility of investigating and analyzing fire and explosion incidents. People involved in investigation/analysis and in rendering opinions as to the origin, cause, responsibility, or prevention of such incidents include on-scene investigators, fire analysts, and other experts with special areas of interest, as well as technicians who analyze the evidence found in the course of the investigation. (The term *fire investigation* is used frequently in NFPA 921 when the context indicates that the relevant text refers to the investigation of either fires or explosions.)

PURPOSE OF NFPA 921

NFPA 921 has been developed by the NFPA process and approved by the American National Standards Institute (ANSI). One of the very important distinctions that must be made is that of a guide versus a standard. The definitions of *guide* and *standard* are taken from NFPA 921 and are used throughout all NFPA documents. It is important to point out that only documents that are adopted into law are officially termed *codes* and *standards,* which differ dramatically from *recommended practices* and *guides,* such as NFPA 921. The NFPA standards process is changing, however, so readers should seek more information at the NFPA website (www.nfpa.org).

guide

A document that is advisory or informative in nature and that contains only nonmandatory provisions. A guide may contain mandatory statements, such as when a guide can be used, but the document as a whole is not suitable for adoption into law.

standard

A document, the main text of which contains only mandatory provisions using the word *shall* to indicate requirements and that is in a form generally suitable for mandatory reference by another standard or code or for adoption into law. Nonmandatory provisions shall be located in an appendix or annex, footnote, or fine print note and are not to be considered a part of the requirements of a standard.

Although NFPA classifies 921 as a guide and not a standard, many courts and professional fire investigation organizations recognize NFPA 921 as the "standard of care" in the profession. The purpose of NFPA 921 is to establish guidelines and recommendations for the safe and systematic investigation or analysis of fire and explosion incidents. It also serves as a model for the advancement and practice of fire and explosion investigation, fire science, technology, and methodology.

The reason for conducting any fire investigation, no matter who is conducting it, is to arrive at the truth in an attempt to prevent an incident's recurrence. Determining what caused the fire, who may be responsible for it, and what factors may have contributed to the fire and its subsequent impact are the objectives.

NFPA 921 is designed to produce a systematic, working framework to ensure an effective fire/explosion investigation and origin/cause analysis. Deviations from these procedures are not necessarily wrong or inferior but do need justification.

Because every fire and explosion incident is in some way different and unique, NFPA 921 is not designed to encompass all the necessary components of a complete investigation or analysis of any individual case. It is up to investigators (depending on their responsibility, as well as the purpose and scope of their investigation) to apply the appropriate recommended procedures in the guide to a particular incident.

Although NFPA 921 was never originally intended as a document to define legal issues involving civil or criminal litigation of fire or explosion incidents, it has become such in the minds and practice of many fire investigation professionals, including attorneys and trial courts.

Relevance of NFPA 921 to Fire Investigators

One of the common issues raised is that NFPA 921 is not an NFPA standard to which fire investigators must adhere when conducting fire investigations. As defined by NFPA, NFPA 921 is a guide and, as such, does not include mandatory language, such as *shall* or *must*. The distinction between standard and guide is important because different jurisdictions have varying interpretations on the importance of NFPA 921. Investigators should be aware of rulings in their local jurisdiction.

However, there is another concern that the fire investigator must consider: the concern that NFPA 921 may be referred to as a standard of care comparable to that in the medical field. It must be stressed that NFPA 921 is not an NFPA standard by NFPA definitions. It might be considered by some as a de facto standard, in that it is a widely used and recognized document that provides guidance to the fire investigator. There are other standards-making bodies that define the word *standard* differently.

Investigator Usage

Investigators do not need to use NFPA 921 all the time at every investigation. Merely because material is covered in this guide does not necessarily mean that its procedures must be carried out in every case. The experienced investigator must use his or her judgment to determine what actions and steps are appropriate at each investigation to ensure that the proper steps are taken to arrive at the truth. If an investigator chooses not to use NFPA 921, however, he or she must be able to explain the method used and the reasons for using it.

Coverage

Since the first edition of NFPA 921 was introduced in 1992, there have been significant advances in fire science as applied to fire investigations. Although every effort is made for NFPA 921 to be as complete as possible, there are certainly areas that are not covered in each edition. However, through the NFPA process, the guide is updated regularly to include new material, as described in the section entitled "Revision Cycles" later in this chapter.

DEVELOPMENT OF NFPA DOCUMENTS

NFPA is the publisher of almost 300 codes, standards, recommended practices, and guides that are related to fire and life safety. All NFPA documents are voluntary documents, which means that they do not have the power of law unless an authority having jurisdiction (AHJ) adopts them. A number of NFPA's codes and standards have been adopted into law in various jurisdictions around the world, including the following:

NFPA 1, *Uniform Fire Code*™
NFPA 54, *National Fuel Gas Code*
NFPA 70, *National Electrical Code*®
NFPA *101*®, *Life Safety Code*®

One of the distinctive features of all NFPA documents is that they are developed through a consensus process, which brings together technical committee volunteers representing varied viewpoints and interests to achieve consensus on the document content. There are more than 200 of these technical committees, and they are governed by a 13-member Standards Council, which is also composed of volunteers.

It is important to point out that it is the technical committees—not NFPA staff—who are responsible for the document content. Contrary to a common misconception, the staff liaisons do not write the material that is contained in the document. Their role is to coordinate the development process, and NFPA's function is to serve as a mechanism for publishing the material that is written or approved by the technical committee volunteers.

Technical Committees

Approximately 6000 volunteers serve on NFPA's technical committees, with selection of these volunteers based on their background or expertise. The committees are balanced to ensure that no single interest has an overriding representation and may, therefore, unfairly influence the outcome of the development process.

NFPA uses the following membership categories to fill and balance a committee:

- Manufacturer
- User
- Installer/maintainer
- Labor representative
- Enforcing authority
- Insurance representative
- Special expert
- Consumer
- Applied research/testing laboratory

Technical committees review and respond to all proposed changes to existing documents. Consensus occurs when a majority of a committee accepts a proposed change to the document it oversees. Acceptance by a committee—and its subsequent recommendation for change—requires at least a two-thirds majority vote by written ballot. Ultimately, committee recommendations are voted on at NFPA's annual meeting, at which the entire Association membership has the opportunity to approve or reject the committee recommendations for change to the documents in the revision cycle for the meeting.

Oversight of Technical Committees

The work of the committees is coordinated by NFPA staff engineers and other staff specialists. NFPA staff members appointed to serve as staff liaisons to the technical committees are responsible for ensuring that the committees follow the rules governing committee activities and for coordinating their meetings. The staff liaisons also record the meeting actions and coordinate the publication of committee reports as well as the final document.

All of the committee meetings and the NFPA annual meeting are open to the public, and anyone, whether an NFPA member or not, can submit proposals to change a document or make comments on proposals the committee accepts. The only step in the process that is limited to NFPA members is the voting that takes place at the annual meeting.

Revision Cycles

All NFPA documents are revised on a staggered basis in two annual revision cycles—a Fall revision cycle and an Annual revision cycle that centers around the annual meeting in June. Typically, individual documents are updated on regular cycles of 3 to 5 years, with anywhere from 20 to 45 documents reporting in a given revision cycle. NFPA 921 has been on a 3-year cycle. The latest revision was issued in 2004.

Most revision cycles have five distinct steps: a call for proposals, the report on proposals, a comment period, the report on comments, the membership vote, and the Standards Council meeting at which the document is issued. Documents in the Fall revision cycle for which an intent to make an amending motion has not been filed are issued on a "consent agenda" in January and are not voted on by the membership at the annual meeting.

Call for Proposals. The first step for any NFPA document that is entering its revision cycle occurs when a call for proposals is issued. Notice that a document is entering its revision cycle, along with the dates for accepting proposed changes, appears in NFPA publications, on the NFPA website (www.nfpa.org), and in various professional and governmental publications that serve parties interested in the subject matter. Anyone, whether a member of NFPA or not, can submit a proposal for the technical committee to consider during a 20-week window. The form for submitting proposals for change includes a section in which the proposer must provide substantiation for the change. The proposal forms can be found at the back of NFPA documents, including NFPA 921, or on the NFPA website.

Tools and techniques are constantly being updated and modified between editions of NFPA 921. Consequently, newer methods that are not covered in the current edition might be used at a fire scene. As new principles and technologies are evaluated by the NFPA process, they may or may not be included in the regular cycle revisions of the document. For this reason, it is important that fire investigation professionals be diligent in making proposals and comments to the document during its revision cycles.

Review and fill out the NFPA proposal form shown in Exhibit 1.1. ◀◀◀

Report on Proposals. After the proposal closing date, the committee meets to act on the proposals, to develop its own proposals, and to prepare its report. Through a series of meetings, the committee determines what material should be added, deleted, or modified and then hammers out the language to be used. If the committee revises or rejects a public proposal (either in whole or part), it must provide a reason for doing so. The committee then votes on its actions relative to each proposal by letter ballot. A two-thirds affirmative vote by the committee is required before an accepted proposal can move forward to the next step.

Call for Comments. All proposals, the submitter's substantiation, and the committee actions pertinent to it are then published in a document called the *Report on Proposals* (ROP). The ROP is published and is available, free of charge, to anyone who requests one—either in hard copy or on CD. In addition, the ROP can be obtained from the NFPA website.

The ROP is circulated for 10 weeks to allow the public to review all of the proposals for change and the related committee actions. The ROP establishes a window of approximately 60 days during which the public can submit comments on proposed changes to the document as published in the ROP.

Report on Comments. After the closing date for public comments, the committee reconvenes to discuss the comments and determine what course of action should be taken in response to each. The committee again votes by letter ballot. The comments and the balloted committee actions are published in a document called the *Report on Comments* (ROC). This report is also publicly available either in hard copy, on CD, or on the NFPA website for a 7-week period.

NFPA Technical Committee Document Proposal Form

Note: All proposals must be received by 5:00 p.m. EST/EDST on the published proposal closing date.

For further information on the standards-making process, please contact Codes and Standards Administration at 617-984-7249.
For technical assistance, please call NFPA at 617-770-3000.

FOR OFFICE USE ONLY
Log #:____________
Date Rec'd____________

Please indicate in which format you wish to receive your ROP/ROC: ☐ **CD ROM** ☐ **paper** ☐ **download**
(*Note*: In choosing the download option you intend to view the ROP/ROC from our Website. No copy will be sent to you.)

Date________ **Name**________________ **Telephone** ________________
Company ________________________________
Address________________ **City**________ **State** ____ **Zip**________
Please indicate organization represented (if any)____________________

1. **a) NFPA Document Title** ____________________________
   b) NFPA No. & Edition ________________ **c) Section/Paragraph** ________________
2. **Proposal Recommends** *(check one)*: ☐ **new text** ☐ **revised text** ☐ **deleted text**
3. **Proposal.** *(Include proposed new or revised wording, or identification of wording to be deleted.)* Note: Proposed text should be in legislative format, that is, use underscore to denote wording to be inserted (inserted wording) and strike-through to denote wording to be deleted (~~deleted wording~~).____________________
4. **Statement of Problem and Substantiation for Proposal.** Note: State the problem that will be resolved by your recommendation. Give the specific reason for your proposal including copies of tests, research papers, fire experience, etc. If more than 200 words, it may be abstracted for publication. ____________________
5. ☐ **This Proposal Is Original Material.** Note: Original material is considered to be the submitter's own idea based on or as a result of his/her own experience, thought, or research and, to the best of his/her knowledge, is not copied from another source.
   ☐ **This Proposal Is Not Original Material; Its Source** ***(if known)*** **Is as Follows:** ________________

I hereby grant NFPA all and full rights in copyright to this proposal, and I understand that I acquire no rights in any publication of NFPA in which this proposal in this or another similar or analogous form is used.

Signature ***(Required)***____________________________

PLEASE USE SEPARATE FORM FOR EACH PROPOSAL • NFPA Fax: (617) 770-3500

Mail to: Secretary, Standards Council • National Fire Protection Association
1 Batterymarch Park • PO Box 9101 • Quincy, MA 02269-9101

EXHIBIT 1.1 Form for Proposals on NFPA Technical Committee Documents

The ROC establishes a closing date for receipt of a Notice of Intent to Make a Motion (NITMAM). Documents in the fall revision cycle that do not receive a notice are issued in January based on a consent agenda. Documents that do receive such a notice are held over to the Annual revision cycle for membership vote at the June meeting.

Membership Vote. Following these steps, the ROP and ROC are then submitted for consideration and vote at NFPA's annual meeting in June. During this meeting, any registered attendee can comment on the document, whether or not that individual is an NFPA member. However, only NFPA members can vote on whether the ROP or ROC actions should be accepted or sent back to the committee for further action.

Standards Council Meeting

Appeals to membership action taken at the Association meeting may be submitted to the Standards Council within 3 weeks following the meeting. The Standards Council meets approximately 5 weeks after the Association meeting to consider the appeals. The council then directs appropriate actions to be taken according to the appeals it accepts and officially issues the documents. The documents become effective 21 days from the date of issuance. Anyone who still has a grievance about a change to the document has an additional 3 weeks to appeal to the NFPA Board of Directors.

Corrections to Documents

Between revisions of the document, errata and Technical Interim Amendments may be issued.

Errata. If a publishing error is found in some part of the document, NFPA can issue an erratum correcting the error. In the next revision of the document, the error is corrected.

Tentative Interim Amendments (TIAs). If there are concept changes that must occur between revisions, a proposal known as a *Tentative Interim Amendment* (TIA) is placed before the committee to address the issue. The committee must conduct a letter ballot on proposed TIAs, and if there is two-thirds agreement, the Standards Council issues the TIA. The TIA then becomes a proposal during the next cycle of the document. Several errata and TIAs have been issued for NFPA 921 since its inception.

Research various TIAs or errata that have been issued for NFPA 921. ◀◀◀

Formal Interpretation of Codes and Standards (FI)

A *Formal Interpretation* (FI) is a mechanism for providing an explanation of the meaning or intent of any specific provision that is included in an issued NFPA code or standard. FIs are processed through the technical committee that is responsible for the document and must be clearly worded to solicit a "Yes" or "No" answer from the committee.

Confirmation from the technical committee for an FI is achieved through a letter ballot. If a three-fourths majority is not achieved, the FI fails, and the item is placed on the committee's next meeting agenda. As an alternative to an FI, many individuals find it faster and more convenient to request an informal interpretation from NFPA staff. This option provides a technical opinion of the NFPA staff liaison assigned to the applicable committee. For more information, contact the staff liaison assigned to the particular NFPA project.

Formal Interpretations of NFPA guides such as NFPA 921, though not unheard of, are unusual, since NFPA guides do not contain mandatory provisions. As of the 2004 edition, no FIs have yet been requested.

NFPA 921 DEFINITIONS

Definitions of terms within a profession or discipline are necessary for a full understanding of the language by which various practitioners communicate. The definitions contained within NFPA 921 are critically important in that they provide a standard, consistent language for all fire investigators to use when conducting a fire investigation and preparing a case. Without consistent language, ambiguities, misunderstandings, and other related problems may arise. These problems can detract from the purpose of the investigation (to find the truth) by creating the appearance of unprofessionalism.

For example, confusion (and hence misuse of the terms) often arises between the concepts of *backdraft* and *flashover.* To avoid any potential communication problems, the investigator should be familiar not only with the definitions as listed in NFPA 921, but also with the reasoning behind those definitions.

For the purpose of this *User's Manual,* we will use the definitions provided in NFPA 921. You may refer to other documents for clarification.

interFIRE TRAINING

The interFIRE VR software is a virtual reality fire investigation training program available to anyone, including the fire service, law enforcement, and the insurance industry. The interFIRE VR program was developed by a team of investigators from NFPA; the Bureau of Alcohol, Tobacco, Firearms and Explosives; the U.S. Fire Administration; and American Re-Insurance. The CD-ROM program offers a comprehensive guide for fire investigation and includes the following features:

- Scenario (virtual reality fire scene investigation)
- An eight-module Tutorial (lessons in fire investigation)
- Resource Center (comprehensive resources for the fire investigator)

The interFIRE VR project has closely followed NFPA 921 throughout its development process. The interFIRE development team recognized that there are a number of different methods of accomplishing specific tasks, but that NFPA 921 is accepted as an established practice. By closely aligning interFIRE VR with NFPA 921, interFIRE would achieve the same level of worldwide acceptance as the NFPA guide.

The interFIRE VR program represents the cutting edge in fire investigations training. Through virtual reality, the user is able to conduct a complete fire investigation—including interviewing witnesses, collecting evidence, and sending the evidence to the laboratory for analysis. The user can undertake a complete scene documentation that includes taking photographs, inspecting all areas of the building to determine potential ignition sources, and determining the extent of the fire damage. Throughout the virtual investigation, the user can refer to the Tutorial and Resource Center sections of the CD-ROM for more information or clarification on specific points.

Once the user has collected all the evidence, interviewed the witnesses, and completed the scene documentation phase, he or she is then given a chance to determine the cause of the fire. The program evaluates the user's determination, and if there is not enough evidence to back it up, the user is then sent back to the fire scene to collect additional information.

A comprehensive website (www.interfire.org) supports interFIRE VR. It features a training calendar, bulletin boards, and a constantly updated Resource Center module that complements the Resource Center module on the CD-ROM. In addition, training programs that further support the concepts outlined in interFIRE are now being developed and are available through the website.

The interFIRE VR program is being used worldwide to help train new fire investigators, even as it provides a realistic learning environment for the seasoned investigator. It is an invaluable tool that joins the latest in computer technology with the depth of information provided by the experts who produced it. Copies of interFIRE VR are available for free from the U.S. Fire Administration and from the Bureau of Alcohol, Tobacco, Firearms and Explosives. Copies can also be ordered through the interFIRE VR website for a minimal charge.

This manual includes a number of tables that provide an interface between the appropriate NFPA 921 chapter and the interFIRE VR training program. The user can apply the interFIRE information or find more information on what NFPA 921 offers. Table 1.1 is the first such table and offers an introduction to the interFIRE program and website.

TABLE 1.1 NFPA 921/interFIRE VR Training

921 Section	Knowledge/Skill	interFIRE Tutorial Student Activity	interFIRE Scenario Student Activity	interFIRE Resource Section Student Activity	www.interFIRE.org Student Activity
Chapter Focus	*Knowledge or skill the investigator needs to research or demonstrate*	Start the program, log in, select *Tutorial.*	Start the program, log in, select *Scenario.* Note: If the student needs to practice the scenario more than one time, simply log in under a different or abbreviated name. This program supports multiple users logged in under different user names.	Start the program, log in, and select *Resource Section* to browse the topic files or to use the Search function to look up specific topics. The easiest way to use the Resource Section is directly from the Tutorial module you are reviewing. Select *Jump to Resource Fire* on the bottom right of the module screen to obtain information related to the module topic.	Start the program, log in, and select *Connect to www.interfire.org* on the bottom right of the menu, or launch your Web browser and type www.interfire.org in the address block and click *Enter.* This website contains the Resource Center and many other powerful features including • New published articles • Search engines for defective products and vehicles • Skills training • Bulletin boards and much more

▶▶▶ QUESTIONS FOR CHAPTER 1

1. Which statement is true about submitting a proposal to the NFPA 921 committee? (*Note:* There may be more than one correct answer.)

A. Only committee members can submit a proposal.

B. Only NFPA members can submit a proposal.

C. There is a proposal form in the back of NFPA 921.

D. Forms are available from the NFPA website at www.nfpa.org.

2. What has been the revision cycle for NFPA 921?

A. 5 years

B. 3 years

C. 2 years

D. 1 year

3. What type of document is NFPA 921?

A. Guide for fire and explosion investigations

B. Standard for fire and explosion investigations

C. Document that must be strictly adhered to in fire and explosion investigations

D. Reference document for laws referencing fire and explosion investigations

4. True or false: Every portion of NFPA 921 is essential to every fire or explosion incident.

A. True

B. False

5. Proper determination of fire origin and cause are essential for what reason(s)? (*Note:* There may be more than one correct answer.)

A. An accurate determination of cause helps to protect life and property.

B. Gathering accurate statistics aids in the development of codes and standards.

C. Accurate origin and cause determinations assist in training programs.

D. Fire chiefs are impressed by proper determination of fire origin and cause.

Basic Methodology

OBJECTIVES

Upon completion of Chapter 2, the user will be able to

- Identify the key elements in the scientific method of fire investigation
- Analyze the data collected
- Understand how to develop a hypothesis

Chapter 4 of NFPA 921 covers Basic Methodology. The investigation of a fire or explosion most often involves the identification, collection, and analysis of data that result from the destruction of materials. The ability to recognize these data, or facts, and then to analyze them properly and objectively is paramount in fire investigation.

It is essential that each investigation be carried out in a consistent manner to ensure that all aspects of any given fire scene are addressed. A systematic approach should be employed to ensure thorough physical evaluation of the scene, careful collection of evidence, and complete documentation of the scene and physical evidence, all of which are required for proper analysis and correct determination of results. The systematic approach recommended by NFPA 921 for analyzing the data is the scientific method.

SCIENTIFIC METHOD

The *scientific method* is an organized, logical method for solving problems. This approach involves recognizing and understanding a particular problem, collecting information (data or facts), analyzing the information, developing hypotheses (theories supported by data), and then testing the hypotheses (either experimentally or cognitively) to determine whether the results are reliable and valid. The scientific method is a means of analysis that has been used and refined in the sciences for many years by researchers. (See Figure 4.3 in NFPA 921, 2004 edition.)

scientific method

Principles and procedures for the systematic pursuit of knowledge involving the recognition and formulation of a problem, the collection of data through observation and experiment, and the formulation and testing of hypotheses. (Source: *Merriam-Webster's Collegiate Dictionary*)

Although they do not recognize it as such, most people use the scientific method every day. Consider the following problem-solving steps suggested by the absence of light in a room.

- *Recognize the need:*
  Light in the room is desired.
- *Define the problem:*
  The light is out.
- *Collect data:*
  Check power to house: Power is "on."
  Check power to circuit: Power is "on."
  Check switch: Switch is "on."
  Check bulb: Filament is broken.
- *Analyze data (inductive reasoning):*
  There is power to the building.
  There is power to the circuit.
  It is unknown whether the switch is working.
  The bulb needs to be replaced.
- *Develop a hypothesis:*
  The bulb is burned out.
- *Test the hypothesis (deductive reasoning):*
  Replace the bulb with a new bulb.
  Turn on the switch.
  The light is "on."
- *Final hypothesis:*
  The bulb was burned out.

The foregoing is a typical scenario in which the systematic process of the scientific method is used to solve a problem for which the answer is unknown. Key elements in the process involve collecting data, and developing and testing hypotheses.

Select an everyday problem or event (for example, a car running out of gas) and identify each of the steps in the scientific method.

FIRE INVESTIGATION AND THE SCIENTIFIC METHOD

Fire investigators have often followed the scientific method or systematic method but have not defined it by name. When questioned on the method used to conduct an investigation, the investigators often said that they document the fire scene from the outside to the inside or from the least damaged to the most damaged.

This is one example of only one part of the scientific method (collecting data), which is the third step in the scientific method. The following are steps used to apply the scientific method to fire investigations.

Recognizing the Need

The first step is to realize that there is a problem to be resolved—something that is often self-evident because the investigator is notified of an incident and asked to determine its origin and cause. The exact nature of the problem that each investigator needs to address depends on the role and responsibility that the investigator fulfills during the investigation.

Defining the Problem

When the investigator arrives on the scene or when the assignment is given, the investigator will have to define the problem to be resolved. Is the problem to determine the origin and cause of the fire? Or is it to determine what role the furnishings may have played in the fire spread? Or is it perhaps to determine the cause of death? It is important to define the problem so that the investigator can apply his or her resources in the most effective and efficient manner.

Collecting the Data

The investigator should then set about gathering the data at the fire scene that will be needed to find the answer to the problem that has been defined. Several chapters in NFPA 921 provide information on data collection. These data can include (but certainly are not limited to) the following:

- Recognition of physical evidence, such as fire patterns
- Collection of materials, such as debris samples, for laboratory evaluation
- Results of laboratory examinations
- Documentation of personal observations, such as witness statements
- Documentation of the fire scene through photographs, sketches, and notes
- Official reports, such as those of fire and police departments
- The documentation or results of prior scene investigations

The data collected using the scientific method that will be later analyzed are empirical data. Empirical data consist of information that can be verified, can be validated as true, or are based on observations or experience.

It is important that investigators properly document and collect the data that will ultimately verify their hypothesis. It is understood that memories fade over time, so proper documentation of interviews must occur. Once the fire scene has been demolished or repaired, the opportunity to collect additional data from the scene may be lost. Although it is impossible to physically collect everything from a fire scene, proper documentation will validate analysis of the data that are collected. Collecting data also includes literature review, pattern analysis, scene documentation, photography and diagramming, evidence recognition and preservation, and review and analysis of others' investigations. Documentation of the fire scene is a topic that is addressed later in this *User's Manual.*

empirical

(1) Originating in or based on observation or experience.

(2) Relying on experience or observation alone often without due regard for system and theory (empirical data).

(3) Capable of being verified or disproved by observation or experiment (empirical laws).

(Source: *Merriam-Webster's Collegiate Dictionary*)

Analyzing the Data

It is now time to employ inductive reasoning, the stage of an investigation when the whole body of evidence (empirical data) and other data will be reviewed. At this point, the evidence is looked at as objectively as possible to evaluate its meaning and to determine a possible hypothesis for the events that are being investigated. Data that are speculative or not related to the fire scene should not be considered.

inductive reasoning

Reasoning from specific observations and experiments to more general hypotheses and theories. (Source: *Funk & Wagnalls Multimedia Encyclopedia,* www.funkandwagnalls.com)

Developing a Hypothesis

Through the analysis of the data, the investigator will develop a hypothesis (or hypotheses) based on the empirical data to solve the problem. The hypothesis is an attempt to answer the *define the problem* section of the scientific method. This task may require the investigator to identify the origin, the cause, the fire spread, the responsibility, or some combination of these as directed by the investigator's assignment. During this stage, the investigator may develop not just a single hypothesis but may have two or more. If a single hypothesis cannot be developed using the scientific method, the investigator may be required to identify another potential hypothesis and list the hypothesis as "unknown" or "more probable."

Testing the Hypothesis

In the next stage, the investigator uses *deductive reasoning* to test the hypothesis that was developed. Through deductive reasoning, the ultimate conclusion is supported, unsupported, or refuted by the complete body of evidence and data.

deductive reasoning

Reasoning from theories to account for specific experimental results. (Source: *Funk & Wagnalls Multimedia Encyclopedia*)

The hypothesis must be supported by the facts. If it is not supported by the facts, the investigator may have to return to one of the earlier steps in the scientific method and repeat the process from that point, to ensure that the hypothesis can be supported when it is subjected to challenge.

Furthermore, it is important to understand that *testing the hypothesis* does not refer only to experimental testing, such as in a laboratory. Testing the hypothesis can be either cognitive or experimental. For example, during the testing and analysis of a hypothesis, the investigator will cognitively test the hypothesis on the basis of his or her knowledge and experience. Cognitive testing is the use of a person's thinking skills and judgment to evaluate the empirical data and challenge the conclusions of the final hypothesis. Other investigators may also provide a forum to test or challenge the investigator's hypothesis.

Many investigators never have their opinions questioned or challenged. However, the critical step is to determine whether a hypothesis is supportable—that is, whether it can withstand reasonable challenge. Testing the hypothesis not only involves the critical examination and expected challenge of others, but also requires critical evaluation by the investigator. This self-challenge requires the investigator to ask questions regarding the hypothesis. In addition to the self-challenge, the investigator should also seek to determine the level of certainty in his or her opinion regarding the hypothesis.

A key question to determine whether a hypothesis is supportable is: What other hypothesis could be supported by the same set of facts? If there are alternative hypotheses that are supportable by the facts, it is likely that the investigator has not gathered enough data. In this case, the investigator should obtain more information and reapply the steps in the scientific method. Occasionally, an investigator cannot develop a hypothesis that can be supported to a reasonable degree of certainty (to a probability or conclusively). Ultimately, the investigator's goal is to find only one hypothesis that is supportable to a probability (more likely than not). In situations where the hypothesis is not supportable to a probability, and is only *suspected* or *possible,* the results should be reflected as such—for example, by stating that the results are *inconclusive.* A hypothesis that is possible is one that is true, but where other hypotheses are also true and none are dominant (or more likely to be true than the others). A hypothesis that is suspected is true, but other hypotheses are also true.

Other types of questions the investigator should use in testing the hypothesis include the following:

- Is there another way to interpret the facts (data)?
- If yes, why is your interpretation more likely true?
- What are the weaknesses in the hypothesis (analysis of the data)?
- What arguments will someone else (an opposing expert) use to refute the hypothesis?
- Are there facts that contradict the hypothesis?
- Is there research that supports the hypothesis?
- Can the hypothesis be proven to someone else?
- Does the hypothesis make sense?

Refer to Chapter 18 in this *User's Manual,* "Failure Analysis and Analytical Tools," for additional methods in testing a hypothesis.

The investigator has to approach every incident with an open mind. There should be no preconceived determination as to what the cause of the fire was, for example. Any preconceived ideas will, consciously or unconsciously influence the investigator's efforts and should be avoided at all costs to maintain a proper level of objectivity during the investigation. The data from the fire or explosion scene should stand on their own. If the hypothesis cannot withstand a reasonable challenge or test, the investigator will need to return to earlier steps in the scientific method, such as collecting additional data, analyzing that data, or redeveloping the hypothesis. If the final hypothesis cannot withstand a reasonable challenge, the hypothesis should be listed as "undetermined."

interFIRE TRAINING

Table 2.1 provides an interface between NFPA 921 and the interFIRE VR program. It brings the user from the NFPA 921 section of interest to the corresponding area in interFIRE. The user can apply the interFIRE information or find more information on what NFPA 921 offers.

TABLE 2.1 NFPA 921/interFIRE VR Training

921 Section	Knowledge/Skill	interFIRE Tutorial Student Activity	interFIRE Scenario Student Activity	interFIRE Resource Section Student Activity	www.interFIRE.org Student Activity

Chapter 4—"Basic Methodology"	*Application of a systematic method for compiling and analylzing factual data in a fire or explosion investigation (4.1)*	Study the interFIRE protocol as a recommended example of a systematic method for fire incident investigation.
		• Study the model protocol in "Process of Fire Investigation."
		• Select *Before-the-Fire* and evaluate three modules:
		  • Guiding Principles
		  • Create a Fire Investigation Plan
		  • Define Official Responsibilities
		
		Activity:
		Brainstorm what local resources and agencies could participate in your fire investigation team and what their roles would be.
		• How does this protocol compare to the way you or your local fire investigation unit conducts fire scene investigations?
		• What are the pros and cons of organizing a fire incident investigation's tactical and evidence collection by a specific protocol?
		• What are the pros and cons of interFIRE's "Team Concept" protocol (an interagency, cross-trained, rapid response unit working under a tactical protocol)?
		• How might adoption of the investigation protocol described in interFIRE have affected the amount/quality of evidence/information collected in your own fire investigations over the past 12 months?
			At a scene, a fire investigator is confronted with many simultaneous imperatives. Should the investigators
			• Interview the fire fighters and police first?
			• Interview civilian witnesses and fire victims first?
			• Protect the fire scene first?
			• Perform an origin and cause examination right away?
			• Begin to document conditions?
			• Assess structural safety?
			
			Activity:
			• To explore the effects of different approaches, try the following:
			1. Start the scenario under a new user name and begin by interviewing the professionals at the scene. Now, write down everything you have learned about the fire from those professionals.
			2. List the questions you want to investigate next and the actions you want to take. Log out, restart the program, and enter a new user name. This time, begin the scenario by examining the scene first. Now, write down everything you have learned about the case from this examination *only* (do not include anything you can't tell from the physical examination). Then, list the questions you want to investigate next and the actions you want to take.
			3. Log out, restart the program, and enter another new user name. This time, begin the scenario by interviewing the civilians at the fire scene. Interview no more than three witnesses. Now, write down everything you have learned about the scene from *only* the witness interviews. Then, list the questions you want to investigate next and the actions you want to take.
			4. Compare your three lists. What were the strengths and weaknesses of each approach? When would you employ one approach over the others?
			
			Discuss these issues:
			• Do you have the right to enter private property without a search warrant? Consent?
			• Do you need help?
			• How do you conduct this investigation without losing potential witnesses and evidence?
				The Resource Section contains vital information about organization of the investigation process and coordination of its function.
				Find this information in
				• *Before the Fire Practices*
				• *Roll-Up Practices*
				• *Preliminary Scene Assessment*
				• *Fire Scene Examination*
				• *Interviewing Witnesses*
				
				Activity:
				Locate and review the following documents:
				• Large Fire Loss Investigation and Management
				• CFI Fire Scene Examination
				• Investigative Fire Contents
				• NFIRS Investigative Forms
				• The Team Concept: Duties and Responsibilities
					All of the reference and photographic information available in the CD-ROM Resource Section is available online at www.interfire.org.
					Additionally, the following "Featured Articles" containing information on *Fire Patterns* are available on the website:
					• *Glass Breakage in Fires*
					• *NIST Fire Dynamics at Cherry Road* (May 2000)
					
					Check the website frequently for additional information on this topic.

▶▶▶ QUESTIONS FOR CHAPTER 2

1. Outline the steps for the scientific method as it relates to fire investigation.
2. True or false: The investigator should consider only facts that can be proven clearly by observation or experiment when analyzing the data.
   A. True
   B. False
3. True or false: Testing the hypothesis is done by using the principle of inductive reasoning.
   A. True
   B. False
4. True or false: Valuable physical evidence recovered at a fire scene should be recognized, documented, and returned to the owner.
   A. True
   B. False
5. What are some benefits of using a systematic approach for fire investigation?

Basic Fire Science

OBJECTIVES

Upon completion of Chapter 3, the user will be able to

- Discuss the importance of fire science
- Discuss how fire dynamics influence fire behavior
- Distinguish between heat and temperature

Chapter 3 of this book explains Chapter 5 of NFPA 921 and is a summary of basic fire science that can be used as a quick reference, but it should not be relied on as the investigator's sole reference in fire science. The investigator should read numerous professional articles, research papers, and books to broaden his or her fire science knowledge. Annex B of NFPA 921 includes a bibliography with specific additional references related to fire chemistry, fire dynamics, and fire science.

Read Chapter 5 of NFPA 921. Read one of the additional references listed in Chapter 2 or Annex B of NFPA 921.

IMPORTANCE OF FIRE SCIENCE

The goal of NFPA 921 is "to provide guidance to investigators that is based on the accepted scientific principles or scientific research." The foundation of the fire investigator's opinions about fire origin and cause is *fire science.* NFPA 921, Chapter 6, "Fire Patterns," states that "one of the major objectives of a fire scene investigation is the recognition, identification, and analysis of fire patterns." To accomplish that objective, the investigator should understand the fire dynamics that produced the fire patterns.

fire science

The body of knowledge concerning the study of fire and related subjects (such as combustion, flame, products of combustion, heat release, heat transfer, fire and explosion chemistry, fire and explosion dynamics, thermodynamics, kinetics, fluid mechanics, fire safety) and their interaction with people, structures, and the environment. (Source: NFPA 921, 2004 edition, Paragraph 3.3.61)

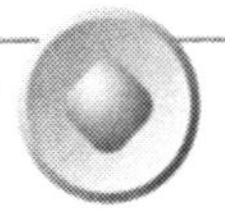

ACTIVITY

Read Chapter 5 of NFPA 921 to understand one way in which fire science knowledge may be used in an investigation.

In addition to understanding fire science, the fire investigator is called on to articulate the principles of fire science in written reports and for civil and criminal proceedings. Therefore, the investigator must understand fire science principles beyond the basic ability to recite them. For example, heat transfer in a fire is not explained merely by the simple principles of radiation, conduction, and convection. *Fire dynamics* involves complex interactions of the principles of chemistry, physics, and engineering.

fire dynamics

The detailed study of how chemistry, fire science, and the engineering disciplines of fluid mechanics and heat transfer interact to influence fire behavior. (Source: NFPA 921, 2004 edition, Paragraph 3.3.56)

PRINCIPLES OF FIRE SCIENCE

The first law of thermodynamics states that energy is neither created nor destroyed but is simply changed in form. Therefore, fire is essentially an energy conversion process.

Energy Transfer

During combustion, complex fuel molecules, typically cellulose-based or hydrocarbon-based, are converted into simpler molecules of carbon dioxide and water. This simple formula illustrates the combustion of methane gas:

$$CH_4 + O_2 = H_2O + CO_2$$

Although it takes an energy input (endothermic) to initiate any combustion process, the conversion releases (exothermic) a greater amount of energy. The energy is released as heat and light. NFPA 921 defines *fire* as "a rapid oxidation process with the evolution of light and heat in varying intensities."

In the 1850s, Michael Faraday presented a series of lectures that were later published as a book entitled *Chemical History of a Candle,* which is still available today. Faraday devised some simple experiments that help to explain fire science principles to young audiences. Investigators can conduct these same types of demonstrations and experiments to broaden their fire knowledge. In fact, many inspectors have already done these experiments during school, training, or life experiences with fire.

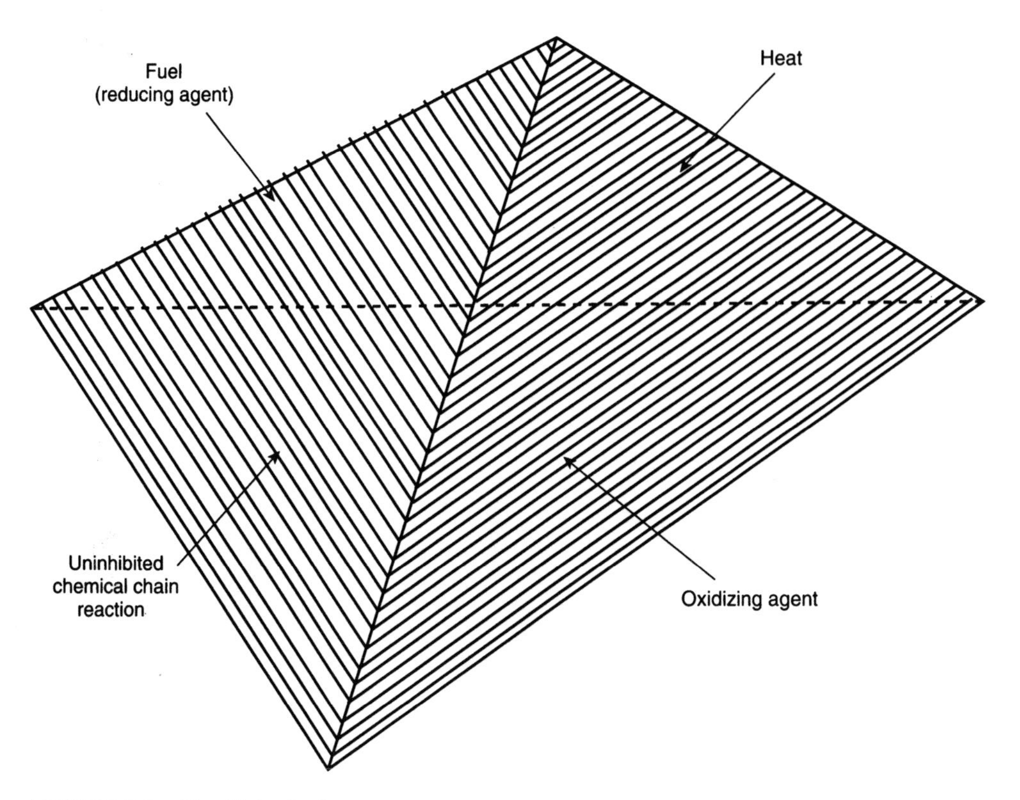

EXHIBIT 3.1 Fire Tetrahedron

ACTIVITY

Design a simple demonstration to show the elements needed for fire. For example, light a candle and explain the elements of the fire tetrahedron. (See Exhibit 3.1.) The *fire tetrahedron* is composed of fuel, heat, oxidizing agent, and an uninhibited chemical chain reaction. Heat is supplied by the introduction of a flame from a match. The match flame heats the wax contained on the wick, turning it to a vapor that ignites in the presence of atmospheric oxygen. Sufficient heat is produced by the candle flame to change the fuel from a solid to a liquid and then to a vapor. Place a jar over the lighted candle. Soon the flame is extinguished as the oxygen is consumed in the enclosure. Gently blow on the flame and disrupt the combustion by cooling the flame, disbursing the fuel vapor, and inhibiting the chain reaction.

Energy Release Rates

The energy released depends on the energy of the fuel and the degree of completeness of the reaction. A partial oxidation reaction releases a percentage of the energy. The heat of combustion, ΔH_c, of a fuel item is the theoretical total amount of energy within the substance or substances that make up the item. The heat of combustion is remarkably similar for many cellulosic and hydrocarbon-based fuels, with an average value of 43 kJ/g. [Joules (J) are a measure of energy. Refer to references for conversion to British thermal units (BTU).] Gasoline and candle wax have similar ΔH_c, but we know that each fuel does not burn in the same manner.

ACTIVITY

Design a simple demonstration to show the three methods of heat transfer. For example, place a metal pan containing water on an electric burner and bring the water to a boil. The metal burner element heats up and transfers heat, through conduction, to the bottom of the metal pan. Now think about the metal of the pan. Would there be a difference if it were copper, tin, or aluminum? Which metal would be the best conductor? How do you know this? The water in the pan heats up through convection. Place your hand near the burner and feel your hand heat up through radiation. ◀◀◀

ACTIVITY 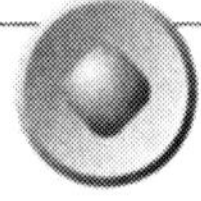

Review the references regarding heat transfer for further explanation of the basic heat transfer methods. Can you explain the processes occurring in the ignition of a match, the burning of a candle, the glowing combustion of charcoal briquettes, the burning of a cigarette, or the sustained burning of wood in a fireplace? These are relatively simple events, common to most people's fire experience. Do you understand them? If you cannot explain these simple fires, how can you investigate and reach a conclusion about a complex fire involving numerous fuels and many compartments? ◀◀◀

What accounts for this difference in the energy release rate? A fire dynamics equation can assist our understanding of the principles governing this process. Examine the following equation:

$$\dot{Q} = \Delta H_c \, \dot{m} A$$

$\dot{Q}$ is the symbol for heat (energy) release rate. Note that the literature refers to this as *heat release rate* (HRR) or *energy release rate* (ERR). The heat release rate is not static but has start, growth, peak, and decay stages. The peak HRR is the most common rate described. The total HRR of a fuel is less important to the fire development than is the peak HRR. In a compartment fire, there is a critical HRR that must be achieved for the fire to transition through flashover to a fully developed stage.

ACTIVITY 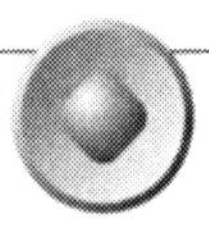

Find a reference (e.g., NFPA's *Fire Protection Handbook,* 2003 edition, or SFPE's *Handbook of Fire Protection Engineering*, 2002 edition) that lists the heat of combustion for different materials. ◀◀◀

ACTIVITY 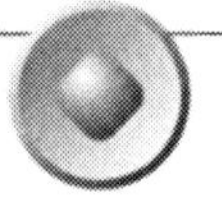

Look at the table of representative peak heat release rates in Chapter 5 of NFPA 921 (Table 5.4.2.1). Note the rate is given in kilowatts (kW). A watt is 1 joule per second. There is a great variance of the HRRs, from the wastebasket with a peak HRR of 4 to 18 kW to the polyurethane sofa with a peak HRR of 3120 kW. ◀◀◀

One of the primary differences in the HRR of objects is the mass loss rate ($\dot{m}$). Going back to the differences between gasoline and candle wax, one can understand that the mass loss rate of gasoline, a fuel in liquid state producing vapors at ambient temperatures, would be greater than that of wax, a solid phase fuel.

The time required for each object to reach its peak HRR also varies. Fire growth within a compartment will be greatly affected by the HRR of the first and subsequent fuel packages that burn. A fuel package that takes a shorter time to reach peak HRR is a greater hazard than one that takes longer. The total HRR of a fuel package is the energy released over the entire burning period and is of less importance than peak HRR and time to reach peak HRR.

ACTIVITY 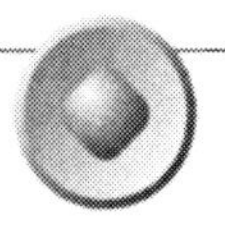

Light a wooden kitchen match. Note that it takes mechanical energy created by the friction of the match head against the rough surface. That energy is sufficient to heat up the chemicals on the match head to the ignition point. The burning match head radiates heat to the wood stick, which pyrolizes and begins to burn. Note the browning of the wood just past the flame. Now try lighting a match using different levels of force and speed as you pass the match head over the striking surface. What do you learn from this? Now try lighting a match on different striking surfaces. What do you learn from this? There are further everyday examples that can be used to explain simple fire dynamics concepts. Use a cigarette to explain heat transfer methods. Blow out a candle and light the vapor being emitted from the wick to show that it is the vapor and not the solid or liquid fuel that burns. ◖◖◖

Many items have been tested to determine their HRR, and this information is available in references, such as NFPA's *Fire Protection Handbook*. An investigator should become familiar with the burning rates of common items.

Energy Sources and Fuel Load

From this discussion we can see that fuels should be thought of as energy sources. How do we apply this knowledge during the fire investigation? During the examination of an area of fire origin, it is important to characterize the type and quantity of fuel. This knowledge will be used as part of the attempt to track the energy release process back to its inception. Ask yourself questions as you do this. What type of fuel was present? Characterize the components of the fuel. Describe the fuel item as fully as possible as to type, size, construction components, fabric, and cushion material.

Which of the following descriptions provides the more useful information about a piece of furniture?

1. The couch was consumed, leaving few remains.
2. Examination of the debris revealed a loveseat of wood frame construction, with metal springs and polyurethane foam cushions covered with a fabric that charred.

Clearly, the second description provides more useful information for reconstructing the fuel load of the compartment and explaining the fire development in the compartment.

Once the fuel load has been determined, the quantity and type of fuel can be analyzed to determine whether they match the fire development as documented by the damage seen.

Mechanisms of Heat Transfer

In further analyzing the area of origin, the investigator looks at the mechanisms of heat transfer. Remember that heat transfer is heat flux, energy transfer per unit time per unit area. For example,

$$\frac{\text{Joules}}{\text{sec} \times \text{m}^2} = \text{kW/m}^2$$

Convection is a critical method of heat transfer to areas and surfaces located above the fire, especially during the initial stages. It is also a mechanism responsible for materials away from the fire being ignited by hot fire gases in later stages.

Radiation dominates as the hot gas layer develops in a compartment. Radiation is a key method of transfer as a compartment approaches flashover and beyond into full compartment involvement.

Thermal Inertia

Ignition of a material is partially dependent on its properties of thermal inertia. We know from experience that it is easier to ignite kindling than logs in a fireplace and easier to ignite crumpled newspaper than charcoal in the barbecue grill. What material properties account for this difference?

Thermal inertia is the product of a material's thermal conductivity (κ), density (ρ), and heat capacity (c). Thermal inertia governs the heating of the surface of a material when it is exposed to a heat flux. The surface of a material with a low thermal inertia, such as polyurethane foam, heats quickly. The surface of a material with a higher thermal inertia, such as wood, heats more slowly.

Ignition

For a fuel to ignite, the surface temperature of the fuel must be raised to its ignition temperature. The quicker the surface heats up, the quicker the ignition. Flaming combustion does not involve the solid or liquid phase of a fuel. What burns is the vapors (gases) generated above the surface of these fuels. Therefore, a fuel must be heated to a point at which it is generating ignitible vapors. It is these vapors, mixed with an appropriate amount of oxygen, that must be raised to a temperature above their minimum ignition temperature for combustion to occur.

To heat a fuel to this point, you generally need an external heat source. Some liquids, such as gasoline, can generate vapors at a temperature as low as 245°F (243°C) and therefore are generating ignitible vapors at normal ambient temperatures. Solid fuel, such as wood, does not generate ignitible vapors under normal conditions and requires its surface to be heated and pyrolized.

To heat a fuel sufficiently to generate ignitible vapors, there are a number of interrelated factors.

- *Form of the fuel:* Is it a solid, liquid, or gas? Does it have a lot of surface area that can absorb heat and a relatively small amount of mass?
- *Amount or mass of the fuel:* How much fuel is present that needs to be heated? The greater the mass, the more heat is needed.
- *Proximity of the fuel to the heat source:* Generally, the closer the fuel to the heat source, the faster the fuel's temperature is raised.
- *Amount of heat being generated:* How much heat is the heat source generating? If it is a relatively small fire, then the fuel needs to be either closer or exposed longer to have its temperature raised.
- *Duration of exposure:* The longer the duration, the more the fuel's temperature is going to be raised. However, some fuels may not need long exposures to have their temperature raised sufficiently to generate combustible vapors.

For example, a single match cannot heat a large block of wood sufficiently to ignite it even though the temperature of the heat source may well be above the minimum ignition temperature of the target fuel. There is too much mass in the wood, and the match generates insufficient heat. Similarly, a solid piece of wood such as a 2 by 4 that is exposed to a brief electrical arc would probably not ignite because the duration of heat exposure is not sufficient.

In the initial stages of a fire, an upholstered chair does not ignite from the radiant heat of a fire in a wastebasket that is several feet away. The fire is too small and the distance too great to raise the temperature of the upholstery sufficiently to

EXHIBIT 3.2
Ignition

generate ignitible vapors and ignite them. However, if the same wastebasket fire is moved to within a foot of the chair, the chair may easily ignite. (See Exhibit 3.2.)

FUEL-CONTROLLED FIRE

In a fuel-controlled fire, there is more than sufficient air in the compartment for the fire to occur. The rate of fire development and spread depends on the fuel that is available. In a typical compartment fire with an open door and window, for instance, there is plenty of oxygen in the initial development stages, so the extent and progress of the fire depend on the nature and quantity of the fuel. The fire is fuel-controlled.

VENTILATION-CONTROLLED FIRE

A ventilation-controlled fire is one in which the rate of development and spread of the fire depends on how much oxygen is available for the fire to propagate. Assume the same compartment as in the above example, but with the door and window tightly closed. The fire reduces the oxygen content in the compartment as it continues to burn. If temperatures in the room remain low, the fire begins to decay despite sufficient fuel available. Even if the window or door is opened, or if a window breaks during the developing fire and the fire develops post-flashover, the fuel can only be burned as rapidly as the existing oxygen allows. There is still plenty of fuel to burn, but there is not enough oxygen for the chain reaction to continue. The fire is ventilation controlled. Most fully developed compartment fires are ventilation-controlled.

It is important that the fire investigator be able to explain how the fire developed. Knowing how the fire developed and what might have controlled the development is key to determining the sequence of events following fire ignition. This means that the investigator must determine the composition of the fuel load, what possible ventilation openings existed, and whether there were any changes in the ventilation during the fire (such as occupant actions, building failures, or fire suppression actions).

FIRE DEVELOPMENT

Fire development describes the process of how a fire grows from its incipient stage and spreads beyond its area of origin. There are five phases of fire development: *ignition, growth, flashover, fully developed,* and *decay.* A number of factors influence fire development, including the fuel load, whether the fire is confined or unconfined, the size of the compartment, the location of the fire in the compartment, and ventilation.

As fire burns, hot gases rise upward from the seat of the fire due to the buoyancy effect. This column of smoke and fire gases is called a *plume.* As the plume moves upward, cooler air is drawn in from around the plume—a phenomenon known as *entrainment.* As the cooler air enters the plume from all directions, it reduces the temperature and increases the mass of the plume as the gases move upward away from the fire. Therefore, the greater the vertical distance from the fire, the cooler the temperature of the plume. Exhibit 3.3 illustrates fire development.

An unconfined fire is not constricted by vertical surfaces such as walls or horizontal surfaces such as ceilings. The fire burns and the gases move without any compartment restrictions. There is essentially no limitation on the amount of air available in an unconfined fire; therefore, it could be considered to be a fuel-controlled fire. Even fires within compartments can burn as unconfined fires until the plume reaches a horizontal surface.

A confined fire is one in which a barrier of some type affects the fire development. This barrier could be either vertical (such as a wall) or horizontal (such as a ceiling). In some fire scenarios, however, even though walls and ceilings may be present, they may be so far from the seat of the fire that they do not play a role in the fire. For example, a relatively small fire that occurs in a large warehouse may not have been influenced by the presence of a ceiling 75 feet (22.9 meters) above it.

When a plume encounters a horizontal barrier, it spreads outward from the center of the plume until it is obstructed by a vertical barrier. While the temperature of the plume away from the fire is generally cooler than the temperature of the plume right above the fire, temperatures can still be significant and capable of igniting combustible materials, particularly within the flame zone of the fire plume. (See Exhibit 3.4.)

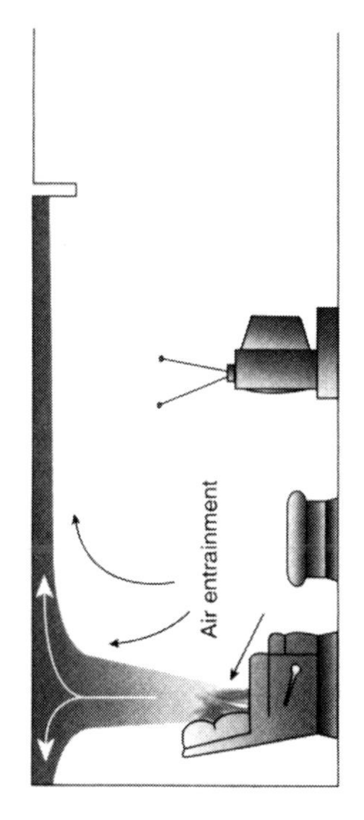

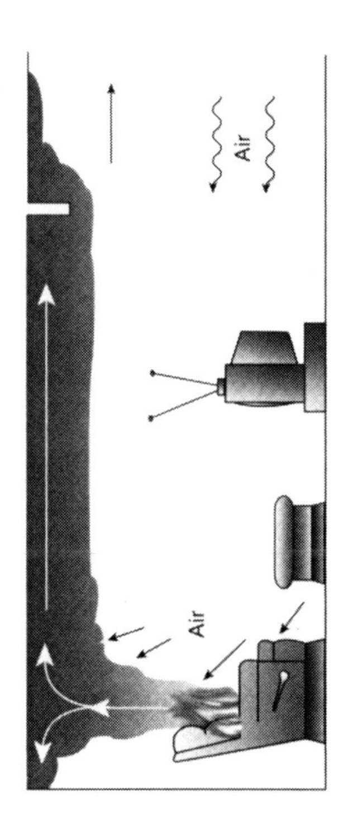

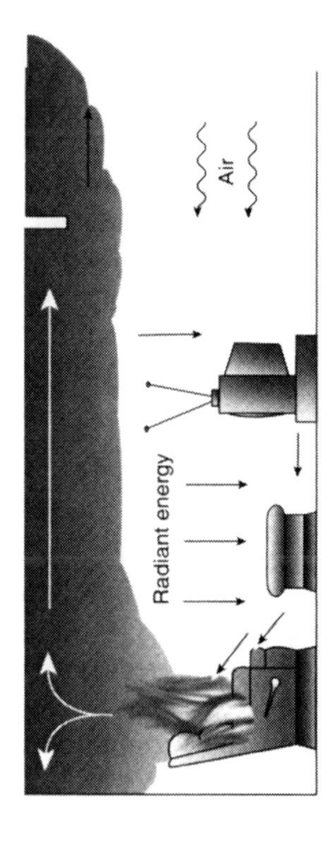

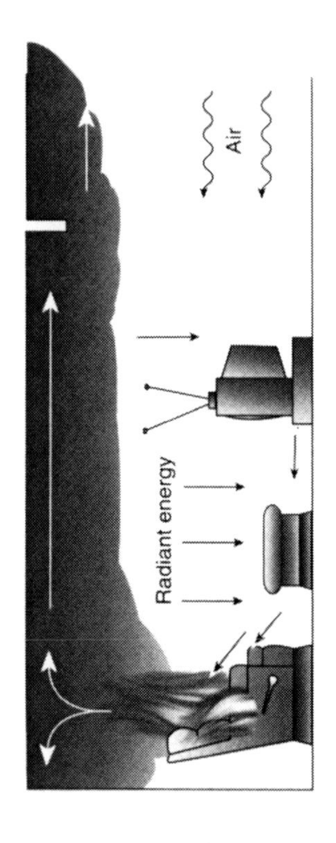

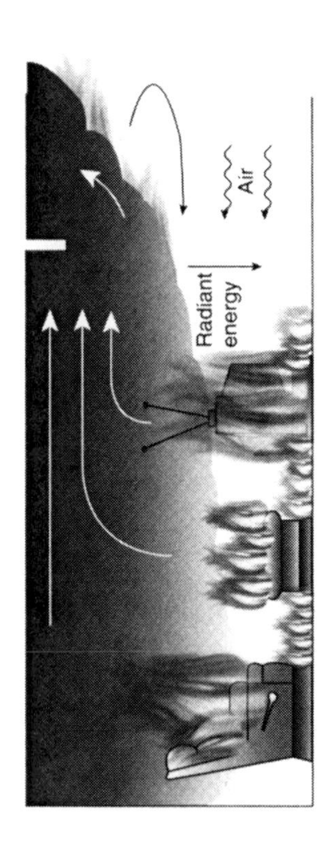

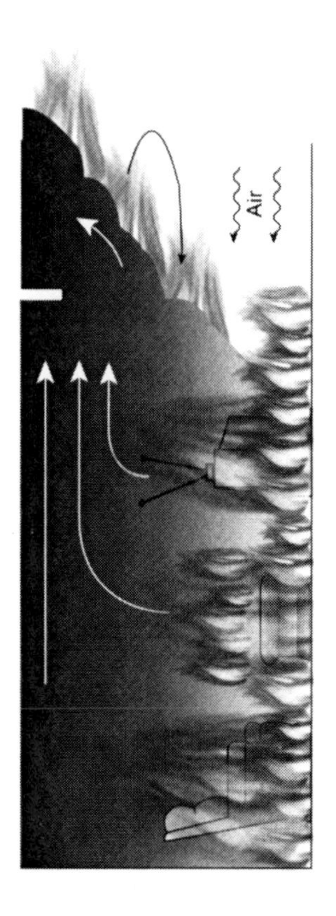

EXHIBIT 3.3 Fire Development

EXHIBIT 3.4
Flame Zone

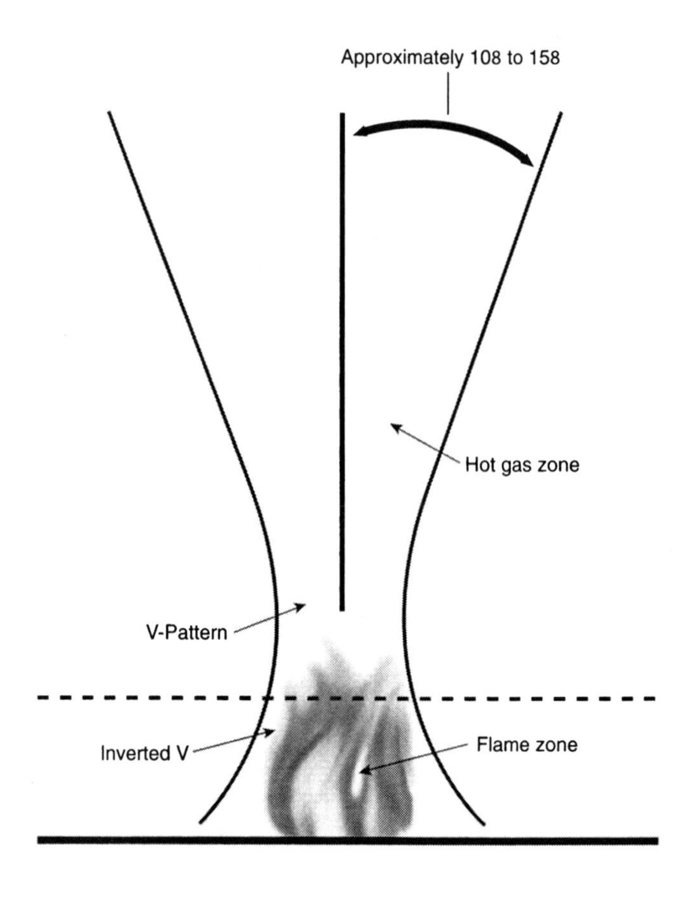

ACTIVITY In NFPA 921, examine Figures 5.5.4.2 through 5.5.4.2.7.

If additional fuel packages are ignited by the initial fire through heat transfer, they can contribute heated gases to the plume or develop plumes of their own. The fire development scenario can become very complex as the fire grows, spreads, and involves other combustible materials.

The reaction of a smoke detector, heat detector, or sprinkler is greatly influenced by its distance from the fire and how the fire is developing and spreading. Theoretically, the plume is going to spread horizontally in 360 degrees when it encounters a horizontal flat barrier. However, additional vertical barriers can influence the reaction time for a detection/suppression device.

FLASHOVER

Flashover is the transition phase of a contained fire in which surfaces exposed to thermal radiation reach ignition temperature almost simultaneously and the fire spreads rapidly throughout the space. As the fire continues to generate hot, buoyant products of combustion, the gas layer begins to deepen and descends to the floor. The gases within the layer are heating objects around them by conduction and convection as well as heating other objects below the gas layer by radiation from the bottom of the layer. When the average gas layer temperature is approximately 1112°F (600°C) and the radiant heat flux to the floor is approximately 20 kW/m^2, most ordinary combustibles in the room are ignited almost simultaneously, and it is likely that the fire will develop to full-room involvement.

As the average temperature of the upper layer approaches 600°C, the underside of the upper hot gas layer can ignite. Fire fighters typically refer to this condition as *flameover* or *rollover.* Flameover is caused by the ignition of the unburned gases in the bottom of the upper gas layer and often precedes flashover.

When interviewing witnesses and fire fighters, the investigator should be sure to get a description of what they saw. The investigator should not rely solely on the witness use of terminology such as flameover and flashover, because these terms are often mistakenly used interchangeably.

Many factors influence when a compartment reaches flashover. (See Table 3.1.)

TABLE 3.1 Factors That Influence Flashover

Factor	*Effect*
The size (volume) of the compartment	The larger the compartment, the more time required to fill the upper layer with hot gases and to raise the temperature of the target fuels.
Height of the ceiling and distance between the fire and the ceiling	The greater the height, the longer it takes for the compartment to reach flashover conditions.
The ventilation (or lack thereof) in the compartment	The growth of the fire that can create flashover is closely tied to the ventilation that is available.
The amount of fuel in the compartment	The amount of fuel includes the initial fuel package, the target fuel packages, and the lining material in the compartment.
The type of fuel in the compartment	The type of fuel determines the rate at which combustible fuels can be generated and how much radiant heat will be required to raise the temperature of the target fuels to a point where sufficient combustible gases are being released to be ignited.
The layout of the room's contents	Target fuels located closer to the initial fire are heated sooner and release combustible gases sooner than those that are remote from the initial fire.
The location of the fire in the compartment	Fires that are located in the center of a compartment have more air entering the plume and cooling it by entrainment, shorter plume heights, and no restrictions to the geometry of the plume. If the same fire is located against a wall, there is 50 percent less entrainment because of the wall barriers. Therefore the following may result: • Longer flames • Faster rise in upper-layer temperatures • Less time to flashover If the fire is located in a corner, the entrainment is reduced by 75 percent. This can result in even longer flames, faster rise in upper-layer temperatures, and a reduced time to flashover than if the fire is located against a wall or in the center of the room. The location of the fire in relation to vertical barriers is important in interpreting the fire development and spread.

FLAME HEIGHT

The height of the flame above the surface of the burning fuel is related to the following:

- Heat release rate (HRR) of the fuel
- Surface area of the burning fuel

This information can be valuable when the investigator is attempting to determine the properties of the fuel. If the investigator knows the flame height (such as from witness statements, examination of the resultant fire patterns, or fire scene photographs), using the formula in NFPA 921, Paragraph 5.5.6.2 makes it possible to estimate the heat release rate. Similarly, if the HRR of the fuel is known, the investigator can calculate the theoretical flame height. This information could then be used to verify findings in the field, witness statements, and so forth.

ACTIVITY

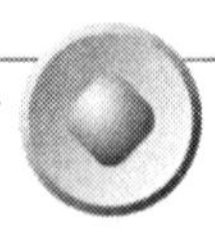

Examine the equation for flame height in NFPA 921, Paragraph 5.5.6.2 and confirm the calculation of 6.9 feet (2.1 meters) for an upholstered chair. In this case, there are no nearby walls ($k = 1$), and the heat release rate is approximately 500 kW.

◗◗◗ QUESTIONS FOR CHAPTER 3

1. What happens to flaming combustion in a compartment when the oxygen content is below 14 percent?
   A. Flaming combustion ceases.
   B. Flaming combustion ceases only in post-flashover high-temperature conditions.
   C. Flaming combustion continues.
   D. Flaming combustion continues in post-flashover high-temperature conditions.
2. Which material has the highest thermal inertia?
   A. Oak boards
   B. Concrete wall
   C. Rigid polystyrene panels
   D. Gypsum plaster
3. Which material's surface heats up the fastest when exposed to a radiant flux?
   A. Oak boards
   B. Concrete wall
   C. Rigid polystyrene panels
   D. Gypsum plaster
4. Which fuel has the lowest ignition temperature?
   A. Kerosene
   B. Gasoline
   C. Methane
   D. Propane
5. Define the ventilation factor and describe its importance.
6. NFPA 921, Paragraph 5.4.1.2, states: "The rate of fire growth as reported by witnesses is not reliable or supported independent evidence of an incendiary fire." Is this a valid statement? Support your answer.
7. In a compartment fire, what radiant energy flux, from the gas layer to the floor, is given to indicate the onset of flashover?
   A. 10 W/cm^2
   B. 2 kW/m^2
   C. 2 W/cm^2
   D. 10 kW/m^2
8. Which fuel burns hotter?
   A. Candle wax
   B. Gasoline
   C. Wood
   D. Kerosene
   E. None of the above
9. Why can the corners of combustible materials ignite earlier than flat portions?
   A. Heat density is less.
   B. There is a higher surface-to-mass ratio.
   C. Angle of recipient radiation is more obtuse.
   D. There is a lower volume-to-mass ratio.

10. What is a flux, as when we talk about an energy flux?

A. Joules per second

B. A unit of measurement per unit of time

C. Joules per square meter

D. A unit of measurement per unit of time per unit of area

11. Compartments A and B have equal volume, but compartment A's ceiling is 3 feet higher than compartment B's. The fire is in the center of the room. HRR is equal in both compartments. Which of the following statements is true?

A. Compartment A would have a higher flame height.

B. Compartment B would have a higher flame height.

C. Compartment B would have quicker flashover.

D. Compartment A would have quicker flashover.

12. What effect does entrainment have on the plume?

A. Decreases the mass of the plume

B. Increases the temperature of the plume

C. Decreases the species composition of the plume gases

D. Increases the cooling of the plume

13. Describe the significance of flashover in a compartment fire.

14. Describe the significance of HRR.

Fire Patterns

▸▸▸ OBJECTIVES

Upon completion of Chapter 4, the user will be able to

- Recognize fire patterns
- Identify the cause of fire patterns
- Analyze fire patterns to produce a hypothesis

Chapter 6 of NFPA 921 provides information relating to the recognition, identification, and analysis of fire patterns. This discussion of fire patterns logically follows the discussion of basic fire science in Chapter 5, because it is impossible to understand fire patterns without a solid knowledge of fire dynamics.

Read Chapter 6 of NFPA 921. Also read the additional references listed in Chapter 2 and Annex B of NFPA 921. ◂◂◂

FIRE PATTERNS DEFINED

Fire patterns are the physical manifestation of the effect of fire on materials. Most fire patterns can be discerned visually, and some can be quantified by measurement.

fire patterns
The physical manifestation of the effect of fire on materials.

In the majority of cases, it doesn't take a fire investigator to determine that a fire has occurred. Even a layperson can easily identify that there has been a fire, just by interpreting visual clues. For instance, people easily identify whether the wood in a fireplace is burned or pristine or whether a discarded cigarette has been smoked. The predominant visual indicator is the presence of burned material.

The role of the fire investigator, however, is more complex than just determining whether a fire has occurred. The fire investigator attempts to recreate the fire development history and then backtracks to identify the origin and the cause of the fire. The primary physical evidence available to the investigator consists of the materials subjected to the fire and the by-products of burning (such as heat, smoke, and soot). It is the understanding of the material response to the effects of fire, coupled with a knowledge of fire dynamics, that enables the investigator to develop and support a hypothesis of a fire's origin and cause.

The interpretation of fire patterns has traditionally been one of the primary processes used in fire investigation. Knowledge about the meaning of the patterns was typically gained through experience and training. In the evolution of fire investigation, there is now greater emphasis on the verification of the fire dynamics that produced the fire patterns. The investigator should maintain current knowledge of the fire science literature and the new research pertaining to fire patterns currently being published.

RECOGNITION OF FIRE PATTERNS

Fire patterns provide data that can be documented and analyzed by the investigator to develop hypotheses. Patterns are produced by the effects of combustion and heat transfer when an affected material undergoes a change that is visually apparent or quantitatively measurable. The changes can include a change in phase, such as a solid melting; a change in chemical composition, such as pyrolysis of wood; deposition of material, such as sooting; deformation, such as distortion of a metal container; physical change, such as spalling; or mass loss of a product, such as the burning of paper. (Spalling, which is illustrated in Exhibit 4.1, will be discussed more fully later in this chapter.) Most fire patterns result from a combination of effects. The properties of the affected material determine the effect produced.

Because effects are dependent on material properties, it is essential to determine the original material that exhibits the fire pattern. At times, identification is difficult and is limited to generic categories such as wood or fabric. If a pattern is critical, the investigator should take a sample of the material for further identification.

The investigator should analyze individual fire patterns within the context of the complexity of all the patterns and is advised not to rely solely or too heavily on only one fire pattern. Seldom is one fire pattern definitive. A pattern analyzed in isolation can lead to an incorrect conclusion. A process of analysis—proceeding from the gross patterns to an area of interest and then to an analysis of the discrete smaller patterns—can provide more meaningful information.

The development stage of the fire affects the fire patterns produced in a compartment fire. Patterns produced during the ignition and early growth stages are primarily influenced by the ignition source and fuel. The development of a gas layer changes patterns as radiation from the layer exerts its effect. Post-flashover conditions of full-room involvement can mask original patterns and change the pattern production to a process more influenced by ventilation than by fuel.

Participate in a fire test or demonstration. Carefully document the fire scene prior to ignition, noting the fuels and compartment conditions. If possible, watch the fire from a safe

EXHIBIT 4.1
Spalling

location or via a remote camera. After the fire, examine the scene again, noting the patterns produced. Remember that a fire demonstration is different from a fire test. A fire test takes place in a controlled environment in which instrumentation is available for measuring HRR, heat flux, and other fire variables, while a fire demonstration is a simulation designed to obtain observable results only. ◖◖◖

Fire pattern analysis should not occur in isolation. Interviews with the owners or occupants can be essential to the investigator's understanding of preexisting conditions (fuel, building construction, occupant activities, etc.), fire suppression activities, and observed fire development. Ignoring relevant information can result in the examination of fire patterns in an area that does not pertain to the task.

Plume-Generated Patterns

Although the patterns left on a single surface are essentially two dimensional, it is important for the investigator to be able to visualize the fire in three dimensions. The pattern that remains on a vertical surface, such as a wall, may be a V-shaped pattern, but the fire plume that created the pattern had a three-dimensional shape. The ability to view a fire pattern and abstractly convert it to the fire dynamic process that produced the pattern is vital to every fire investigation.

Exhibit 4.2 shows a V-shaped pattern on an inside wall, and Exhibit 4.3 shows an exterior V-shaped pattern.

EXHIBIT 4.2 V-Shaped Pattern (Interior)

EXHIBIT 4.3 V-Shaped Pattern (Exterior)

Light a candle. Take a 3 by 5 index card, hold it above the flame, and notice the pattern produced. Move the card into the plume, bisecting the plume. Remove the card before it ignites and observe the pattern. Hold the card perpendicular to the flame, and move it close enough so that the card begins to scorch. Note the shape of the pattern. (Remember that small-scale fire demonstrations, such as those with a candle, are illustrative but do not usually correlate to the complexity of large-scale fires.) ◀◀◀

Ventilation-Generated Patterns

Ventilation can affect the resulting fire patterns in several ways. Increased ventilation, or airflow, can raise the temperature and increase the rate of combustion, which in turn often results in greater burning of materials than in cases not affected by ventilation. This result is often measurable in wood, appearing as greater depth of the char layer.

Ventilation can also affect the resulting fire patterns by the development of hot fire gases. The heated gases rise until they are obstructed by a horizontal surface and then move laterally. The gases can be hot enough to char or alter material, leaving behind a pattern as they contact the material.

Hot Gas Layer-Generated Patterns

Radiant heat from the gas layer also plays a role in creating patterns in a room. As the hot gas layer builds up at the top of the compartment, heat is radiated outward from this layer, including downward radiation toward any combustible fuel load. These conditions can lead to flashover. The investigator can observe these patterns on both horizontal and vertical surfaces, including furniture that may be below the hot gas layer.

The hot gas layer on a wall may create a distinct horizontal line of demarcation that indicates where the gas layer stopped descending. (See Exhibit 4.4.)

EXHIBIT 4.4 Hot Gas Layer–Generated Pattern

Fire gases fill the room from the top downward, creating patterns as they pass over vertical surfaces.

Patterns Generated by Full-Room Involvement

Patterns created by full-room involvement can usually be found on all exposed surfaces throughout the room, all the way down to the floor level, as shown in Exhibit 4.5. Traditional fire patterns, such as V-patterns, may be more difficult to document and analyze if the fire continues to burn for a long time.

Lines of Demarcation

A line of demarcation indicates the height of the hot gas layer and the direction of travel of the hot gas layer. It can also indicate whether the fire was fast or slow in developing. Exhibit 4.6 shows lines of demarcation.

Surface Effect

The extent of the pattern depends on the nature of the surface. Generally, rougher surfaces experience greater damage than smooth surfaces (of the same material), due to greater surface-to-mass ratio and turbulence created by the rough material.

EXHIBIT 4.5 Full-Room Involvement

EXHIBIT 4.6 Lines of Demarcation

Fire Penetration of a Horizontal Surface

Penetrations of horizontal surfaces can come from above or below. The investigator can determine the direction of fire flow by inspecting the sides of the hole. For instance, in the top portion of Exhibit 4.7, the fire direction is from below. In the bottom portion, the fire direction is from above. Exhibit 4.8 shows the penetration of a deck from above, and Exhibit 4.9 shows the same damage from below.

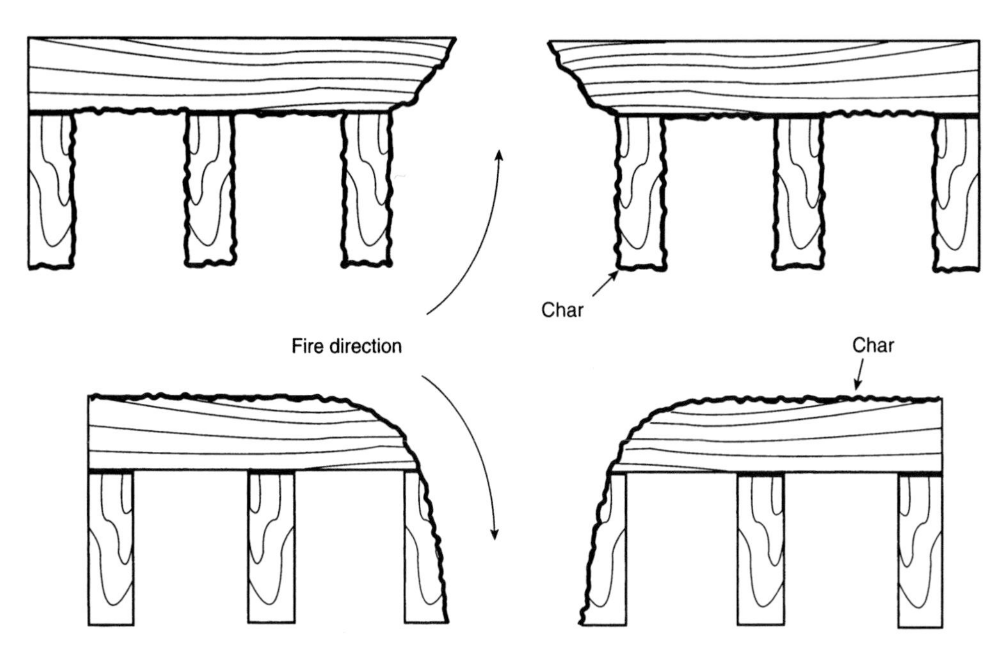

EXHIBIT 4.7 Fire Penetration of a Horizontal Surface

EXHIBIT 4.8 Fire Penetration from Above

EXHIBIT 4.9 Fire Penetration from Below

Patterns Detected in Fire Victims' Injuries

Although there may be a natural instinct to quickly remove a dead victim from a fire scene, the body should not be moved until it has been analyzed and documented. As with any other object in the fire, the body could have effects and contain fire patterns that can assist in the analysis.

Beveling

Beveling is an indicator of fire direction on wood wall studs. The bevel leans toward the direction of travel, as is shown in the top portion of Exhibit 4.10. Note how the beveling changes when the heat source is directed toward the narrow surface of the wall stud, as is shown in the bottom portion of Exhibit 4.10.

OTHER FIRE PATTERN CONSIDERATIONS

Char

Char is the blackened, pyrolized, carbonaceous material that occurs when solid fuels are exposed to heat. The rate of charring is not a reliable indicator for estimating the time of burning. Likewise, depth of char is not a reliable method for determining burn times, although it can be a reliable method to indicate fire spread.

Spalling

The presence or absence of *spalling*—physical change in the burned object—is not indicative of the presence of an ignitible liquid. Spalling is typically caused by the expansion of the moisture within the concrete from the heat of the fire. (Refer again to Exhibit 4.1.)

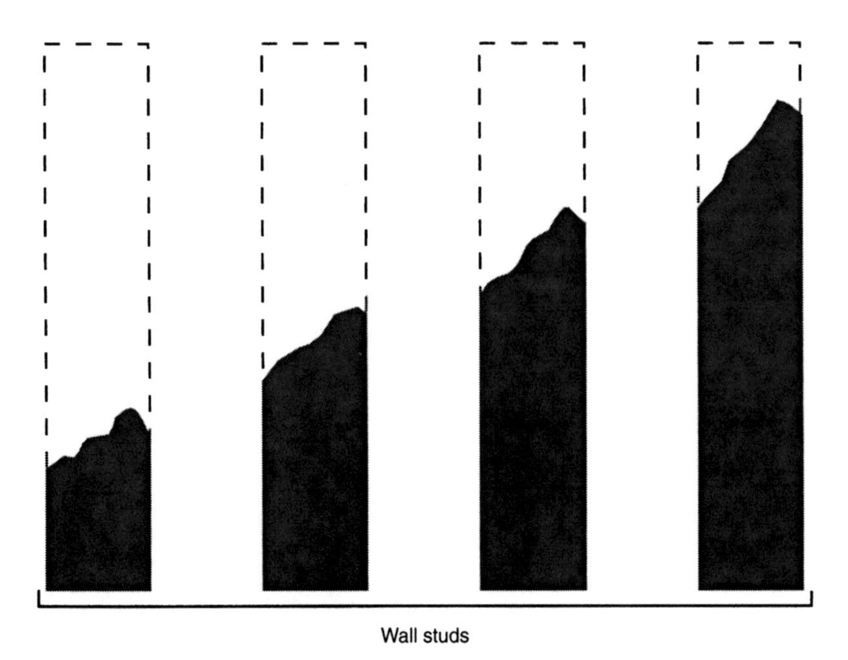

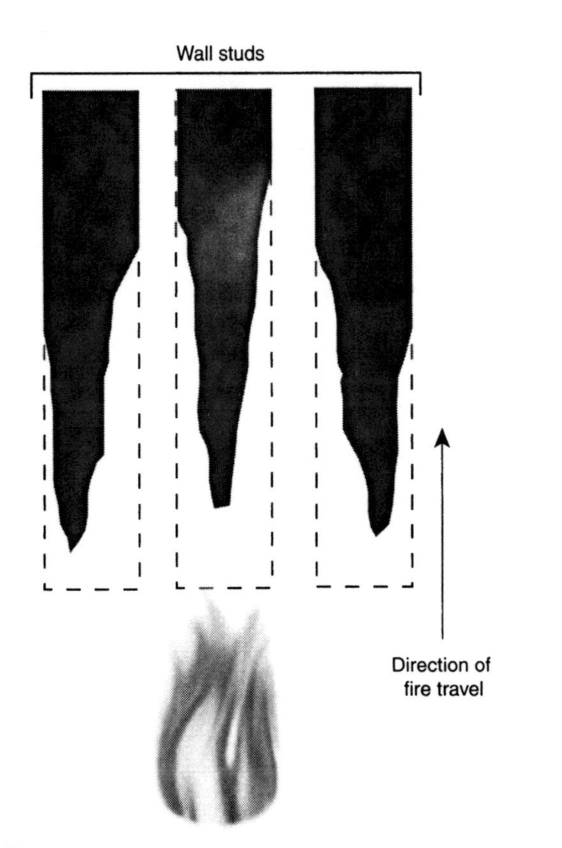

EXHIBIT 4.10 Beveling

Oxidation

Oxidation is a physical change in appearance of a material resulting from the combination of oxygen with metal, rock, or soil in the presence of high temperature. Oxidation indicates an area of metal that was exposed to intense heat.

Melting

Melting is a function of duration and intensity of heat transfer to the material. It is possible to make some general assumptions about the fire environment temperatures based on which materials were melted. Melting temperatures listed in the literature are not exact, as these temperatures can vary with heat flux. The effective fire temperature in a structural fire rarely exceeds 1900°F (1040°C) for extended periods of time. Exhibit 4.11 shows melted plastic, and Exhibit 4.12 shows melted copper.

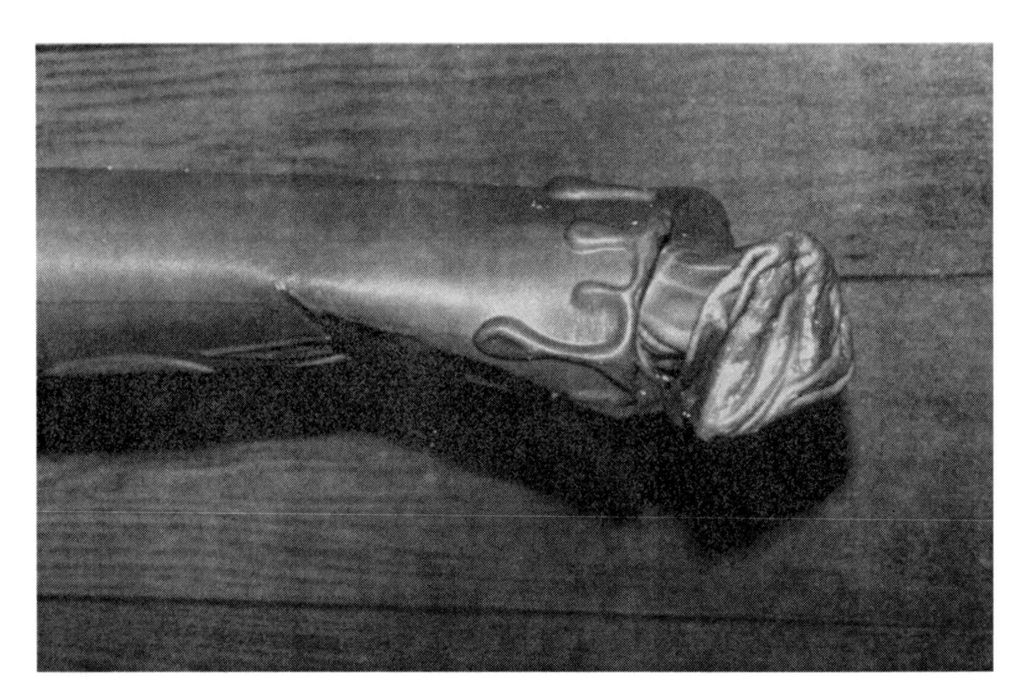

EXHIBIT 4.11 Melted Plastic

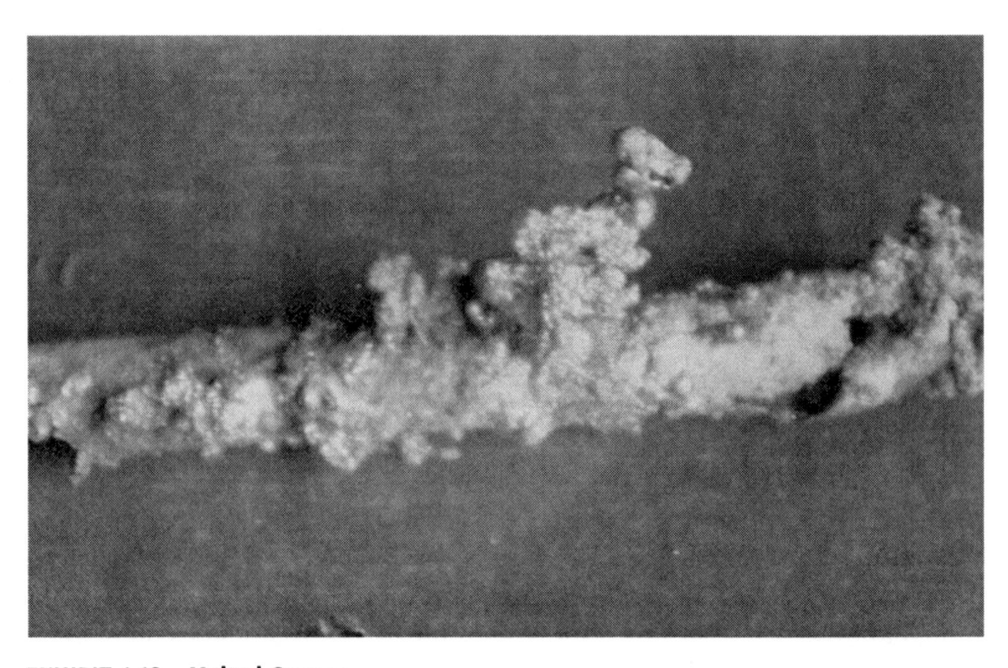

EXHIBIT 4.12 Melted Copper

ACTIVITY

Light a candle. Suspend a piece of aluminum wire between two pliers and place it in the flame. What happens? Take a thin piece of copper wire and repeat the experiment. What happens? Take a piece of thin steel wire and repeat the experiment. What happens? Repeat the experiment with a piece of copper pipe. What happens? ◖◖◖

Alloying

Remember that material interactions can affect melting temperatures, especially with metals. A common example is the interaction of aluminum and copper that results in a lower melting temperature for the copper. The interaction of metals is called *alloying*.

Soot

Soot—the elemental carbon produced during incomplete combustion—can result from the burning of fuel that contains carbon. Soot can be deposited where flames touch surfaces and by settling on surfaces, as shown in Exhibit 4.13.

ACTIVITY

Light a candle. Note that no soot is emitted from the flame. The soot is consumed in the upper portions of the flame, producing the yellow luminescence. Interrupt the complete combustion by placing a metal spoon into the flame. What happens? Why? ◖◖◖

Clean Burn

A *clean burn* occurs on noncombustible surfaces such as gypsum wallboard (remaining after the combustible surface has been burned away), masonry, and concrete. (See Exhibit 4.14.)

EXHIBIT 4.13 Condensates on Walls

EXHIBIT 4.14
Clean Burn

ACTIVITY

Take the sooted spoon from the previous activity and place it above the tip of the flame. What happens to the soot?

Calcination

Calcination occurs in plaster or gypsum wall surfaces when the chemically bound water is driven out of the gypsum by the heat of the fire. Exhibit 4.15 shows the calcination of drywall. The rate of calcination is not a reliable indicator for estimating the time of burning. Likewise, depth of calcination is not a reliable method for determining burn times, although it can be a reliable method for indicating fire spread.

Window Glass and Light Bulbs

Variables that affect the reaction of glass when it is exposed to fire include heat flux, duration of exposure, direct flame impingement, type of glass, thickness of glass, glass edge quality, and glass assembly.

It is important to distinguish between glass that has been mechanically broken and glass that is broken by the fire. Fire suppression activities can either deliberately or unintentionally cause mechanical breakage of glass, particularly

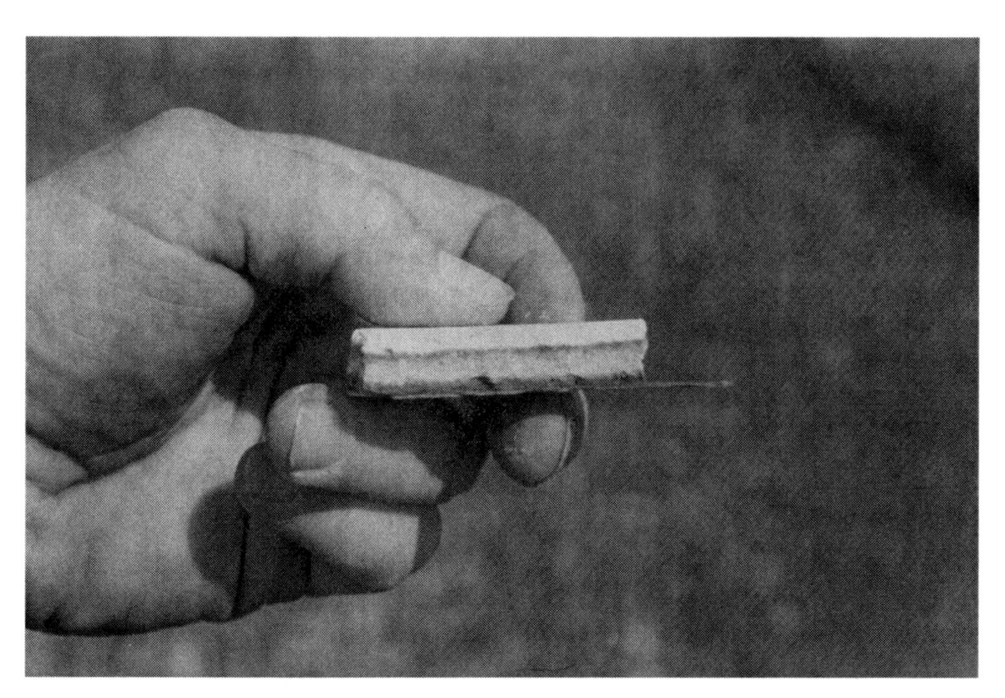

EXHIBIT 4.15 Calcination

window glass. The subsequent venting of smoke, hot gases, and flames through such mechanically broken glass can alter the remaining indicators. In this case, as others, interviews and documentation of fire suppression activities are important in the analysis of the fire patterns

Heat can also have an effect on the glass of a light bulb. The investigator can examine the distortion of a light bulb, for instance, to determine the direction of heat impingement. Bulbs over 25 watts expand toward the heat, and bulbs that are 25 watts or less pull inward on the side of the heat. The bulbs in Exhibit 4.16 demonstrate this phenomenon. The bulb in the middle is greater than 25 watts, and the bulb on the right is less than 25 watts.

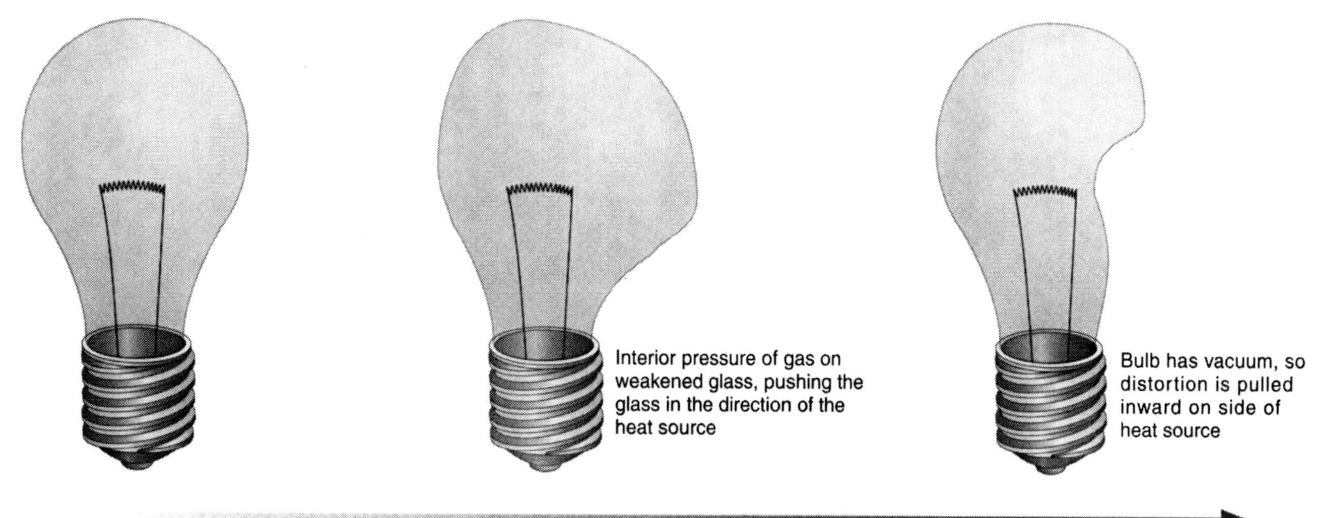

EXHIBIT 4.16 Damage to Light Bulbs, Indicating Direction of Heat Source

ACTIVITY

At a fire scene, note the difference in appearance of the windows that were broken by fire fighters from those broken by the fire.

Furniture Springs

The loss of strength in furniture springs is a function of the duration and intensity of heat exposure. Springs can lose strength by short-term heating at high temperatures or by long-term heating at moderate temperatures of around 750°F (400°C). The presence of a load on the springs at the time of the fire may also contribute to loss of strength. Exhibit 4.17 shows the damage to mattress springs from heat exposure.

Heat Shadowing

Heat shadowing is caused by an object blocking the travel of radiated heat, convected heat, or direct flame to a surface, thus creating a discontinuous pattern on that surface that relates to the interruption of that heat transfer.

Protected Areas

When an object is physically in contact with another material in such a way that the material shields the object from the effects of heat transfer, combustion, or deposition, then the object is said to be in a *protected area.* The investigator will find that protected areas are very useful in reconstructing the fire scene. Exhibit 4.18 shows the protected area beneath an ottoman.

EXHIBIT 4.17 Mattress Springs

EXHIBIT 4.18
Protected Area

Pattern Location

It is important to remember that the fire is a three-dimensional event, and therefore fire effects can be three dimensional. The fire investigator needs to examine all the affected surfaces, including surfaces that are not readily apparent. The scene examination includes looking for large-scale patterns such as aerial views of the roof, and small-scale patterns such as the heat effects on the insulation of wiring.

Exhibit 4.19 shows burn patterns on the floor caused by the pouring and ignition of gasoline. Note the sharp contrast in the damage lines.

It is important to examine all areas in the building or near the fire to determine whether there are other areas with fire patterns. Fire patterns that are separate from the main body of the fire can be attributable to a number of factors, including fire extension, fire brands, drop down, separate areas of origin, or even previous fires.

Chapter 6 in NFPA 921 offers an extensive discussion of pattern location. As you read that material, it should become apparent why the fire investigator also needs a good understanding of building construction.

Flashover and full-room involvement can produce relatively uniform burning on surfaces. Some exposed surfaces may have little or no damage due to ventilation effects and the location of furnishings. Additional care must be taken in evaluating patterns.

EXHIBIT 4.19
Ignitible Liquid Burn Patterns

ACTIVITY

Read additional reference material on building construction that relates the materials and methodology of construction to fire pattern development. ◀◀◀

Pattern Geometry

The shape and type of a pattern can provide clues to the investigator about the location of the heat source and the fire dynamics that occurred. It is again important to relate the patterns observed to the fire's development. Post-flashover fires can alter the initial patterns from fuel-controlled to ventilation-controlled. Sustained burning from collapse or lack of suppression can mask many of the initial patterns. Pattern analysis cannot occur in a vacuum devoid of the understanding of fire dynamics, building construction, and fuels.

QUESTIONS FOR CHAPTER 4

1. Why is it important for the investigator to recognize fire patterns?
2. True or false: Determining which pattern was produced at the point of origin becomes more difficult as the size and duration of the fire increase.
   A. True
   B. False
3. What damage pattern can be created on the surface when a plume is truncated by a wall surface?
   A. V-shape
   B. U-shape
   C. Inverted U-shape
   D. Hourglass
4. How do ventilation-generated patterns affect an investigation?
5. List the two basic types of fire patterns and give a brief description of how they are produced. (See Chapter 6 in NFPA 921.)
6. True or false: Spalling of concrete at a fire scene is a positive indicator of a liquid accelerant.
   A. True
   B. False
7. What is the difference between charring and spalling?
8. True or false: Analysis of the depth of charring is more reliable for evaluating fire spread rather than for the establishment of specific burn times or intensity of heat from adjacent burning materials.
   A. True
   B. False
9. List the key variables that affect the validity of a depth of calcinations analysis.
10. True or false: The angle of the borders of the V-pattern indicates the speed of fire growth.
    A. True
    B. False

Building Systems

OBJECTIVES

Upon completion of Chapter 5, the user will be able to

- Identify construction types in commercial and residential structures
- Explain fire behavior as it relates to the method of construction
- Discuss the structural integrity of construction assemblies during a fire

If a builder build a house for someone, and not construct it properly, and the house which he built falls in and kills its owner, then that builder shall be put to death.

—*Hammurabi, King of Babylon*

The design, materials, and construction methods of a structure play a role in the development and movement of any fire that may occur within it. The impact can be either positive, containing the fire within a compartment, or negative, allowing the fire to spread from the area of origin. The fire investigator must have an understanding of building construction and building systems in order to properly analyze fire growth and movement as well as the fire patterns that remain after a fire. Computer models have been developed to help investigators analyze fire incidents. Some of these mathematical, engineering, and graphic models are described in Chapter 18, "Failure Analysis and Analytical Tools," of this *User's Manual.*

This chapter offers guidance on building construction and building systems. However, many existing fire protection systems—such as sprinkler systems, smoke control systems, and alarm systems—are not mentioned in Chapter 7 of NFPA 921. An investigator must have a basic understanding of these systems, too, in order to conduct a thorough fire investigation. See NFPA's *Fire Protection Handbook,* 2003 edition, for a basic discussion of these fire protection systems and their functionality.

BUILDING SYSTEMS OVERVIEW

Modern design considerations and construction features are a direct result of the analysis of fires, often catastrophic fires, some involving conflagrations that occurred at the turn of the 20th century. More than 50 percent of the modern codes are usually related to fire protection. (See NFPA's *Fundamentals of Fire Protection,* 2004 edition, for more details.)

One of the basic principles of modern building construction is the concept of confinement of fire by means of compartmentation. According to *Fundamentals of Fire Protection,* one primary cause of severe fire damage to structures is failure to contain or confine the fire. Basic principles of construction provide that fire should be contained to the room of origin, area of origin (often called *fire area*), or structure of origin.

When a fire does move from the room of origin, it is most often through open doors or construction defects rather than system failures. Failure to manually close fire doors (or impeding their automatic closure), failure to subdivide open spaces by the use of draftstops, and penetration of rated assemblies such as walls and ceilings for building or utility services provide ample methods for fire movement.

However, it should be noted that fires originating in interstitial spaces are not afforded the same level of protection as a fire that develops within a compartment. (See Paragraph 7.2.2.5 of NFPA 921.) *Interstitial spaces* are associated with the spaces between the building frame and interior walls and the exterior façade and with spaces between ceilings and the bottom face to the floor or deck above. Fires that originate in these concealed locations may develop undetected and thus move freely and rapidly without barriers to stop their spread.

DESIGN, CONSTRUCTION, AND STRUCTURAL ELEMENTS

A building's characteristics that affect the development, spread, and control of a fire include its type of construction, the integrity of its structural elements under a fire load, and its fire protection and other building systems. The strength of a building and the number of doors, windows, and other openings it contains affect the amount and type of damage it sustains, from smoke damage to explosion damage. Factors that influence the origin, development, and spread of a fire include the interior layout, interior finish materials, and building services and utilities. Exhibit 5.1 shows these components in a typical single-family dwelling.

Building Design

In a building fire, fire spread and development are largely effects of radiant and/or convective heating. In compartment fires, the following factors significantly affect fire spread:

- Room size
- Room lining material
- Shape
- Ceiling height
- Placement and area of doors and windows

Limiting the fire spread to a specific area through compartmentation is a primary objective of fire safety design.

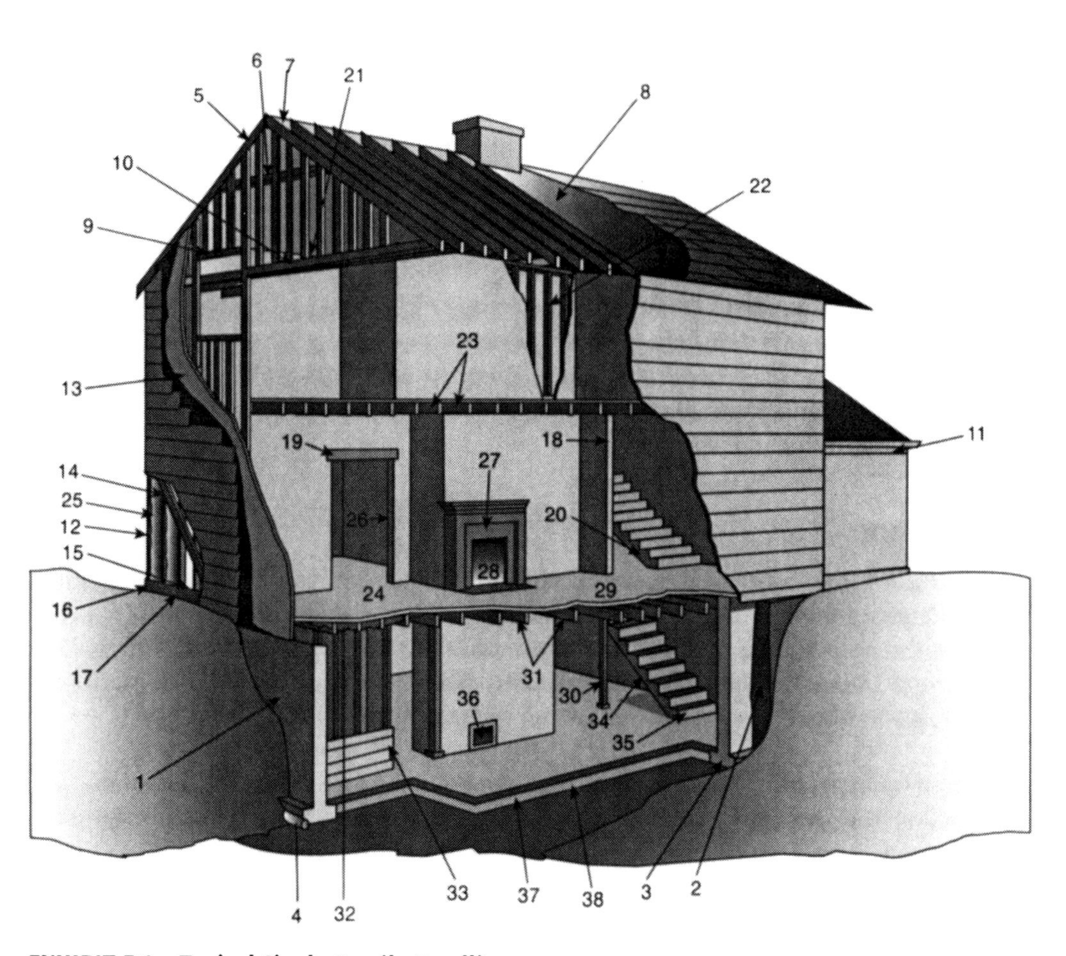

1. Foundation wall
2. Frost wall
3. Wall footing
4. Perimeter drain
5. Rafter
6. Collar beam
7. Ridge board
8. Roof sheathing
9. Window header
10. Attic joist
11. Box beam
12. Exterior wall stud
13. Wall sheathing
14. Corner bracing
15. Exterior wall plate
16. Box sill
17. Sill
18. Wall stinger
19. Header
20. Stair partition casing
21. Attic insulation
22. Partition studs
23. Second floor joists
24. Finish flooring
25. Wall insulation
26. Cripple stud
27. Damper control
28. Ash door
29. Hearth
30. Post or column
31. First floor joist
32. Subfloor
33. Basement partition
34. Stair stringer
35. Tread and riser
36. Cleanout door
37. Concrete floor slab
38. Granual fill

EXHIBIT 5.1 Typical Single-Family Dwelling

Building Loads. A building is designed to support a specific amount of weight. This weight is classified as either a live load or a dead load. A *live load* is described as a load that can move. Examples of live loads include people, wind, water, and snow. *Dead loads* are constant and immobile, such as floor slabs, structural columns, and roof coverings. Dead loads include the weight of the structure, structural members, and building components.

There are certain conditions under which a building will no longer be in balance or capable of supporting its loads. For instance, additional loads applied above and beyond the structure's design parameters may create instability. Examples of such conditions include the following:

- Extreme snow loads
- Extreme wind loads
- Additional contents, such as stock
- Additional mechanical components, such as HVAC and elevator equipment
- Large congregation of people in a limited area
- Water from fire-fighting operations

Structural Changes. Structural changes can occur either deliberately—during building renovations, for example—or during a fire. As the fire grows and progresses, it may damage structural support elements so that they are no longer capable of supporting their designed loads.

Geometry of the Space. The geometry of the room—its height, width, and specifically the distances of the walls from each other—is an important factor in fire development. Given a similar fuel package and similar ventilation points, a smaller room reaches flashover in less time than a larger room. Flashover is an important mechanism for fire spread beyond the room of origin.

Planned Design Versus "As-Built" Conditions. The original plans and designs for a structure may not match what was finally constructed. Sometimes, plans are revised during or after construction to reflect the true "as-built" conditions of the building. It is important for the investigator to determine the accuracy of any plans in relation to the actual structure.

Occupancy

The investigator should determine whether there were any changes in occupancy type during the life of the building. A change in type of occupant can result in the introduction of fuel loads for which the building's fire protection system was not designed. (See Chapter 18, "Failure Analysis and Analytical Tools," for additional discussion of occupancy and how it can affect fire spread in a building.)

Compartmentation. The *compartmentation* of a structure refers to its subdivision into separate sections or units. A common mechanism of fire spread outside the compartment of origin is horizontal movement through openings such as doors, windows, and unprotected openings in walls. Vertical openings such as stairways, utility chases, and shafts also contribute to fire spread outside the compartment of origin.

compartmentation

Design features of a building that limit fire growth to the room or building section of origin. These features include but are not limited to fire walls and fire doors.

Concealed and Interstitial Spaces. Concealed and interstitial spaces exist in most buildings. They can provide a mechanism for fire spread beyond the compartment of origin. Interstitial spaces generally lack firestops. In high-rise buildings, these spaces are generally located between the interior and exterior walls and between the ceiling and the underside of the floor above. Concealed spaces are another major concern for the fire investigator. Failure to account for concealed spaces can lead to erroneous interpretation of fire patterns. In modern construction, concealed spaces are often equipped with fire sprinklers, early detection devices, or firestops. Modern building codes sometimes allow concealed spaces if they are constructed of noncombustible material. These "rated" concealed spaces become a code enforcement problem, as well as a fire spread problem, because they are often used for the general storage of combustible items.

Contact a building inspector from a local building department or a construction supervisor at a local construction site, and arrange for a tour of a building under construction. As an alternative, visit the site of a building demolition. Identify concealed spaces, interstitial spaces, and vertical and horizontal avenues for fire spread. Locate the major structural members and determine how the building carries its load. ❮❮❮

BUILDING CONSTRUCTION

There are many construction types employed throughout the world. This *User's Manual* addresses the most common types. The investigator should determine the type of construction based on the structural elements of the building. The following discussion of construction types includes a list of features typical of each. (Note that a wood wall stud is commonly called a 2 by 4, reflecting the *nominal* lumber measurement. The actual dimension of the 2 by 4 is 1¾ by 3½ inches, which is the *dimensional* lumber measurement.)

Ordinary Construction

In ordinary construction, exterior walls are masonry or other noncombustible material. In frame construction, exterior walls and load-bearing components are wood. The characteristics of ordinary construction versus frame construction are shown in Table 5.1.

Wood Frame Construction

Wood frame construction is generally associated with lightweight commercial construction or with residential construction. Buildings utilizing wood frame are of limited size, with their floor joists and vertical supports generally spaced 16 inches on center. Vertical supports can be 2 by 4 or 2 by 6 inch nominal wood members. Such buildings offer little fire resistance. They may be sheathed with a fire-resistive material such as gypsum board, lath and plaster, or mineral tiles. Even non–fire-rated sheathing can provide some fire resistance. It is important to remember that a fire-rated wall may not necessarily stop heat from being conducted through the wall to underlying members or to combustible materials on the other side of the wall.

Platform Frame. Platform frame is the most common construction type currently used. (See Exhibit 5.2.) In this method, walls are placed on top of platforms or floors, and thus the platforms become an effective firestop for floor-to-floor vertical fire spread. However, barriers that are combustible can be overcome in time. Areas of concern with this type of construction are concealed spaces in the soffits (the horizontal undersides of the eaves or cornice) and vertical openings for utilities.

TABLE 5.1 Ordinary Construction Versus Frame Construction

	Ordinary Construction	*Frame Construction*
Exterior walls	Masonry or noncombustible	Wood
Interior walls	Wood assemblies Platform or braced frame assembly	Wood assemblies Platform or braced frame assembly

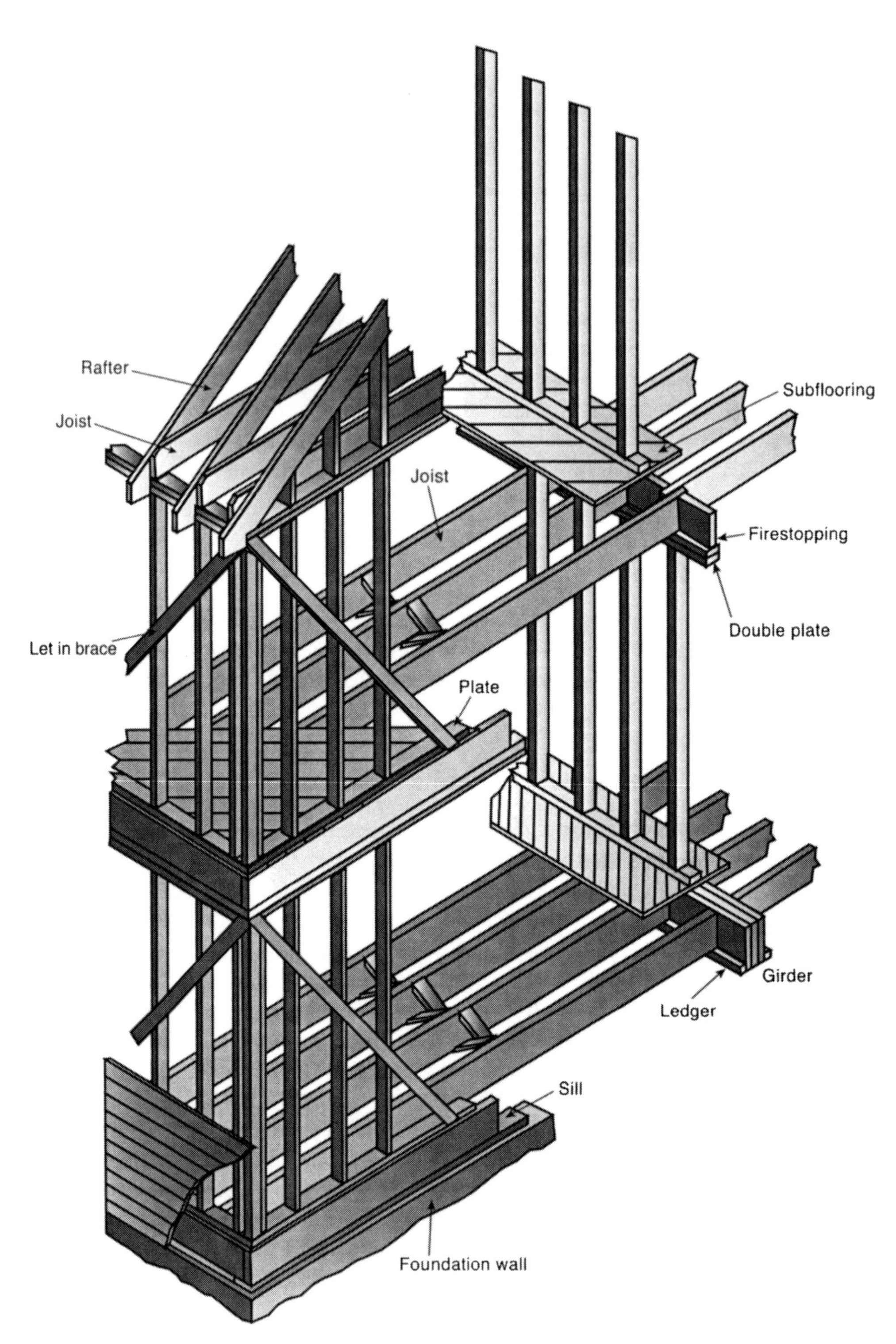

EXHIBIT 5.2 Platform Frame Construction

Balloon Frame. In balloon frame construction, shown in Exhibit 5.3, the exterior wall studs extend from the foundation to the roofline. Because of this feature, building codes have called for the installation of firestopping in the channels created by the studs. If firestops are properly installed, the fire performance of the building can be similar to that of platform frame construction. However, lack of firestopping can allow uninhibited vertical fire spread or fire ignition by fall down

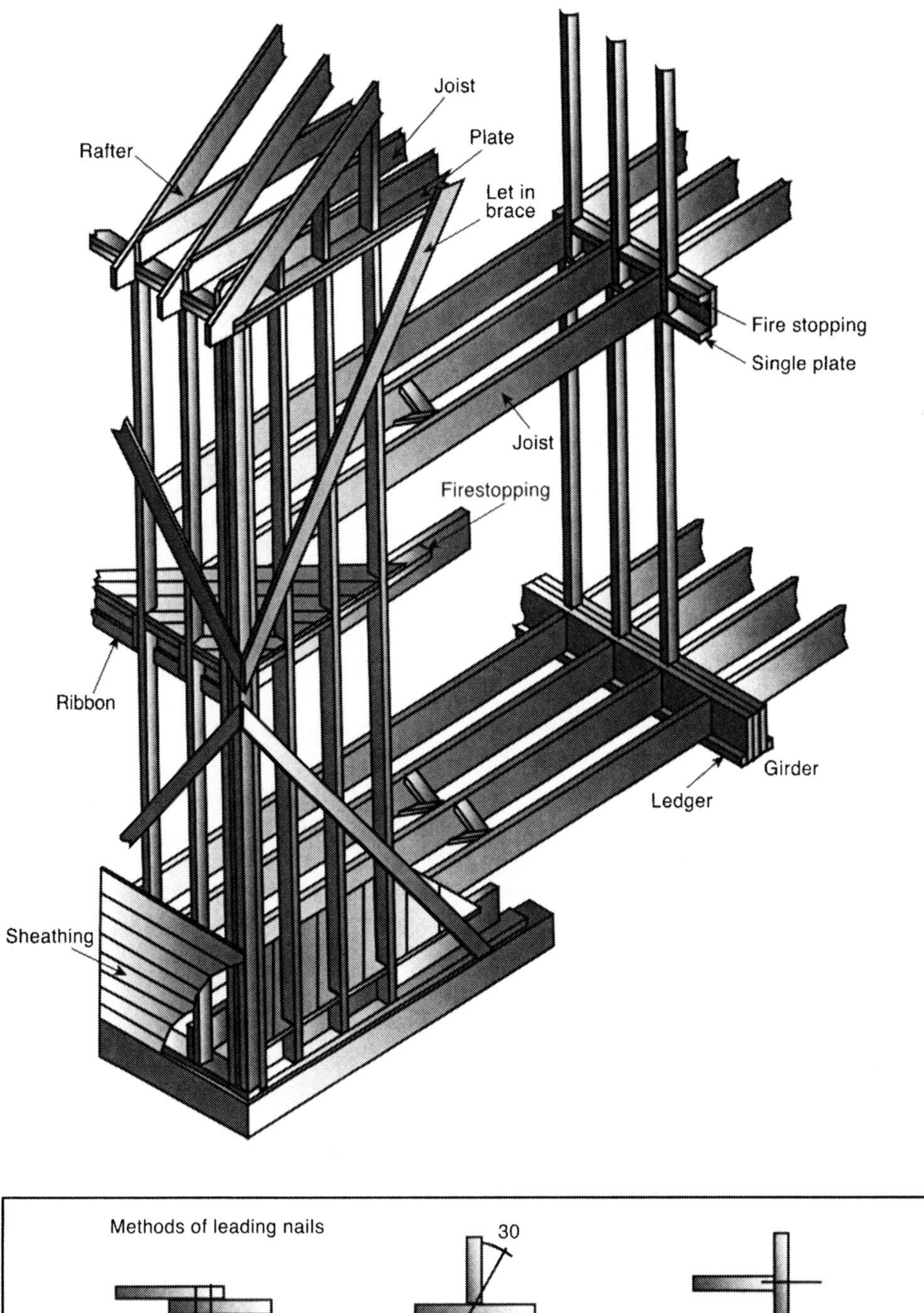

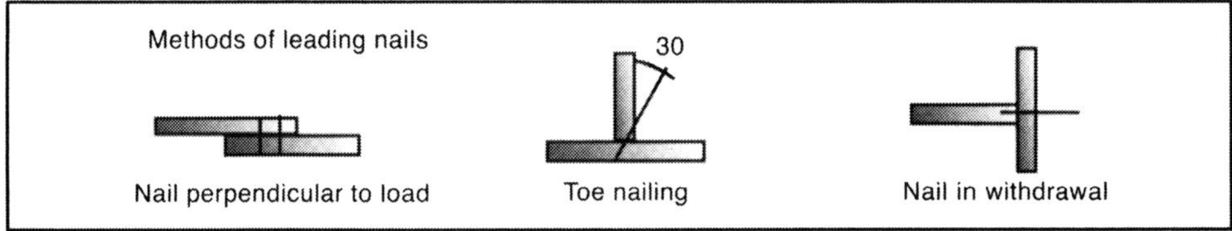

EXHIBIT 5.3 Balloon Frame Construction

from the attic area. The investigator must identify all avenues of fire travel in this type of construction. Fire can break out in an area remote from the point of origin because of the ability to spread through vertical channels.

Plank-and-Beam. In plank-and-beam construction, larger beams such as 4 by 10 or 5 by 12 inch are used. The beams are more widely spaced at 4 or 6 feet on center, and the beams are supported by posts. The floor decking has a minimal thickness of 2 inches and is generally of the tongue and groove type, which helps limit fire spread. This construction type has a limited amount of concealed spaces and the exterior finish has no structural value.

Post-and-Frame. Post-and-frame construction is similar to plank-and-beam construction. An example is typical barn construction.

Heavy Timber. Heavy timber construction employs structural members that are of unprotected wood, the smallest dimension being 6 or 8 inches. Floor assemblies are constructed of 2-inch-thick tongue and groove end-matched lumber, and no concealed spaces are permitted. When heavy timber construction is used, building codes require the bearing walls to have a 2-hour rating.

Alternative Residential Construction

In addition to wood frame site-built construction, other forms and materials are being used for residential construction, such as manufactured housing and steel-framed construction.

Manufactured Housing. Manufactured housing employs a construction technique whereby the structure is built in one or more sections. These sections are then transported to the building site and assembled on the site. Another type of manufactured housing builds the structure on a steel frame equipped with wheels that allow it to be transported easily to the site. This type of manufactured housing is often referred to as a mobile home.

Steel-Framed Residential Construction. Steel-framed residential construction is becoming more common. It has characteristics similar to those of wood frame construction and is noncombustible. However, steel framing can lose its structural capacity during extreme exposure to heat; exposed steel beams and joists can fail during flashover in a period as short as 3 minutes.

Manufactured Wood and Laminated Beams. Laminated beams are structural elements that have the same characteristics as solid wood beams. They are composed of many wood planks that are glued or *laminated* together to form one solid beam and are generally for interior use only. They will behave like heavy timber until failure. Effects of weathering decrease their load-bearing ability. The investigator should document the size of individual members as well as the overall size of the beam.

Wood I-beams have a smaller dimension than floor joists and therefore can burn through and fail sooner than dimensional lumber. Openings in the web for utilities may reduce the structural integrity of the web. I-beams must be protected by gypsum board to help delay collapse under fire conditions.

Wood trusses are similar to other trusses in design. Individual members are fastened together using nails, staples, or metal gusset plates (gang nail plates) or wooden gusset plates. Failure can occur from gusset plates failing even before wood members are burned through. Failure of one truss can place additional loads on adjacent trusses.

Building Materials

Building materials can have an impact on a fire's ignition and growth. Table 5.2 includes some of the factors that should be considered. Orientation, position, and placement of materials make a difference in how the materials react under

TABLE 5.2 Building Materials

Characteristics	*Influencing Factors*
Ignitability	Minimum ignition temperature Minimum ignition energy Time/temperature relationship for ignition
Flammability	Heat of combustion Average and peak heat release rate Time to peak heat release rate Mass loss rate Air entrainment
Thermal inertia	Reaction to heating Ease of ignition
Thermal conductivity	Good conduction versus poor conduction
Toxicity	Quantity and type of gases produced by a material while burning
Physical state and heat resistance	Temperature at which a material changes phase Amount of heat required to ignite the material in its different phases
Attitude, position, and placement	Different burning characteristics exhibited by the materials, depending on whether they are vertical or horizontal Flame spread ratings, which can be obtained through the Steiner Tunnel Test (see NFPA 921 Chapter 7, Paragraph 7.2.3.7.2)

fire conditions. For instance, carpet is generally placed horizontally on the floor. When carpet is placed vertically on the wall, the flame spread of the carpet is greatly increased.

Rating of Wood Frame Assemblies

A fire rating is accomplished by covering the wall with a noncombustible finish, commonly gypsum board. A fire-rated assembly may conduct heat through to the underlying combustible members. Field installations can vary from those used in testing and evaluating the fire integrity. Therefore, a fire-rated wall in the field may fail sooner than expected based on its fire rating.

Noncombustible Construction

Noncombustible construction is used primarily in commercial and industrial storage and in high-rise construction. The building materials used in this type of construction do not add to the fuel load. Columns and walls may be built of materials that are strong in compression but weak in tension. Examples of these materials are brick, stone, cast iron, or nonreinforced concrete.

Noncombustible construction of the elastic range deforms and then assumes its original shape with no loss of strength after the load is removed. Plastic range noncombustible construction is permanently deformed but may continue to bear the load.

TABLE 5.3 Metal Construction Building Material Features

Material	*Factors*
Steel	Can conduct electricity Good conductor of heat Loses its ability to carry a load well below temperatures encountered in a fire Can distort, buckle, or collapse as a result of fire exposure Amount of distortion depends on factors such as heat of the fire, duration of exposure, physical configuration, and composition of the steel
Aluminum	Can be used as a curtain wall or as a siding Melts well below temperatures found in typical fires Can conduct electricity
Wrought iron	Found in older buildings Largely used for decorative, not structural, elements Can withstand greater temperatures than steel Columns can distort when exposed to fire Can shatter or break instead of deforming

Metal Construction. Metal is commonly found exposed in unfinished spaces and may fail in a period as short as 3 minutes during flashover. Fire-rating tests are not necessarily indicative of how the member will perform under fire conditions. Table 5.3 describes some of the features of metal construction.

Concrete or Masonry Construction. Concrete and masonry constructions have inherent fire resistance due to their mass, high density, and low thermal conductivity. They are strong under compression but weak under tension. Structural failure often occurs at connection points.

CONSTRUCTION ASSEMBLIES

Floor, Ceiling, and Roof Assemblies

Floor, ceiling, and roof assemblies are of particular concern to the fire fighter as well as the fire investigator. These assemblies are among the first to fail when structural elements are exposed to fire conditions. Types of failures include collapse, deflection, distortion, heat transmission, and fire penetration.

Factors that can affect the failure of floor, ceiling, and roof assemblies include type of structural element, protection from the elements, span, load, and beam spacing. Added live loads, such as water during fire-fighting operations, can contribute to failure. Penetrations in assemblies for utilities are common. Although penetrations are required to be sealed in a fire-rated assembly, often they are not. Floor assemblies are tested for fire spread from below, not from above. Roof stability can be a critical factor during fire-fighting operations and fire dynamics.

Potential collapse should be a vital concern to the fire investigator on a fire scene. Structural elements are not intended to maintain their strength when subjected to fire conditions. The protection afforded to structural elements, such as gypsum wallboard, is intended to prevent the immediate involvement of the structural elements. However, once this protection fails and the structural elements are affected, their ability to maintain their load-bearing capabilities is diminished.

Walls

Walls serve as barriers to fire spread and can be made to a wide variety of standards and in a wide variety of types. They may be fire-rated or non–fire-rated, load-bearing or non–load-bearing. Penetrations in these assemblies for utilities are common and are required to be sealed in a fire-rated assembly. Fire barrier walls or firewalls using gypsum board will use a type X gypsum wallboard. Non–fire-rated walls can provide varying levels of fire resistance.

Doors

Doors can serve as a critical factor to limiting fire spread throughout the structure. Fire doors may be constructed of solid wood, steel, or steel with an insulated core. They can be rated or nonrated and must be closed to provide an effective barrier to smoke, heat, and fire. Any door installed in a fire-rated wall assembly should be installed as a fire-rated door assembly. The rating of the door will be dependent on the rating of the fire wall and will be less than the wall system rating. A fire-rated door assembly must have the following components rated as part of that system:

- Hinges
- Closures
- Latching devices
- Glazing

Access the attic space in a friend's or your own home or apartment. What can you determine about the design and construction? Photograph and document design factors, construction techniques, and floor, ceiling, and roof assemblies. ◖◖◖

Fireplaces

Fireplaces are a common area of origin for many residential fires. Most old masonry fireplaces were not constructed to withstand constant use, such as being an auxiliary heat source for the home. The most common construction flaws in masonry fireplaces are not having adequate brick thickness between the smoke chamber and the combustible framing members and not having enough firestopping material to adequately protect the framing members. Cracked flue liners and creosote buildup from lack of maintenance are among the major contributors to fires originating from the fireplace assembly. The investigator should have a basic understanding of fireplace construction in order to perform a thorough investigation. See Exhibit 5.4 for a list of some of the problems associated with chimneys.

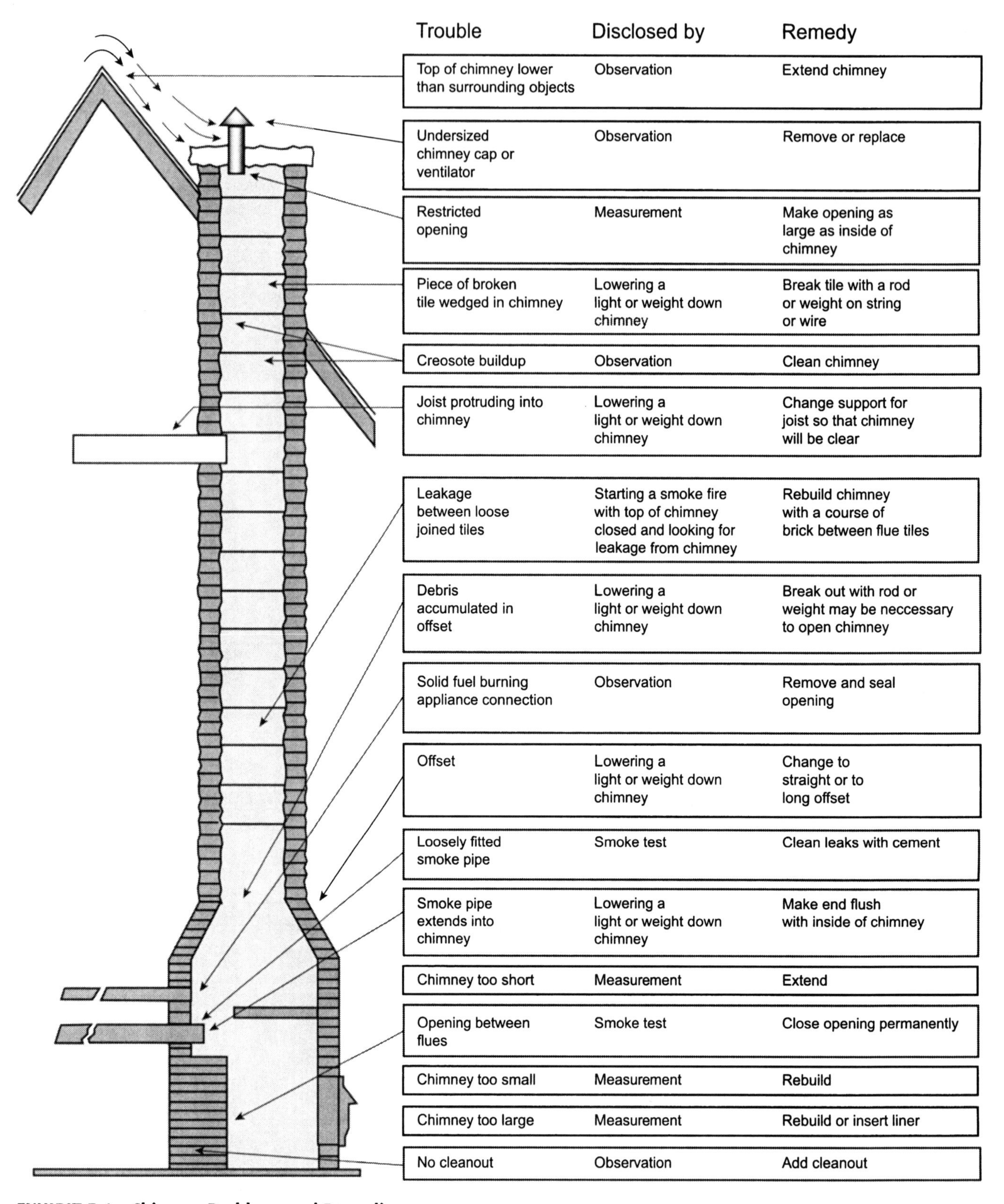

Trouble	Disclosed by	Remedy
Top of chimney lower than surrounding objects	Observation	Extend chimney
Undersized chimney cap or ventilator	Observation	Remove or replace
Restricted opening	Measurement	Make opening as large as inside of chimney
Piece of broken tile wedged in chimney	Lowering a light or weight down chimney	Break tile with a rod or weight on string or wire
Creosote buildup	Observation	Clean chimney
Joist protruding into chimney	Lowering a light or weight down chimney	Change support for joist so that chimney will be clear
Leakage between loose joined tiles	Starting a smoke fire with top of chimney closed and looking for leakage from chimney	Rebuild chimney with a course of brick between flue tiles
Debris accumulated in offset	Lowering a light or weight down chimney	Break out with rod or weight may be neccessary to open chimney
Solid fuel burning appliance connection	Observation	Remove and seal opening
Offset	Lowering a light or weight down chimney	Change to straight or to long offset
Loosely fitted smoke pipe	Smoke test	Clean leaks with cement
Smoke pipe extends into chimney	Lowering a light or weight down chimney	Make end flush with inside of chimney
Chimney too short	Measurement	Extend
Opening between flues	Smoke test	Close opening permanently
Chimney too small	Measurement	Rebuild
Chimney too large	Measurement	Rebuild or insert liner
No cleanout	Observation	Add cleanout

EXHIBIT 5.4 Chimney Problems and Remedies

◗◗◗ QUESTIONS FOR CHAPTER 5

1. What is the goal of compartmentation as it applies to building design?
   - **A.** Containing a fire in an enclosed storage space
   - **B.** Limiting a fire to its area of origin
   - **C.** Restricting ceiling height for large enclosed areas
   - **D.** Ensuring the erection of fire walls
2. Define the following terms:
   - **A.** Live load
   - **B.** Dead load
3. Define the following terms:
   - **A.** Nominal lumber
   - **B.** Dimensional lumber
4. Wood frame construction is generally associated with what type of facility?
   - **A.** Warehouse
   - **B.** Residential or lightweight commercial construction
   - **C.** Office buildings
   - **D.** Manufactured homes
5. Which of the following descriptions is not true about platform frame construction?
   - **A.** Is the most common construction type
   - **B.** Creates barriers to limit vertical fire spread
   - **C.** Does not have any concealed spaces
   - **D.** Places walls on top of platforms
6. Which of the following descriptions is not true about a wood I-beam?
   - **A.** Has smaller dimension than traditional lumber floor joist
   - **B.** Has openings in the web for utilities that reduce the structural integrity
   - **C.** May fail sooner than traditional lumber joist if exposed to fire
   - **D.** Does not need to be protected by gypsum board
7. Which description is not true regarding heavy timber construction?
   - **A.** Smallest dimension of the wood timbers is 6 or 8 inches.
   - **B.** No concealed spaces are permitted.
   - **C.** Bearing walls are required to have a minimum 1-hour rating.
   - **D.** Structural members are unprotected wood.
8. True or false: It is acceptable for laminated wood beams to be used on the exterior as well as the interior of a building.
   - **A.** True
   - **B.** False
9. Which of the following descriptions is not true of wood trusses?
   - **A.** Members are fastened together with nails, staples, or wooden or metal gusset plates.
   - **B.** Gusset plates can fail before the wooden members when exposed to fire.
   - **C.** Failure of one truss does not affect adjacent trusses.
   - **D.** Their design is similar to other truss designs.

10. Which of the following descriptions is true of floor assemblies?

A. Live loads such as water from fire-fighting operations have no effect.

B. Penetrations for utilities are not required to be firestopped (sealed).

C. Floor assemblies are tested for fire spread from above and not from below.

D. Floor assemblies are tested for fire spread from below and not from above.

11. Which of the following descriptions is true of door assemblies?

A. Doors are not a factor in limiting flame spread.

B. Fire doors are required to be equipped with hinges and latches, at a minimum.

C. A fire-rated door must be used when installed in a fire-rated wall assembly.

D. Fire doors may be constructed of steel only.

12. Which of the following descriptions is true of wall assemblies?

A. Non–fire-rated walls do not provide any level of fire resistance.

B. Penetrations in fire-rated assemblies are not required to be sealed in a fire-rated assembly.

C. A fire wall must be a load-bearing wall.

D. Penetrations in a fire-rated wall must be sealed in a fire-rated assembly.

13. Which of the following descriptions is not true of concrete or masonry construction?

A. Low thermal conductivity

B. Strong in compression loads

C. Strong in tension loads

D. Failure often occurs at connection points

Electricity and Fire

▸▸▸ OBJECTIVES

Upon completion of Chapter 6, the user will be able to

- Define 120/240-volt, single-phase electrical systems
- Describe the systems most common in residential and commercial structures
- Identify characteristics of electrical equipment to determine whether damage is the result of electrical activity or a result of the fire

This chapter is based on Chapter 8 of NFPA 921 and provides an introduction to basic electricity and electrical systems. Knowledge of electricity and electrical systems is necessary to identify whether observed damage is the result of electrical activity or fire. If the investigator is not qualified to perform an analysis of the electrical equipment, a qualified individual should be contacted to assist. A word of caution: Before analyzing an electrical circuit, the investigator must ensure that the power to the system has been disconnected.

ACTIVITY

Read NFPA 921, Sections 8.1 through 8.3 ("Introduction," "Basic Electricity," and "Building Electrical Systems"). ◂◂◂

BASIC ELECTRICITY

Parallels can be drawn between hydraulics (Exhibit 6.1) and electricity (Exhibit 6.2) to explain basic electricity. The components shown in Exhibits 6.1 and 6.2 are also explained in Table 6.1.

Hydraulic and electrical systems are not perfectly parallel, of course. First of all, when making the comparison, it is important to envision the hydraulic system as a closed system that recirculates water back to the source, not an open system that discharges water, such as a fire hose. NFPA 921's section on basic electricity contains graphics to help explain simple electrical circuits and electricity. Table 6.1 compares a hydraulic system to an electrical system.

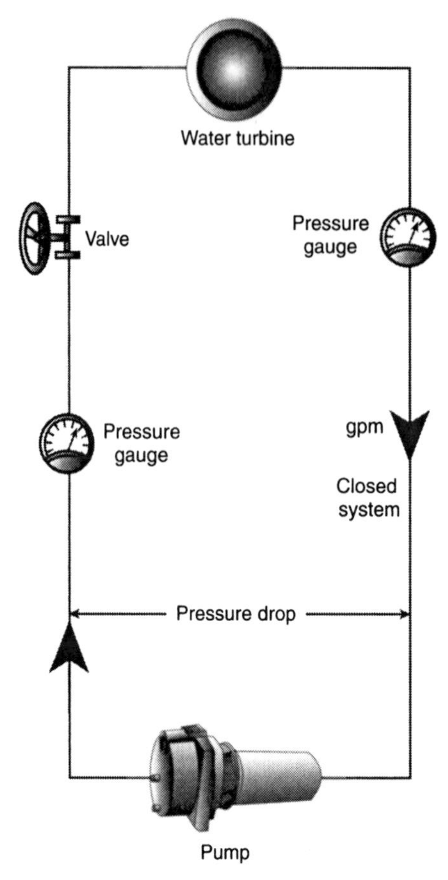

EXHIBIT 6.1
Closed Hydraulic System

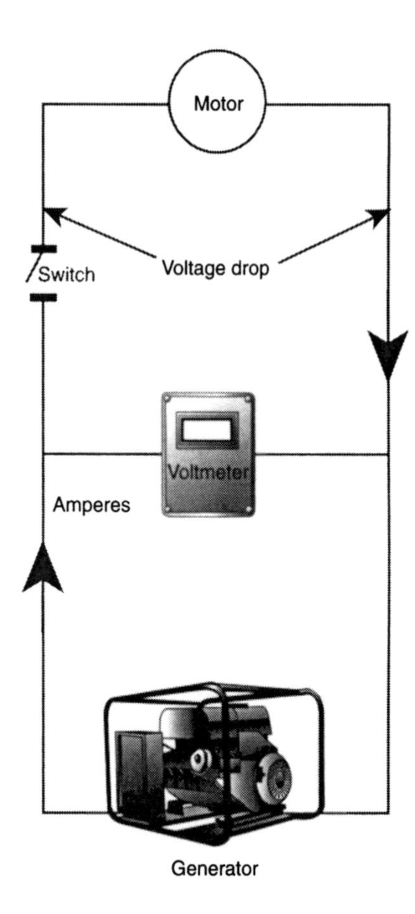

EXHIBIT 6.2
Electrical Flow

Formula Used to Calculate Electrical Values

Ohm's Law. A basic law of electricity is called *Ohm's Law.* It defines the relationship between voltage, current, and resistance. If two of these three values are known, it is possible to determine the third. NFPA 921 introduces Ohm's Law and the Ohm's Law Wheel. The values in Ohm's Law are shown in Table 6.2. Note that voltage is often represented in equations by the letter "E" but abbreviated by the letter "V." Likewise, current appears as "I" in equations, but current in amperes is abbreviated as "A."

Ohm's Law is

$$\text{Voltage} = \text{current} \times \text{resistance}\ (E = I \times R) \qquad \text{Volts} = \text{amperes (amps)} \times \text{ohms}$$

By using this law, it is possible to rearrange the values to determine current and resistance, as follows:

$$\text{Current} = \frac{\text{voltage}}{\text{resistance}} \quad \left(I = \frac{E}{R}\right) \qquad \text{Amps} = \frac{\text{volts}}{\text{ohms}}$$

$$\text{Resistance} = \frac{\text{voltage}}{\text{current}} \quad \left(R = \frac{E}{I}\right) \qquad \text{Ohms} = \frac{\text{volts}}{\text{amps}}$$

TABLE 6.1 Hydraulic System Comparison to Electrical System

Hydraulics	*Electrical*
Pump Creates the force that moves the water	**Generator/Battery** Creates the force that moves the electrons
Pressure Measured in pounds per square inch (psi) Measured by a pressure gauge	**Voltage (*E*)** Measured in volts (*V*) Measured by a voltmeter
Water Moves through the pipes and does the work	**Electrons** Move through the wires and do the work
Flow Measured in gallons per minute (gpm) Measured by a flowmeter	**Current (*I*)** Measured in amperes or amps (*A*) Measured by an ammeter
Valve Controls the flow of the water: • Open (water flowing) • Closed (no water flowing)	**Switch** Controls the flow of the electricity: • Off (open circuit; no electricity flowing) • On (closed circuit; electriticy flowing)
Friction The resistance of the pipe or hose to the water moving through it Measured in pounds per square inch	**Resistance (*R*)** The opposition of the wire to the electrons moving through it Measured in ohms (Ω)
Friction Loss Amount of pressure lost between two points in a pipe layout	**Voltage Drop** Amount of voltage drop between two points in a circuit
Pipe or Hose Size Measured in inches, inside diameter: • Larger pipe = greater flow • Smaller pipe = lower flow	**Conductor Size** Given by AWG or wire gauge size: • Larger wire = greater current • Smaller wire = lower current

One of the most useful formulas in Ohm's Law is the measurement of resistance. Once the investigator knows the resistance, he or she can determine the current for that particular appliance. Resistance is measured with a special meter, by taking a reading across the two spades of the male plug of an appliance. The result is the resistance of that appliance in ohms. Then the investigator can

TABLE 6.2 Ohm's Law Values

Value	*Symbol*	*Units*
Voltage	*E*	Volts
Current	*I*	Amperes (amps)
Resistance	*R*	Ohms
Power	*P*	Watts

Note the difference between symbols and units as well as the relationship between the unit values when analyzing electrical circuits.

divide the voltage by the ohms as indicated in the formula above to determine the amps or current used by that particular appliance.

The investigator can determine the current of all the appliances on a particular circuit by using this method. The investigator can also determine what the resistance should be for a particular appliance, as the current or amps are often printed on the appliance's nameplate. The investigator divides the voltage by this amperage to determine what the resistance should be, as indicated in Table 6.2.

Power

Electrical power refers to the rate of doing work in an electrical circuit, such as in a hair dryer, electric motor, or light bulb. Electrical power is measured in watts. A term that most people are familiar with is *horsepower,* when determining the rate of doing work for some mechanical objects.

The symbol for power is P. The formula for determining power is

$$\text{Power} = \text{voltage} \times \text{current} \qquad (P = E \times I)$$

Relationship Between Voltage, Current, Resistance, and Power

It is important for the investigator to understand the relationship between voltage, current, resistance, and power because it will help him or her to understand the potential of electricity to be a fire cause. For example, these values will allow the investigator to determine *how much* of the current or energy was used by an appliance, whether a circuit was overloaded, or whether the overcurrent protection was properly matched to the power needs of the circuit.

The investigator should be able to calculate the total *power requirements* of a circuit by inspecting the equipment, determining the power requirements of each, and then totaling all of these values. The labels on the appliances provide some information regarding the current or energy used. Often these values are given in amperage, watts, or voltage.

Resistance in an electrical fault may determine whether the fault can continue. If the fault in a circuit is of a high resistance, there will be lower amperage draw and the circuit may not trip. For example, if a fault develops across a carbon path and there is 100 ohms of resistance (amps = volts/ohms), the *current draw* for that particular fault would be 1.2 amps and the circuit protection may not trip. However, if a fault occurs where there is very little resistance, such as 0.2 ohms, there is the potential of 600 amps flowing through that fault, and the circuit protection should immediately open.

Exhibit 6.3 illustrates the relationship among voltage, current, resistance, and power.

ACTIVITY

Calculate the total current rating in amperes of the appliances on your kitchen counter. How many circuits supply these appliances?

Overload Situations

If the power needs in a circuit exceed the circuit's capacity, then an overload situation may occur.

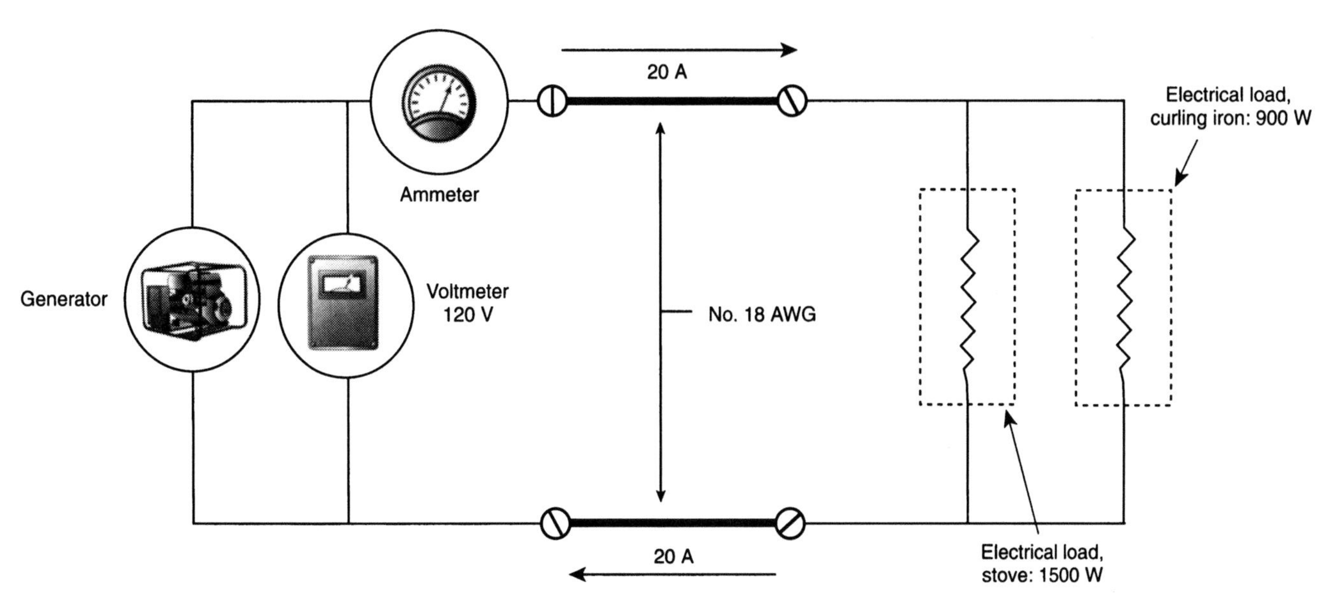

EXHIBIT 6.3 Voltage, Current, Resistance, and Power

overload

Operation of equipment in excess of normal, full-load rating or of a conductor in excess of rated ampacity, which, when it persists for a sufficient length of time, would cause damage or dangerous overheating. A fault, such as a short circuit or ground fault, is not an overload. (Source: NFPA 921, 2004 edition, Paragraphs 3.3.110 and A.3.3.110)

An overload can cause the various components in a circuit to overheat. If the heat has sufficient temperature, enough duration, and proximity to a fuel, ignition may occur. However, all three factors must be considered when attempting to determine whether an overloaded circuit was the contributing factor to the fire. The investigator must examine the circuit and the appliances on that circuit to determine whether an overload occurred. Overloading of the wiring in a house by only a few amps will not cause overheating, as there is a safety factor in most electrical circuits. Overloading a circuit to overheating usually requires a great amount of overcurrent and time, something that circuit protection is designed to prevent. It is important to check the circuit protection if an overload is suspected.

Problems can occur in household wiring under normal loads—such as a high-resistance connection at an outlet, electrical switch, or other mechanical connection where there is enough resistance to prevent exceeding the circuit protection. This type of connection may cause heating at the connection.

Wire Gauge

The diameter of wire is commonly measured in sizes given by the American Wire Gauge (AWG). The smaller the AWG number, the larger the wire diameter. Common household wiring is 14 and 12 gauge (AWG). There are also electrical circuits in the home that use larger amounts of electrical currents and are served with larger wires. Appliances such as an electric range, dryer, or water heater, for example, are often served with 6, 8, or 10 gauge (AWG) wire.

The size of the conductor gauge is usually stamped on the insulation. The number of conductors on multiconductor cable and the insulation type are also stamped on the cable insulation. Often residential wiring will be designated "NM 12/2 with ground," which means it is nonmetallic cable, 12 AWG, and with either two conductors and a ground conductor or a hot conductor, a neutral conductor, and a ground wire.

Ampacity

Amperage is the amount of current flow measured in amperes (*amps* or *A*). Exhibit 6.2 indicates that amperage is similar to the flow of water in gallons per minute. *Ampacity* is the current, in amperes, that a conductor can carry continuously under the conditions of use without exceeding its temperature rating.

Alternating Current and Direct Current

Alternating Current. It is important to understand that most electrical current used in buildings is alternating current, in which the electrons flow out from the electrical source, such as the electrical panel, and then flow back into the electrical source—thus the term *alternating current* (ac). The current flowing out from the source and then back is considered one cycle. One cycle includes both a positive and a negative component, going from 120 V positive to 120 V negative during the cycle. There is no current flowing, or zero voltage, in some part of that cycle. This change in voltage over the cycle is why it is often difficult to sustain an arc with ordinary household alternating current.

Alternating current flows from the source to ground and then reverses back to the source. In the United States, the frequency—measured in hertz (Hz) or cycles per second—is 60 Hz (60 cycles per second). The alternating current follows a path that can be described as a *sine wave.* (See Exhibit 6.4.) The ac voltage changes rapidly from positive to negative.

In the early days of electricity usage, when ac systems were just beginning to become prominent, engineers devised a mathematical computation to equate the voltage level of an ac system to that of a more familiar direct current (dc) system. The name of the computation is *root mean squared* (RMS). For an ac voltage source with a frequency of 60 Hz, the formula works out to the peak voltage divided by the square root of 2 (or 1.707), as follows:

$$V_{\text{rms}} = \frac{V_{\text{peak}}}{\sqrt{2}}$$

The common 120-volt ac system in most homes has a peak of about 170 volts. For unusual voltage signals that contain frequencies in addition to 60 Hz, the computation is rather complicated.

Direct Current. *Direct current* (dc) flows from the source to ground only. This type of current is found in a variety of installations and devices, including motor vehicles, appliances, and control systems. The voltage is mainly constant, so the direct current has no voltage sine wave. The measure of voltage to ground is a straight line.

Read NFPA 921, Section 8.4, "Service Equipment," through Section 8.8, "Outlets and Devices." ◀◀◀

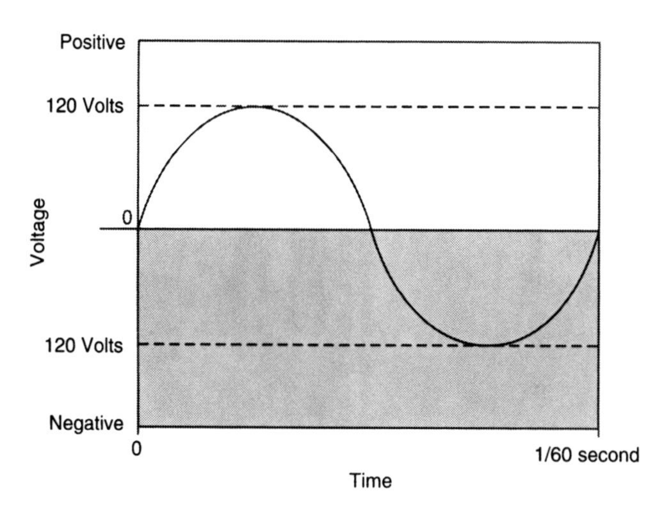

EXHIBIT 6.4
Alternating Current Sine Wave of One Cycle

Review several residential electrical systems while accompanied by the local electrical inspector or an electrician. It may be most beneficial to inspect these systems while the home is under construction, so that the electrical components can be examined before they are covered by the finish materials and electrical plates. ◀◀◀

Single-Phase Service

A single-phase system requires three conductors: two insulated conductors and a noninsulated conductor. The insulated conductors, called *hot legs,* carry the alternating current in opposite directions (120 times per second). They go back and forth at the same time but in different directions. The noninsulated conductor is the grounded conductor. This single-phase system is frequently found in residential occupancies (single-family houses) and small commercial buildings.

A single-phase system can be delivered to the structure either overhead or underground. Wiring coming in from an overhead pole is called a *service drop.* In a triplex drop, the hot conductors may be wrapped around the neutral, as shown in Exhibit 6.5. Wiring coming in underground is called a *service lateral.* The voltage between either one of the hot conductors and the ground is 120 V. The voltage between the two hot conductors is 240 V. The three conductors in a service entrance are multistranded and of a larger gauge than is typically found in branch circuits. An example of 240 volts between the hot conductors is a sine wave, as shown in Exhibit 6.6.

Three-Phase System

In a three-phase system, electricity is supplied on three conductors. The sine wave of the three conductors is out of phase, so that only one conductor has current at its highest voltage potential while the other two conductors are at some other position in the sine wave. Three-phase systems usually require four conductors: three insulated conductors carrying the current and one bare conductor that is neutral and grounded. Three-phase systems are frequently found in industrial occupancies such as large commercial buildings or large multifamily dwellings. The voltage between the hot conductors can be 480 V, 240 V, or 208 V. The voltage between one of the hot conductors and the neutral can be 277 V, 208 V, or 120 V. A common configuration in large commercial and industrial occupancies is 277/480 V. (See Exhibit 6.7.)

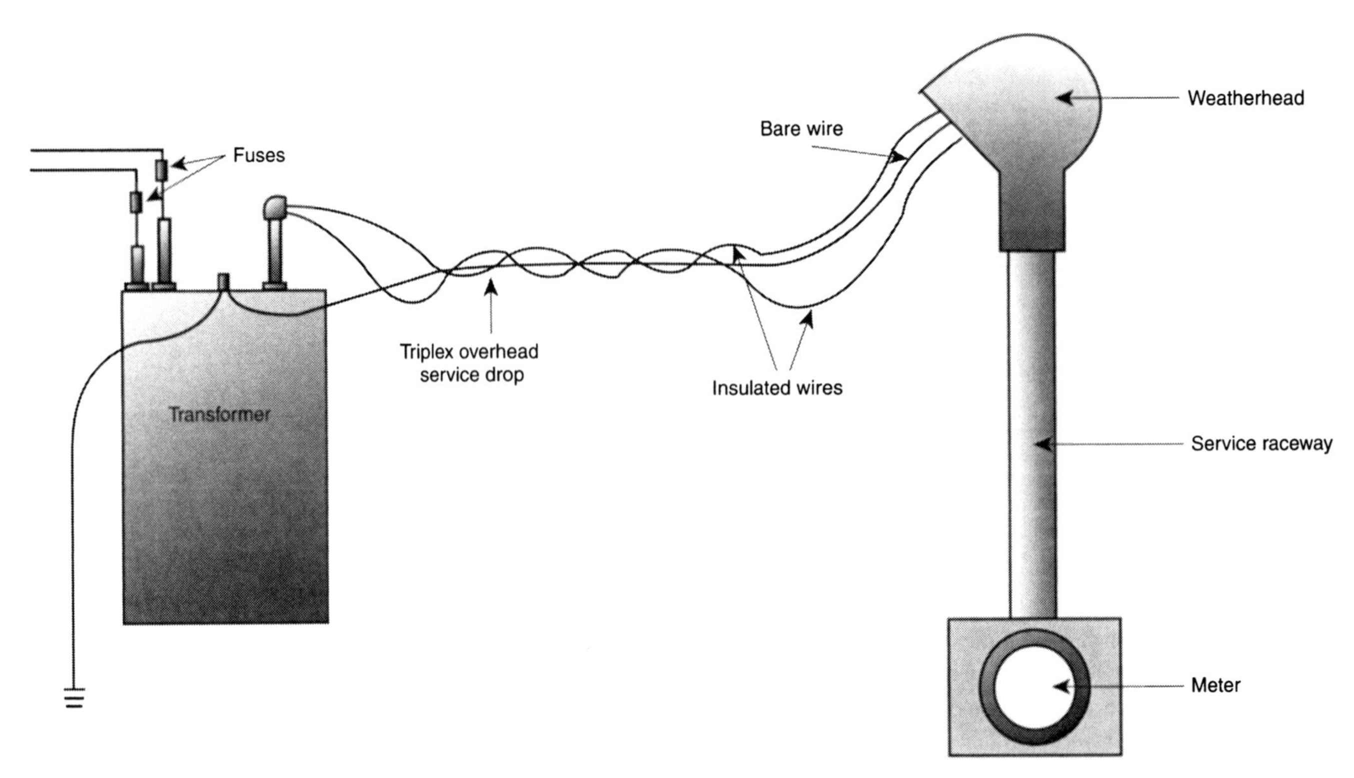

EXHIBIT 6.5 Triplex Overhead Service Drop

In large occupancies, there may be more than one service entrance for electrical power. In some occupancies with very high electrical demands, the service entrance may have very high voltage (4000 V, for example), and transformers inside the occupancy will reduce, or "step down," the voltage. The voltage differential between conductors on this type of system is the maximum difference between the positive and negative phase.

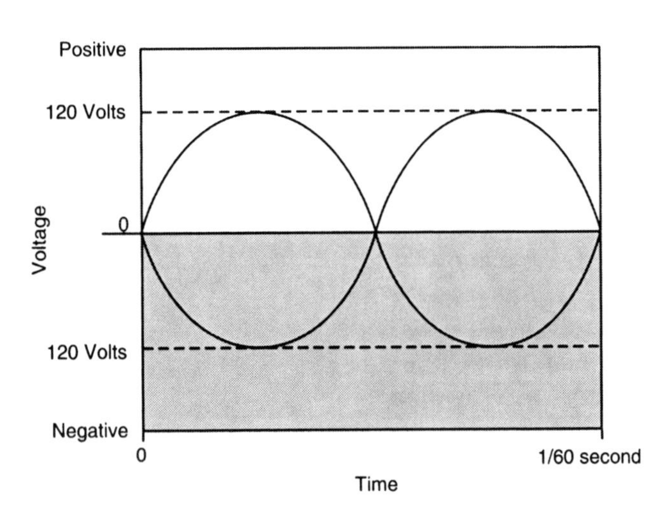

EXHIBIT 6.6
Alternating Current Sine Wave of 120-Volt Conductors

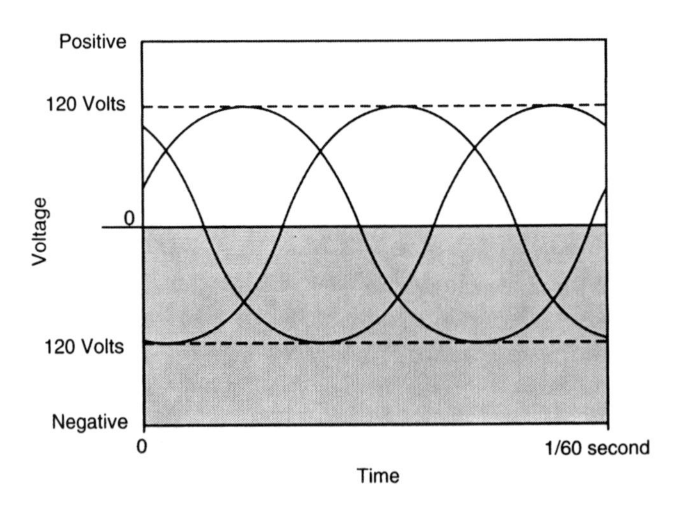

EXHIBIT 6.7
Alternating Current Sine Wave of Three Cycles

BUILDING ELECTRICAL SYSTEMS

This section describes the components of a building's electrical system. The systems described are for residential and small commercial occupancies. Larger occupancies may have more complex systems.

Service Entrance (Meter and Base)

The service entrance is the point where the electrical service enters a building. It is often the transition point between the municipal power and the private owner's electrical distribution service.

A service entrance consists of the following components:

- *Weatherhead:* The point where service entrance cables connect to the structure, which is designed to keep water out of the conduit that carries the wires. An underground service entrance might not have a weatherhead, as the conductors enter from underground directly into the panel.
- *Meter base:* The component where the service cables come in through the weatherhead and go down the conduit to connect to an electric meter that measures the amount of electricity being used.
- *Meter:* A watt-hour meter that plugs into the meter base to measure the flow of electricity.

NFPA 921 provides figures of these three components and an overhead service entrance.

The service equipment is most often located close to where the cables enter the structure. The NFPA's *National Electrical Code*® (*NEC*®) provides for the maximum distance from the location of the main disconnects to the point where the service cables enter the structure.

The *service equipment* includes the main breaker and overcurrent protection (fuses and circuit breakers). The *main disconnect* provides the mechanism for shutting off power and provides protection using overcurrent protection devices. There may be more than one main disconnect in some electrical services. The main electrical distribution panel (Exhibit 6.8) also distributes the power to different circuits inside the structure (Exhibit 6.9). This may be referred to as the circuit breaker panel board, the distribution center, or the load center.

Grounding

Grounding is the mechanism for making an electrical connection between the system and the earth (ground). If a hot conductor should contact a grounded component such as the metal case of an appliance or water pipes, the unimpeded electrical current would flow to ground and the overcurrent protection devices would open, causing the electricity to cease flowing through the circuit. The investigator should be aware that electrical systems have changed over the years and that different requirements may have been in place at the time of installation. If the ground was not in place and if the hot conductor came into contact with a metal component of the system that was not grounded, such as a conduit pipe, then this metal component could become electrically charged. If something electrically conductive and connected to the ground comes into contact with this ungrounded circuit, it may become the ground path to carry the current to ground. This scenario

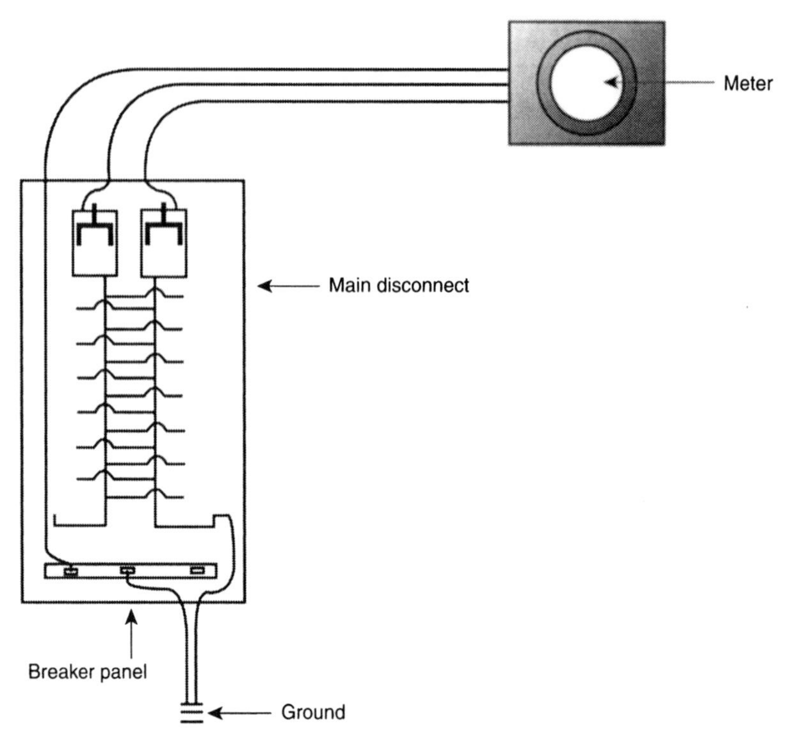

EXHIBIT 6.8 Main Electrical Distribution Panel

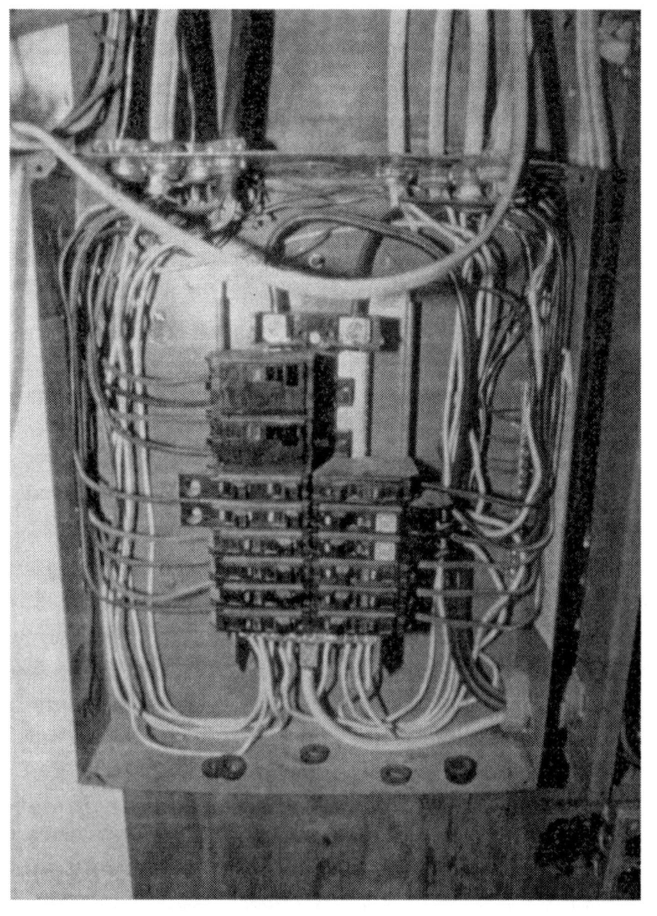

EXHIBIT 6.9 Electrical Panel

may occur when, for instance, someone touches the conduit that is not grounded and that person becomes the ground path.

Using current *NEC* requirements, there are two methods of creating a ground:

1. Connecting the overcurrent panel (circuit breaker or fuse panel) to a metal cold water pipe that extends at least 10 feet (3.048 meters) into the soil.
2. Connecting the overcurrent to a grounding electrode that may include a galvanized steel rod or pipe and/or a copper rod that is at least 8 feet (2.438 meters) long and must be driven into the soil.

Whichever method is used, there must be a secure bond between the panel and the pipe.

There must be a secure bond to the ground, which is often accomplished by connecting the grounding block in the fuse or breaker cabinet to the grounding rod or pipe with a copper or aluminum conductor and using proper connecting clamps to connect the conductors to the ground. Because electricity flows to ground and follows paths (including people) that allow that flow, it is important to have a good grounding system.

Overcurrent Protection

Overcurrent protection is a current-limiting device that may limit the damage that can occur to a circuit and the devices/appliances on the circuit. The objective is to stop the flow of electricity in the circuit when an abnormal condition is detected. The circuit protection device can be renewable or nonrenewable. Most circuit breakers are renewable, meaning that they can be used again to protect the circuit after they have been reset. The protective devices have two current ratings: regular current rating and interrupting current rating.

- *Regular current rating:* The amount of current that the circuit carries under normal conditions. Above this value, the device opens, stopping the flow of current. Common values found are 15 A (15 amps), 20 A (20 amps), 30 A (30 amps), and 50 A (50 amps).
- *Interrupting current rating:* The maximum amount of current the device is capable of interrupting.

There are several types of circuit protection devices. The most common are fuses and circuit breakers. Table 6.3 provides examples of various circuit protection devices.

Circuit breakers react to heat and overcurrent. (See Exhibit 6.10.) Whereas it may take several minutes for the circuit breaker to trip for a small increase in current flow over its rating, it reacts quickly to a large current flow over the rating of the breaker. This phenomenon is termed the *time current curve* of the breaker or fuse.

The time current curve (also known as the *characteristic trip curve*) for breakers and fuses relates to the amount of time required for a device to interrupt at a specific level of current. The more current there is, the faster it will trip. For example, *The American Electrician's Handbook* states that a typical 100-A breaker will take about 30 minutes to trip at 135 A, and 10 seconds to trip at 500 A. Ground-fault circuit interrupter (GFCI) breakers respond the same as a regular circuit breaker to overcurrent and short circuits, but they also respond to an

TABLE 6.3 Types of Circuit Protection Devices

Type	Application	Amps	Notes
Plug (also referred to as Edison)	Older application; not used in new installations	30 A (amps) or less	Possible to overfuse (put a 30 A fuse in a circuit that would require a 15 A, for example) because fuses with different ratings are all interchangeable Possible to bypass the fuse by using a penny to complete the circuit
Type S	New installations		Designed to make bypassing more difficult Can be used in older installations (Edison bases) with an adapter
Time-delay fuses	Allow short duration overcurrent May be up to six times normal current		
Cartridge fuses	Fast action or time-delay Single use or replaceable element Can be used for high-current loads (water heaters, ranges, etc.) Greater than 100 A are found in commercial or industrial occupancies	Circuits greater than 30 A	

imbalance of the current between the energized and nonenergized conductors. Arc-fault circuit interrupter (AFCI) breakers, which monitor the circuit for some abnormal conditions, are now required in many new residential occupancies.

Table 6.4 provides a list of the types of renewable circuit protection devices.

Branch Circuits

Branch circuits are the circuits that distribute the electricity from the circuit breaker or fuse panel throughout the occupancy. Each circuit should have its own overcurrent protection device (either a fuse or a circuit breaker). Table 6.5 lists the types of conductors in a typical branch circuit.

Conductors

There are many sizes of conductors. The AWG number refers to the wire diameter. The larger the AWG number, the smaller the wire diameter. For instance, 15-A circuits usually have 14 AWG diameter wire, while 20-A circuits usually have 12 AWG. Larger circuits usually have 10 or 8 AWG. The size of the conductor is related to the ampacity of the protective device. The type of insulation and bundling of conductors also affect required conductor size. Conductors may be larger than required but cannot be smaller than required. Undersized conductors have more resistance to larger current flow and therefore can generate more heat.

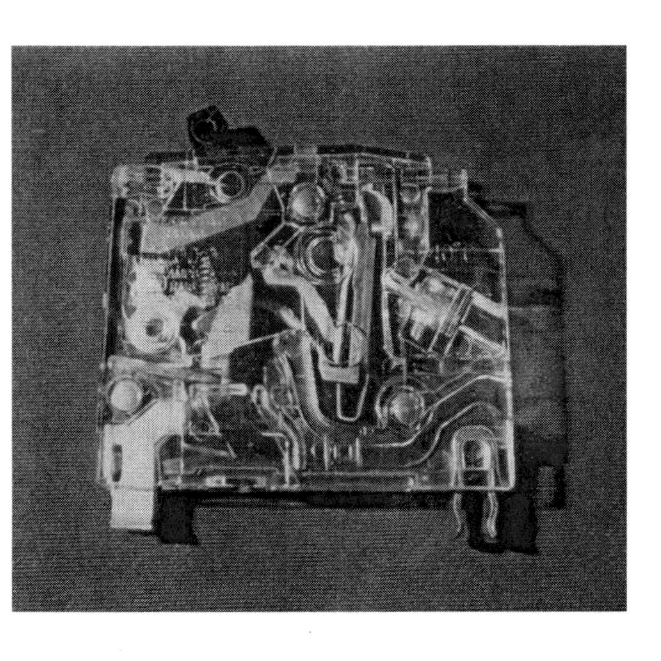

EXHIBIT 6.10
Circuit Breaker

TABLE 6.4 Types of Renewable Protection Devices

Type	*Amps*	*Notes*
Main breakers	100 A to 200 A (residential)	Interrupt the circuit at the main disconnect panel Pair of breakers Handles fastened together
Branch circuit breakers	15 A or 20 A (general lighting and receptacle circuits) 30 A, 40 A, or 50 A (ranges, water heaters, etc.)	Switch Used on the branch circuits that distribute electricity throughout the structure Rating generally imprinted on the handle Cannot be manually tripped (can be placed in the "off" position manually); can be tripped by physical impact or external heat Will trip even if the handle is locked in the "on" position May trip to the "off" position Have body of phenolic plastic Body does not melt or sustain combustion Can be destroyed by fire impingement Do not rely on visual inspection to determine the status of the breaker
Ground-fault circuit interrupter breakers (GFCI)		Monitor the electrical flow for abnormal conditions Used in locations where person might become grounded while using appliances, such as • Bathroom • Bedrooms • Kitchen • Patio/garage

TABLE 6.5 Conductors in a Typical Branch Circuit

Type of Conductor	*Function*	*Grounding Protection*	*Description*	*Notes*
Ungrounded	Hot	Attached to the protective device	Carries the current	All circuits must have this conductor.
Grounded	Neutral	Attached to the grounding block	Carries the current	All circuits must have this conductor.
Grounding	Ground	Attached to ground	Provides protection and allows the current to go to ground	Not required to make a circuit function. Grounding may be provided through metallic conduit instead of a separate conductor.

ACTIVITY 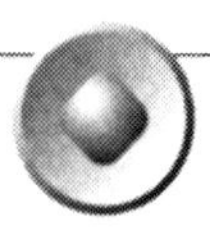

NFPA's *National Electrical Code,* Article 310, offers tables that list the amount of current that conductors may carry. Review this information with the assistance of someone knowledgeable, such as your local electrical inspector or an electrical contractor. Determine the amount of current that conductors are allowed to carry. ◀◀◀

There are three types of conductors: copper conductors, aluminum conductors, and copper-clad aluminum conductors.

Copper Conductors. Pure copper is used in copper conductors. Their melting temperature is approximately 1980°F (1080°C). However, surface melting may occur below 1980°F (1080°C) because formation of copper oxide on the surface can leave an unmelted core. When fire attacks a copper conductor, the conductor is heated in a larger area or along its length (as opposed to an electrical activity that is usually confined to a smaller defined area where the heat is focused to a small area). Because the fire attacks the conductor from the exterior, the investigator will observe surface melting and formation of copper oxide. When they melt by heat, copper conductors form pointed ends, globules, and thinned areas. In fires, bare copper conductors that have lost their insulation oxidize, and the surface is blackened with cupric oxide. The surface may additionally have no oxide or may be covered with reddish cuprous oxide. Refer to NFPA 921, Figure 8.10.1, as a guide for interpreting damage.

A copper conductor's melting temperature may be affected by "alloying," or *utetic melting,* which occurs when other metals such as aluminum or zinc come into contact with the copper and change its melting temperature, often lowering it. This frequently happens during a fire when metals drip or splash onto or come into contact with the copper conductors. The investigator can often, but not always, determine that this has occurred by noting a color change on the copper, which may be a brass color. This is not a cause of the fire but a result of the fire.

Exhibit 6.11 shows various forms of conductor damage.

Aluminum Conductors. Primarily used as service-entrance conductors on new installations, aluminum conductors may be used as branch circuits on some older

EXHIBIT 6.11
Conductor Damage

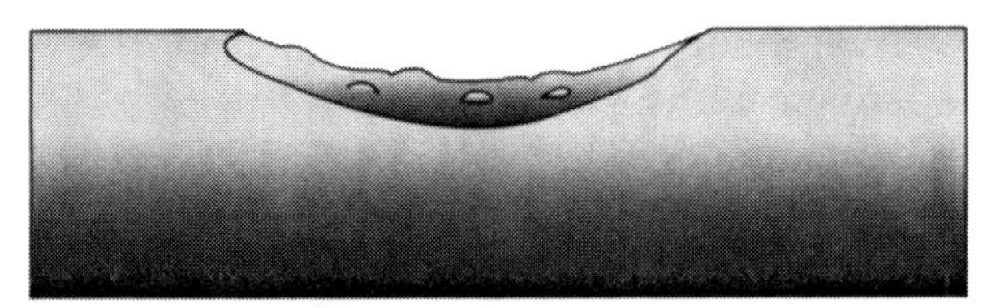

Melt from brief arcing through char or contact short at about 50–200 amperes

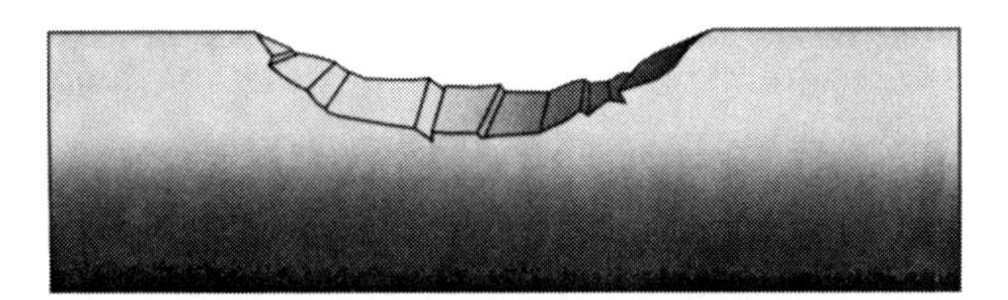

Scrape shows marks from whatever gouged it

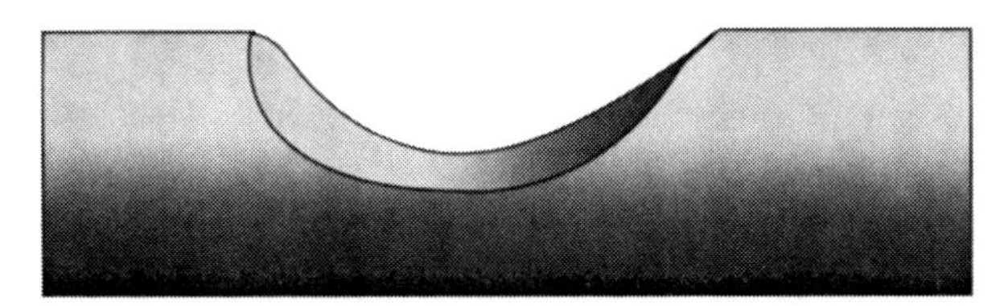

Contact short circuit melts a spot in conductor

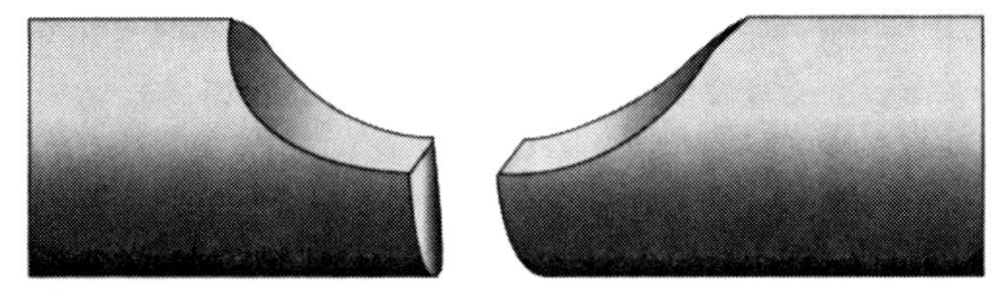

Conductor breaks easily when moved

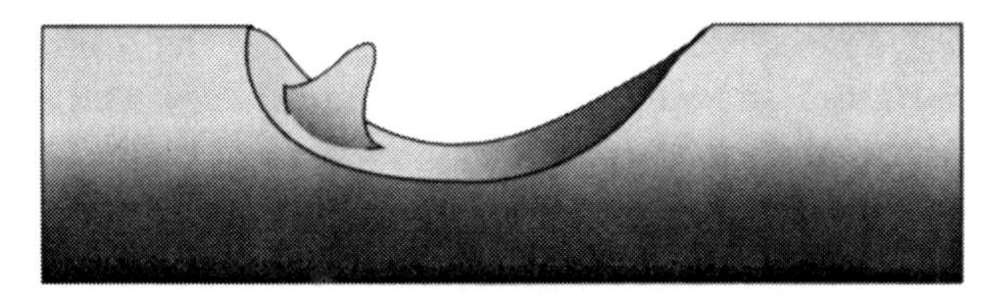

Projection of very porous copper is sometimes formed (not a bead)

installations. An aluminum conductor's conductivity is lower than copper and must be two AWG sizes larger than copper for the same ampacity (10 AWG aluminum = 12 AWG copper). Aluminum in its pure form is used in conductors. Its melting temperature is approximately 1200°F (650°C). Aluminum oxide may form on the surface, but it does not mix with the aluminum. Therefore, surface melting does not occur. Aluminum will melt all the way through the wire and can flow through the skin of aluminum oxide. The heat of the fire often melts aluminum conductors and the original electrical activity is destroyed.

Copper-Clad Aluminum. A less common type of conductor is copper-clad aluminum. As the name implies, the aluminum conductor is encapsulated in copper. Its melting characteristics are similar to aluminum.

Insulation. Insulation on the conductors is used to ensure that the current does not take unwanted paths. Insulation includes any material that can be applied readily to conductors, does not conduct electricity, and retains its properties for an extended time at elevated temperatures. Air can be an insulator for high voltages when the conductors are kept separate. An arc can occur when dust, pollution, or products of combustion contaminate air or the insulation. A code on the insulation identifies it and lists its temperature rating. The coding is listed in Table 310.13 of NFPA 70, *National Electrical Code.*

The color of the conductor indicates the function it provides: Green means it is a grounding conductor; white or light gray means it is a grounded conductor (neutral); and any color except green, white, or light gray means it is an ungrounded or hot conductor. In 120-V circuits, the ungrounded conductor (hot leg) is often black. In 240-V circuits with nonmetallic cable, the two hot legs are black and red. In certain applications, white or gray conductors could be hot when used in three-way or four-way switches and therefore could be energized. The investigator is cautioned to treat all wires as being energized until they are proven otherwise.

Polyvinyl chloride (PVC), a common insulator, can become brittle with age or heat and may give off hydrogen chloride in a fire.

Rubber was a common insulating material until the 1950s. It can become brittle with age, and once it becomes brittle, it can easily be broken off the conductor. Rubber chars when exposed to fire and leaves an ash.

Other materials used as conductors include polyethylene and related polyolefins. These materials are commonly used on large circuits in residential applications. Nylon jackets may be used to increase thermal stability.

Outlets and Devices

Circuits will terminate at or connect to switches, receptacles, or appliances. Switches are used to control the flow of current to receptacles and/or appliances and will have either screw terminals, push-in terminals, or both. Back-fed (push-in terminal) receptacles are those in which the conductor is not attached to the screw terminal but is pushed in the back of the receptacle. Because of their small surface contact area, some push-ins have created problems such as resistance heating. As was previously discussed, terminals or connection points are the most likely areas for resistance-type heating to occur because of faulty connections.

Exhibit 6.12 illustrates an outlet damaged by fire.

There are two types of 120-V receptacles: duplex receptacles on 15-A and 20-A circuits (two outlets per receptacle) and single receptacles on 30-A or greater circuits. Receptacles must now be polarized in that the blade of the cord can only be plugged into the larger opening on the receptacle, the neutral leg. Receptacles also must be grounded in new installations. They may have screw terminals, push-in terminals, or both. Hot conductors (usually black) should be attached to the brass screw. Neutral conductors (white) should be attached to the colorless screw. The green screw is for ground.

GFCI (ground-fault circuit interrupter) outlets are used in bathrooms, kitchens, or other locations. These outlets monitor the amount of current going in and out and trip if the amounts are not close to equal. They are intended to trip current flow before there is a large enough current flow to ground to hurt a person.

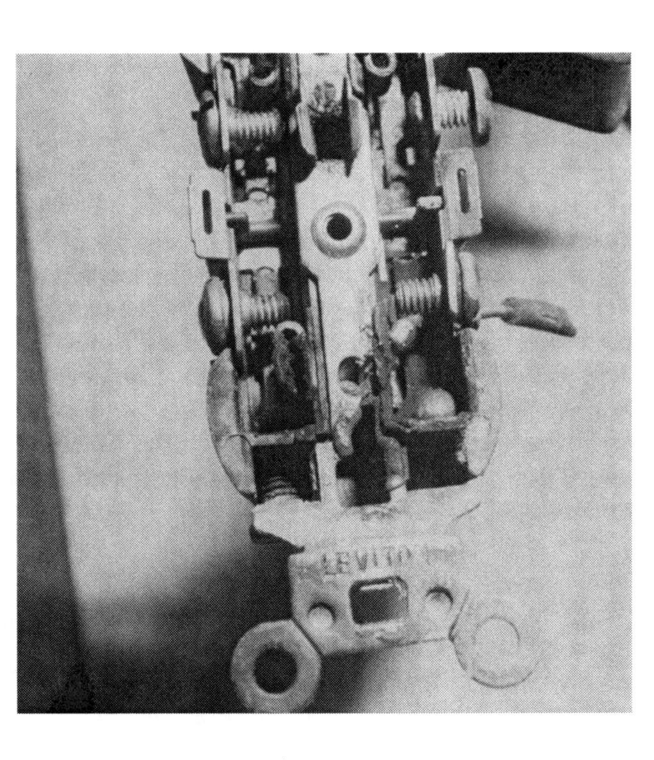

EXHIBIT 6.12
Fire Damaged Outlet

ACTIVITY

Read NFPA 921, Section 8.9, "Ignition by Electrical Energy." ◀◀◀

IGNITION BY ELECTRICAL ENERGY

The following conditions must exist for ignition from an electrical source:

- Wiring must be energized from some source (power must be on).
- Sufficient heat and temperature must be produced (enough energy to ignite the first fuel; see Chapter 16, "Fire Cause Determination").
- Combustible material must be a material that can be ignited by the energy produced by the electrical failure (to raise temperature to ignite).
- The heat source and the combustible fuel must be close enough for a sufficient period of time for the combustible material to generate combustible vapors; a simple arc does not produce enough energy to ignite most ordinary combustibles. (See Chapter 16.)

  Some methods that may generate heat include

- Resistance heating such as occurs when current flows through something that provides high resistance to current flow, such as a carbon
- Short circuit, or the current taking an unintended path, such as through part of an appliance
- Ground fault or short circuits, in which the current flows to ground instead of the intended circuit path
- Parting arcs such as occur when wires or switches are separated and have current flowing, which will create a momentary arc
- Excessive current, as when there is more current flowing than an appliance

or conductor is intended to carry and may overheat (carrying 50 A instead of 8 A)

- Light bulb overlamping and close proximity to combustibles, heaters, or cooking equipment

Electricity flows from a higher voltage to a lower voltage, such as to ground. Electricity takes any path that allows this flow. If this path is not in the intended flow (e.g., an electrical circuit), it is considered a *short circuit.* When electricity flows through most materials, including wires and conductors, heat is produced because of the inherent resistance of the material it passes through. The heat is dissipated under normal conditions and with the safety factors calculated into the size of the wire. Sometimes this heating is desirable, such as in the heating element on a stove. Other times, of course, it is not desirable, such as current flowing through an undersized conductor.

Heat-Producing Devices

Some devices are designed to generate heat. When this is the case, the design of the devices also allows for the safe dissipation of the heat. However, in cases where the heat is not allowed to dissipate and where there is a combustible fuel nearby, ignition may be possible. For example, if a cloth or piece of clothing is placed over a halogen lamp, the heat cannot safely dissipate, and the material may begin to produce combustible vapors that are then ignited.

In other cases, the device may malfunction, allowing the heating elements to become hotter than they are designed to be, causing a fire. An example could be a deep fat fryer whose controls fail, allowing the grease to heat up to its ignition temperature.

In some cases, an arsonist may deliberately use a heat-producing device to cause a fire, either by sabotaging its safety features or by placing it in proximity to a combustible fuel load.

Poor connections are another heat producer. They can allow heating at the point of the poor connection and also allow the formation of an oxide interface at the poor connection. There may also be heating caused by the resistance of the interface, called *resistance heating.* The circuit is functional, but there is increased resistance at this point. A hot spot can develop that can glow. If there are combustible fuels in proximity to this spot, they can possibly ignite. Often connections are contained within a box or appliance, which can reduce the chances of igniting a combustible. In currents of 15 to 20 A, power of up to 40 W can be produced at a poor connection with temperatures exceeding 1000°F (538°C).

Overcurrent and Overload

In a properly designed circuit, *overcurrent* is the momentary excessive flow of current through a conductor. For overcurrent to occur, either a fault must occur (such as a short circuit) or too many loads must be placed on the circuit (an excessive number of appliances, for example). Circuits can handle these types of momentary overcurrents.

overcurrent
A momentary excess current flow, such as a refrigerator motor starting up.

As discussed earlier in this chapter, *overload* is the persistent excessive flow of current through a conductor. Factors that influence the potential for fire include the magnitude of the overcurrent and the duration of the overcurrent condition. Overcurrents that are large and that persist (overload) can cause the conductor to melt the insulation or cause the surrounding material to heat. Overloads can occur in instances in which an undersized extension cord is used. The cord can be heated in excess of its temperature rating. (See Exhibit 6.13.) Overloads that cause fires are uncommon in properly sized circuits with operating protective devices. In Exhibit 6.14, the overprotection of the meter has been bypassed by a bolt and a socket tool. This would create an overload condition.

Arcs

Arcs are high-temperature discharges across a gap. The temperature of the arc can be several thousand degrees depending on current, voltage drop, and the metal involved. For the spontaneous initiation of an arc to occur across an air gap, there must be a relatively large voltage potential (above the normal 120/240-V systems). In 120/240-V systems, arcs do not occur spontaneously under normal conditions. Arcs can occur when the conductors separate while current is flowing, even in 120/240-V systems. Arcs may not be competent ignition sources. Arcing is brief because of the cycling of alternating current. The voltage difference is zero to ½ cycle. Arcing is localized. Adjacent combustible fuel that does not have a low surface-to-mass ratio and that cannot be heated sufficiently to sustain combustion (such as a wood 2 by 4, for example) will likely not ignite. If the fuel has a high surface-to-mass ratio (such as a gas or vapor), combustion may occur. Fuels that may ignite from an arc include cotton batting, dust, tissue paper, gases and vapors, and lint. Arcs are generally of short duration, and there is often insufficient energy available to ignite solid fuels under normal conditions. Note that when a conductor-to-conductor fault occurs—for example, cutting an energized extension cord

EXHIBIT 6.13 Electrical Cord Fire Damage

EXHIBIT 6.14 Overprotection Bypassed

with metal snips—there is low resistance and high current flow, causing the circuit protection to open.

High-Voltage Arcs. High-voltage arcs can occur in 120/240-V systems when there is an accidental contact between the power company's high-voltage distribution system and the system on the premises. This situation may cause a momentary arc that exceeds the safety limits of the devices or circuits on the system. Fire may occur if there is combustible fuel in the area of the arc path. Lightning can also cause high-voltage arcing in the building's electrical system.

Parting Arcs. Parting arcs are brief discharges created when the electrical path is opened while energized. Examples include opening a switch, pulling a plug, and motors with brushes. Once the conductor separates, the heating stops because the current flow has stopped. Parting arcs occur in arc welding, which is often dc voltage, so the arc continues.

Short-circuit or ground fault may cause metals to melt at the initial point of contact. An arc occurs as the metals part, and the arc is quenched immediately. It may throw sparks.

Arc Tracking. Arc tracking has been observed in high-voltage systems and has been demonstrated experimentally in 120/240-V systems. It may occur on surfaces of noncombustible materials and lead to the development of a path of electrical current through this path over time. The arc path or arc tracking will start small and may continue to grow in size. The phenomenon begins when the surfaces become contaminated with salts, conductive dusts, and liquids. This contamination causes degradation of the base material. Arc tracking can also occur when a surface is contaminated with water that has itself been contaminated with material such as dirt, dusts, salts, or mineral deposits. Tap water and fire-fighting water may contain contaminants. Often, the arcing will create sufficient heat to dry the wet path and stop the flow of current. If the water is replenished, thereby reestablishing the current, deposits of metals or corrosion can form. Arc tracking is more pronounced in dc current flows.

Sparks. Sparks are metal particles thrown out by arcs. When a high-current ground fault occurs (low resistance), hundreds or thousands of amperes may be flowing. The energy is sufficient to melt the metal and throw out sparks. Protective devices generally open almost immediately, so the event should only happen once. If the metal involved is copper or steel, the particles cool as they fly through the air. If the metal involved is aluminum, the particles may burn as they fly through the air and may ignite nearby combustibles. In branch circuits, sparks are generally not competent ignition sources for ordinary combustibles. The size of the spark particle is important because it determines the total heat content.

High-Resistance Faults

High-resistance faults are long-lived events. The current flow through the fault is not sufficient to trip the protective devices, so (at least in the initial stage) the current keeps flowing. These types of faults develop more heat, depending on how long the fault continues, and may be capable of igniting combustibles. It is difficult to find evidence of a high-resistance fault after a fire because the ensuing fire and heat destroy the evidence of the original failure. Arcing through charred insulation of conductors is a high-resistance fault that does not open the circuit protection. Multiple events with sustained arcing can occur under these conditions.

Read NFPA 921, Section 8.10, "Interpreting Damage to Electrical Systems." ◀◀◀

INTERPRETING DAMAGE TO ELECTRICAL SYSTEMS

Arc Mapping

Arc mapping is often an important part of a fire investigation, both for large-scale and small-scale events. It can be helpful in defining the general area of origin in the structure, and the investigator can use it to find the origin within an appliance. Arc mapping requires a careful examination of the electrical circuits to identify arcing that occurred within the circuit. Mapping those arcs with each of the circuits can help to identify an area that needs additional examination.

Downstream electrically from the severance point of a conductor, power is stopped, and therefore additional electrical damage cannot occur. The conductors and insulation will likely remain downstream from the severed conductors. Upstream from the first severance point, the circuit may remain energized if the circuit protection does not open, and the circuit may continue to arc upstream. On conductors with multiple arcing, the first arcing most likely occurs farthest from the power source, and additional arcing will be toward the power source. The investigator should attempt to find as much of the conductor as possible to identify first arcing. This will identify the first failure point on the conductor. The identification of the first points of failures on conductors will help to identify where the fire first attacked the circuits and possibly the area of origin. Branch circuits may also have visible holes in the conduit or metal panels.

Short Circuit and Ground Fault Parting Arcs

A short circuit will occur when current passes from an energized conductor to a grounded conductor or to a metal object that transfers the current to ground if there is almost zero resistance in the current. A surge of current results. Melting occurs at the point of contact, creating a gap and a parting arc. A parting arc melts metal only at the point of contact; adjacent areas are unmelted. Thus the investigator can reason that if the adjacent areas are melted, the fire or a subsequent event probably caused it. It may be difficult to identify the area of the initial parting arc if subsequent melting occurred.

Stranded conductors used in lamps and appliance cords may have several conductors notched or severed, or all strands may be severed and fused together. The investigator must determine where the insulation failed or was removed and how the conductors came into contact with each other.

There are many ways to interpret the damage to electrical conductors, including arcing through char, parting arcing, and so forth. Examples of this damage are shown in Table 6.6.

Arcing Through Char

Insulation exposed to fire may char. Char may be conductive and allow sporadic arcing. This arcing may leave surface melting at spots and melt through the conductor, depending on the duration and repetition of the arcing. (See Exhibit

TABLE 6.6 Types of Conductor Damage

Mode of Damage	*Effects*	*Result*	*Cause of Fire?*
Arcing through char	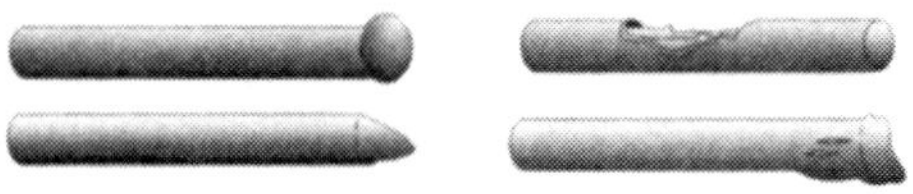	Direct fire heating	No, always a result of fire
Parting arcing	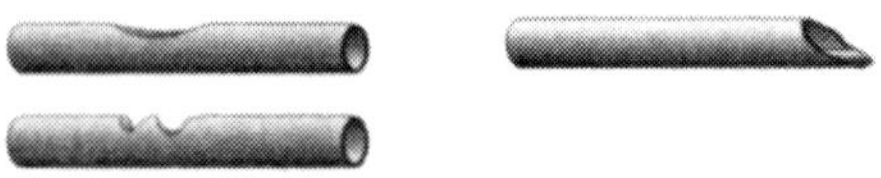	Heating at about 400°F (250°C) but no direct fire	Usually not
Overcurrent		Short circuit or failure in a device plus failure of overcurrent protection	Yes, but also may be a result of fire
Fire	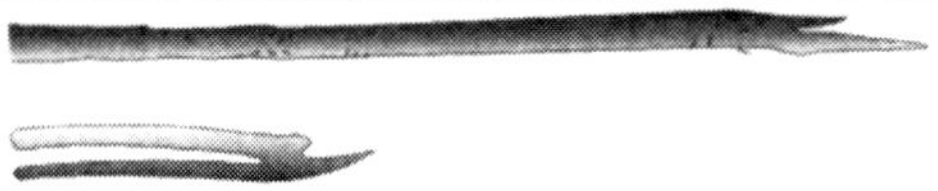	Cable exposed to existing fire	N/A
Heating connection		Connection not tight	Yes
Mechanical	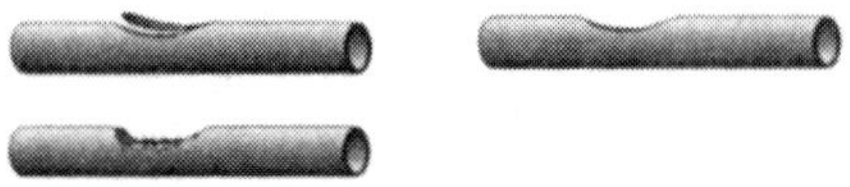	Scraping or gouging by something	No
Alloying		Melted aluminum on the wire	No

EXHIBIT 6.15 Arcing Through Char

6.15.) Often the investigator will notice multiple points of arcing. Several inches of conductor may be destroyed. Ends of conductors may be severed and have beads on the ends; the beads may weld two conductors together. In conduit, holes may be melted in the conduit at one point or along several inches. Exhibit 6.16 shows holes melted in a fluorescent ballast. Table 6.7 provides some general indicators to help determine whether the damage to the conductor is from the fire, arcing, or overload. This damage, by itself, does not necessarily indicate whether it was or was not the cause of a fire.

EXHIBIT 6.16 Holes in a Fluorescent Ballast

TABLE 6.7 Potential Indicators on Conductors

Beads	*Globules*
Localized heating Distinct line of demarcation	Nonlocalized heating, such as overload or fire melting

Service-entrance conductors have no overcurrent protection, and several feet of conductor can be partly melted or destroyed if electrical activity occurs. Arcing can continue on service-entrance conductors until the power is cut.

Overheating Connections

Poor connections are likely places for overheating. The probable causes are loose connections and/or the presence of oxides at the point of connection, creating resistance. Poor connections can often be verified by color changes at the point of the connection. At the point of the poor connection, portions of the metal connection may be found deformed or destroyed. As shown in Exhibit 6.17, contact at small spots causes heating and oxidation of metals, pitting at points of glowing, and greater oxidation of metals than at other terminals.

Overload

Overloads, as explained earlier, are overcurrents that are large enough and persistent enough to cause damage and create danger of a fire. This situation occurs when currents exceed the rated ampacity. The amount of damage depends on the degree and duration of the overcurrent. The most likely place for an overload to occur is on stranded cords such as extension cords for large-draw appliances (e.g., air conditioners, space heaters, and refrigerators). This is unlikely to happen on circuits with proper overcurrent protection. The effects of an overload cause internal heating on the conductor, which occurs along the entire length of the conductor (except where it has been thinned), and may cause sleeving. If the overloading is severe, the conductor can ignite fuels in the vicinity and melt the conductor. Once the conductor melts, the current stops flowing, and heating stops. Overcurrent melting is not necessarily indicative of ignition, however, as the fire itself may cause overcurrent to the circuit. Effects of overcurrent may be seen as sleeving of the insulation on the conductor. This overheating and evidence of sleeving are usually observed along the whole length of the circuit of the cord, as the overload occurs along the whole length of the conductor. There may be areas of limited overloading in appliances and short conductor paths.

In Exhibit 6.18, overload brings the conductor to melting. The first point to sever stops current and heating. Other neatly melted spots freeze as offsets.

EXHIBIT 6.17
Poor Connections

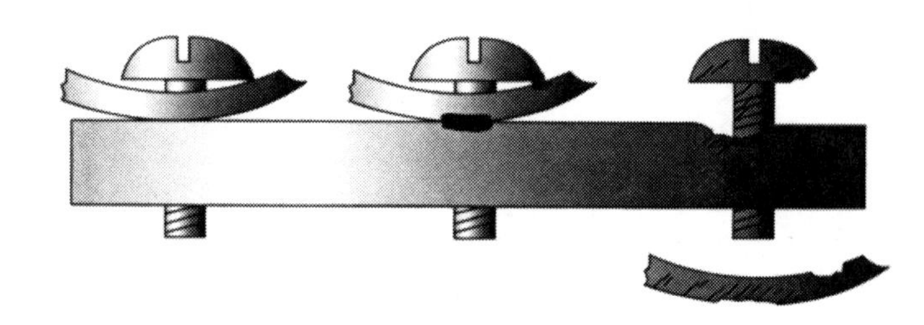

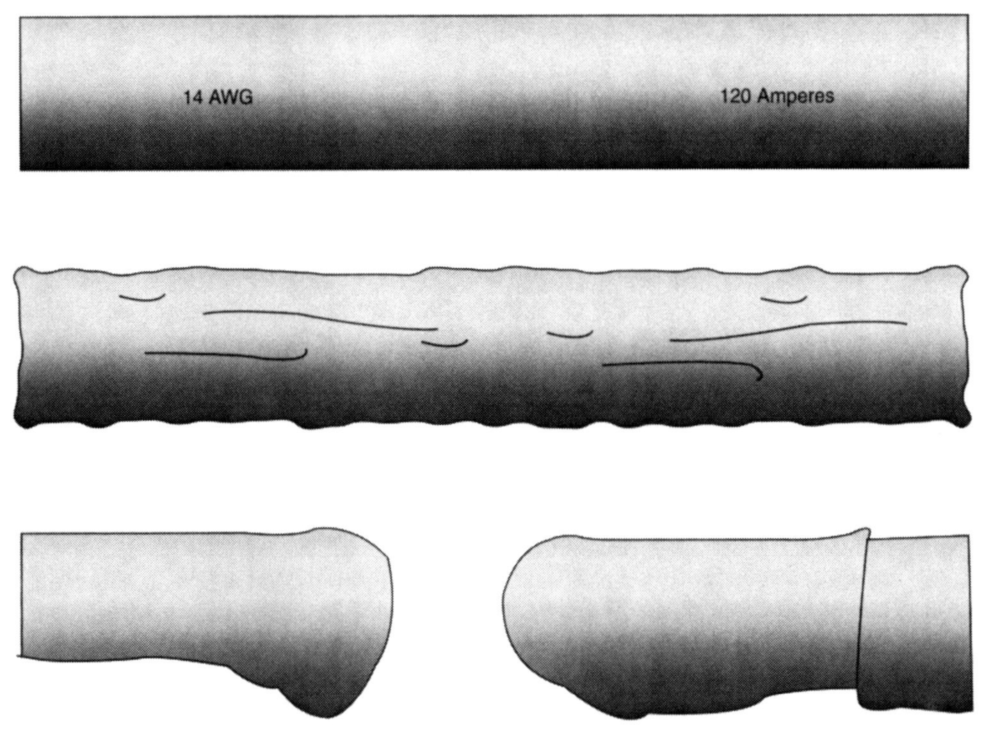

EXHIBIT 6.18 Overload

The investigator could observe some effects on wires that are not caused by electricity. These effects, which include conductor surface colors (dark red to black oxidation, green, or blue) are of no value in determining cause but are always present in fires.

Melting by Fire

Solid copper conductors first become blistered and distorted on the surface. Some copper flows on the surface, and some hanging drops form. Further melting allows thin areas to form necking and drops, and the surface of the conductor becomes smooth. Resolidified copper forms globules that are irregular in shape and size and are often tapered and may be pointed. No distinct line of demarcation is present. Electrical damage is usually very localized and defined, as the electrical energy to melt the conductor is very localized.

Stranded conductors melted by fire become stiff as they reach melting temperature. Further heating melts the individual strands together, and the surface becomes irregular, showing individual strands. Continued heating creates conditions similar to those in solid conductors. In large-gauge stranded conductors, the individual strands may fuse, separate, and have beadlike globules.

Aluminum conductors have low melting temperatures. They melt in any fire and solidify into irregular shapes. They are generally of little help in determining a fire's cause.

Alloying

Alloying is the combining of different metals of different melting temperatures, such as aluminum dripping onto the surface of a copper conductor and sticking lightly to the surface. If heating continues, the melted aluminum can penetrate the copper's surface and form an alloy. The alloy has a melting point lower than that of pure copper or pure aluminum. The investigator observes an alloy spot as either a rough gray area or shiny silvery area. Copper/aluminum alloy is brittle and may break easily. If the alloy drips off during the fire, the investigator notes a pit lined with the alloy. Chemical analysis can verify the existence of an alloy.

alloy

A solid or liquid mixture of two or more metals, or of one or more metals with certain nonmetallic elements, caused by fusing the components. The properties of alloys are often greatly different from those of the component metals. (Source: *Hawley's Condensed Chemical Dictionary*)

Mechanical Gouges

Mechanical gouges can be distinguished from arcing marks by microscopic examination. They usually show scratch marks, dents in the insulation, and deformation of the conductors. They do not show fused surfaces caused by electrical energy.

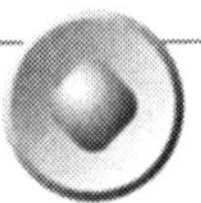

Review electrical damage of some known failures with someone who has knowledge in interpreting the damage. Try to contact a local forensic electrical expert who is working on a fire investigation and ask that person to show you examples of damage caused by fire and damage caused by electrical phenomena.

Considerations and Cautions

Investigators need to be aware of the hypotheses that have been and may be used to describe electrical failures. The following discussion provides some considerations and cautions on some of these hypotheses.

Undersized Conductors. Conductor sizing is an important safety factor in the design of circuit ampacity. However, applying 20 A to a 15-A circuit will likely not cause overheating. Likewise, overcurrent may cause increased heating but not necessarily fire ignition. The investigator must carefully evaluate circuits that have more current flow than the code allows. An undersized conductor in and of itself is not indicative of the fire cause. For example, a 20-A current flow in a 14 gauge conductor will not cause the wire to heat sufficiently to cause ignition.

Nicked or Stretched Conductors. Nicked or stretched conductors that reduce a conductor's diameter in a localized area usually cannot be said to cause heating at that location under normal load conditions. Additional heating at this reduced cross section is negligible. It is not possible to stretch copper conductors sufficiently to reduce the cross-sectional area.

Deteriorated Insulation. Insulation deteriorates with age and heating. Rubber deteriorates more quickly than thermoplastic. It may become brittle and will crack when bent. Cracks do not allow for conduction of electricity unless conductive solution should enter the cracks.

Mechanical damage and vibration also deteriorate insulation. In Exhibit 6.19, the conductor presses against a metal edge of a grounded device with no

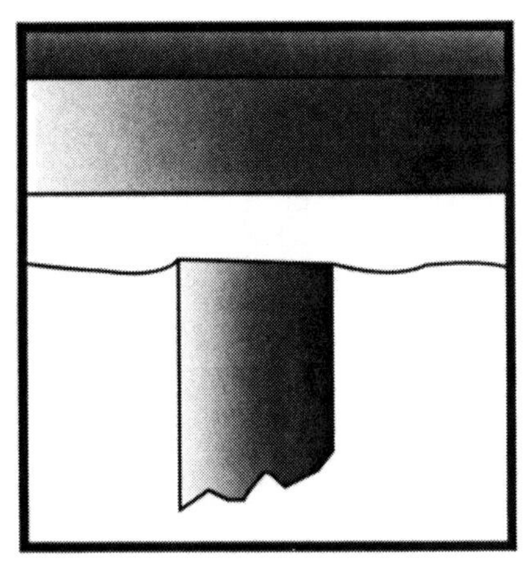
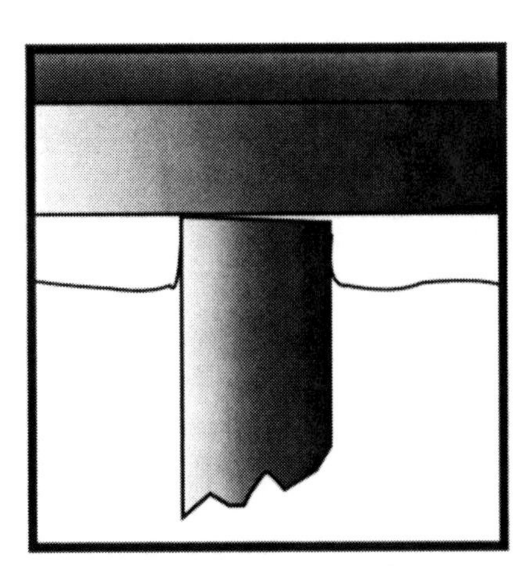
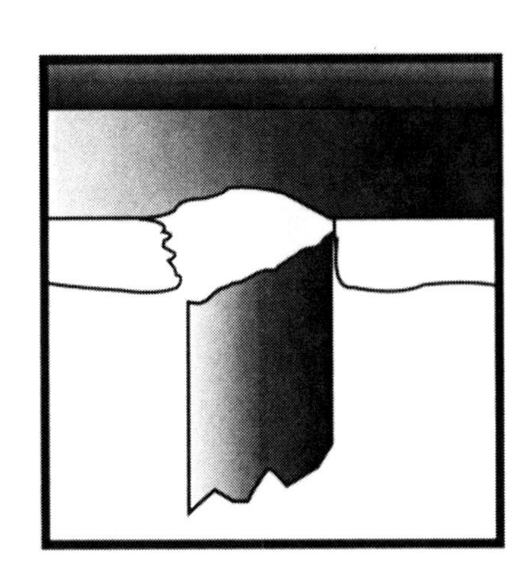

EXHIBIT 6.19 Ground Fault from Deteriorated Insulation

grommet (left), vibration lets the edge cut through insulation (middle), and a ground fault melts the spot at contact (right).

Overdriven or Misdriven Staple. Properly driven staples cannot be overdriven into the conductor. Misdriven staples at an angle can cut across several conductors, creating a short circuit. This occurrence should be evident following the fire at the point where the staple was misdriven. Overcurrent protection should stop the flow of electricity and stop any heating. A parting arc from this event would not be sufficient to ignite the insulation or supporting wood. If a leg of the staple penetrates the insulation and contacts an energized conductor and a grounded conductor, a short circuit will occur. There are some who suggest that if the staple severs an energized conductor and a high-resistance fault develops, a heating connection may be formed.

Short Circuit. Short circuits or conductor-to-conductor faults generally cannot generate sufficient heat to ignite combustibles nearby before the overcurrent protection activates. If the overcurrent protection does not function, then an overload may occur, which, along with arcing, may generate sufficient heat to ignite nearby material.

Beaded Conductors. A beaded conductor is not indicative, in and of itself, of the cause of a fire. Beading may be the result of the fire damaging the wire and then creating a failure.

Read the remainder of NFPA 921, Chapter 8, from Section 8.12, "Static Electricity," to the end of the chapter. ◀◀◀

STATIC ELECTRICITY

Static electricity is a stationary charge caused by the movement of one object in relation to another. Examples of situations that can create static electricity include walking across a carpet, conveyor belts moving over rollers, and flowing liquids.

static electricity

The electrical charging of materials through physical contact and separation and the various effects that result from the positive and negative electrical charges formed by this process. (Source: NFPA 921, 2004 edition, Paragraph 8.12.1.1)

Static electricity can be generated when two surfaces pass over each other and a sudden recombination of the separated positive and negative charges creates an electric arc. A static charge may build up. It is difficult to absolutely prevent static electricity. Conditions must be ideal for it to be a competent ignition source, and it may be a competent ignition source if the discharge occurs in an ignitible atmosphere.

Static can be generated by movement of liquids in relation to other objects and can occur in flow through pipes, mixing, pouring, pumping, spraying, filtering, and agitating. Static may accumulate in the liquid, particularly with liquid hydrocarbons. If sufficient static builds up, an arc may occur, and if there is a flammable vapor–air mixture, ignition results. Lower liquid conductivity implies a higher ability to create and hold a charge. Surface charge on the surface of the liquid is of the greatest concern.

Relaxation time is described as the amount of time for a charge to dissipate. It can range from several seconds to several minutes and is dependent on the conductivity of the liquid, the rate at which the liquid is being introduced into a tank, and the manner in which the liquid is being introduced into the tank. If the electrical potential between the liquid and the metal tank shell should become sufficient, the air may ionize, and an arc may form between the liquid surface and the tank shell. However, an arc between the surface and the shell is less likely than an arc between the surface and a projection into the tank. Projections are known as spark promoters. Bonding or grounding can remove these static charges. If the tank or container is ungrounded, an arc may occur between the tank and a nearby object.

Switch loading occurs when a liquid is introduced into a tank that held a liquid of different properties. Static discharge can ignite the vapors from the more flammable liquid.

Spraying operations can produce significant static charges on the surfaces being sprayed and on an ungrounded spraying nozzle or gun. If the material being sprayed is flammable, then ignition can occur. High-pressure spraying operations have a greater potential for generating static than low-pressure spraying operations do.

Static can build up when a flowing gas vapor is contaminated with metallic oxides, scale particles, dust, and liquid droplets or spray. When a contaminated gas is directed against an ungrounded object, a static buildup may occur. If the static charge is sufficient, then an arc may occur.

The minimum electrical charge required to ignite a dust cloud is 10 to 100 millijoules (mJ). This charge is below the amount of energy in a static arc from the human body. The human body can accumulate charges in atmospheres less than 50 percent relative humidity. The charge can be as high as several thousand volts.

Charges can build up when layers of clothing are separated, when layers of clothing are moved away from the body, or when layers of clothing are removed entirely. This is common when the layers are of dissimilar fabric. Ignition sources can occur in ignitible atmospheres from synthetic garments and from removing outer garments.

Some energy is lost in heating the electrodes. The energy required to create an arc on flat plane electrodes is 350 V for a gap of 0.1 millimeter and 4500 V for a gap of 1 millimeter. The minimum amount of energy required to be incendive is 1500 V. Dusts and fibers require 10 to 100 times more energy to ignite than gases and vapors do.

Controlling Accumulations of Static Electricity

Static charges can be removed or dissipated through humidification and by bonding and grounding.

Humidification refers to the moisture present in a material. The more moisture there is in a material, the more conductive it will be and the less likely it will accumulate static charge. The moisture content is related to the relative humidity of the surrounding air. In atmospheres with high relative humidity, that is, humidity of 50 percent or more, materials and air reach equilibrium, contain sufficient moisture to be adequately conductive, and do not have significant static electricity accumulation. In atmospheres with low relative humidity, that is, humidity of 30 percent or less, materials dry out and become good insulators, and static accumulations are likely. Conductivity of the air itself is not changed by humidity.

Bonding refers to the electrical connection of two or more conductive objects in such a way as to reduce the electrical potential differences between them. *Grounding* is the process of electrically connecting an object to ground. It reduces the electrical potential differences between objects and the earth. Objects such as pipes and tanks that are embedded in the earth are naturally bonded. The grounding and bonding should be tested to determine their effectiveness.

Conditions Necessary for Static Arc Ignition

Five conditions must be present for static arc ignition:

1. A means of static charge generation
2. A means of accumulating and maintaining the charge
3. A static electric discharge arc with sufficient energy
4. A fuel source
5. Coexistence of the arc and fuel source

Investigating Static Electric Ignitions

Investigating static electric ignitions often requires the gathering of circumstantial evidence. The five conditions listed above must exist. The investigator must determine how the static electricity was generated by identifying the materials involved and determining the materials' conductivity, motion, contact, and separation. The investigator must identify how the charge was able to accumulate through grounding, bonding, or conductivity of the material. The investigator must determine the relative humidity, identify the potential location of the static arc, determine whether the arc contained sufficient energy to be a competent ignition source, and determine whether the arc and the fuel coexisted.

Lightning

Lightning bolt characteristics include the following:

- Core energy plasma of ½ to ¾ inch (1.27 to 1.9 centimeters) diameter
- Surrounded by a 4-inch-thick (10.2-centimeter) channel of superheated ionized air

- Average 24,000 A but can exceed 200,000 A
- Potentials that range up to 15,000,000 V

Lightning often strikes the tallest object and follows it as a path into the ground. Lightning can enter a structure in four ways: striking a metal object on top of structure, striking the structure, striking a nearby tall structure and moving horizontally to the building, or striking overhead conductors and being conducted into the building via the conductors. Lightning strikes carry high electrical potentials, and extremely high heat energy and temperatures are generated. The damage may be displaced or may explode building features. The investigator may observe damage resulting from a lighting strike as damage to the structure or to the electrical system. Investigators must pay particular attention to any point where the building object may be grounded.

ACTIVITY 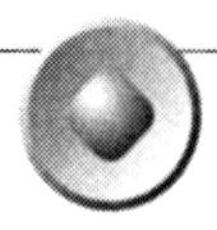

Examine and trace the wiring for a circuit breaker panel as seen in the diagram and photograph in Figure 8.6.1.1(b) in NFPA 921. Review the color code on the wiring to indicate the ground, neutral, and main connections. *Caution:* Make sure the electricity is off before doing this activity. ◀◀◀

interFIRE Training

Table 6.8 provides an interface between NFPA 921 and the interFIRE VR training program. It brings the user from the NFPA 921 section of interest to the corresponding area in interFIRE. The user can apply the interFIRE information or find more information on what NFPA 921 offers.

TABLE 6.8 NFPA 921/interFIRE VR Training

921 Section	Knowledge/Skill	*interFIRE Tutorial Student Activity*	*interFIRE Scenario Student Activity*	*interFIRE Resource Section Student Activity*	*www.interFIRE.org Student Activity*
Chapter 8—“Electricity and Fire”	*Develop the knowledge to analyze electrical systems and equipment (8.1)*	Review the following Tutorials containing information on examination and documentation of electrical systems in fire damaged buildings: • Size Up the Structure • Document the Scene • Sources of Ignition • Elimination of Sources of Ignition • Use Non–Forensic Labs and Technical Experts Activity Review NFPA 921 Chapter 8, “Electricity and Fire.” Write a summary/engage in a class discussion describing a method for systematic examination and documentation of electrical supply and utilization equipment in accordance with this chapter. Discuss the pros/cons of including a qualified electrical inspector/electrical engineer on your fire investigation team.	Using your mouse, locate the service drop on the exterior of 5 Canal Street, examine the raceway, the meter enclosure, and the service penetration for signs of electrical activity. Locate the breaker box. Examine the installation and service wiring closely using the mouse and *Shift* key to zoom in and the *Ctrl* key to zoom back. Trace wiring along the ceiling. Is there evidence of electrical faulting . . . a fire origin in the basement? Examine all rooms, closets, and corridors for evidence of fire origin. In the living room (AO?), examine each electrical fixture, appliance, and circuit for signs of electrical activity. Point the cursor to an appliance you wish to examine, then click on words highlighted (blue) to view its components. You may perform a 360-degree examination of the entire space heater and the outlet behind the couch. Examine each appliance, fixture, and service cord in this room. Did you find evidence that any were a source of ignition for this fire? Activity Locate and examine the following: • Lamp, service cord, and outlet near the staircase • Wall outlet, service cords, and lights near to the couch • The electric space heater, service cord, extension cord, and outlet • Remainders of a TV • A cabinet stereo, service cord, and outlet For each item, write out your reasons for eliminating it as a cause. If you cannot eliminate an item, state why.	Open the Resource File and click *Search.* • Type in “accidental causes” and click *Go.* Open the file *Accidental Causes: Photo catalogue of common*. Review the photos of overheated circuit and electrical arc. • Click on *Search* and type in “electrical” to review articles and abstracts in this topic.	All of the reference and photographic information available in the CD-ROM Resource Section is available online at www.interFIRE.org. Enter “Electricity and Fire” in the Search engine at the top of the Web splash page and click *Go* to access over 20 articles on the topic. Click on *Recall Alerts* to access the Consumer Product Safety Commission (CPSC) website. Click *Recalls/News* then *Find by Product Type* to search for a product category. Click on a product category that has caused a fire in your jurisdiction and see if the model responsible is listed as a recall. There are many other features on this website that provide information on electricity and fires. Explore the site. Check the website frequently for additional information on this topic.

QUESTIONS FOR CHAPTER 6

1. What unit of measurement is used to define the flow of electricity?
   - A. Volts
   - B. Current
   - C. Amperes
   - D. Ohms
2. How is the flow of electrons best described for alternating current?
   - A. One direction on the circuit
   - B. Toward ground
   - C. Back and forth in the circuit
   - D. In a conductor
3. By using the Ohm's Law Wheel, it is possible to determine the amount of current that flows in a particular item (NFPA 921, Figure 8.2.7). How many amps are in a heater that has 10 ohms of resistance in a normal 120-V circuit?
   - A. 0.083
   - B. 0.83
   - C. 1.2
   - D. 12.0
4. What is the wattage for a hair dryer that draws 8.3 A in a home with 120-V service?
   - A. 830
   - B. 1000
   - C. 1200
   - D. 1500
5. The current flowing out from an electrical panel is 12.5 A. It then flows through an electrical load of 1500 W. What is the amperage of the remaining current flow of the return circuit?
   - A. 0
   - B. 12.5
   - C. 15
   - D. 20
6. Most single-family residences in the United States have single-phase service with three wires entering the main disconnects. What is the voltage between the two hot wires of the service entrance cables?
   - A. 240 A
   - B. 60 V
   - C. 120 V
   - D. 240 V
7. The third noninsulated wire on service-entrance cable that enters a residential structure is often called the *neutral wire.* In residential wiring the neutral wire is connected to the neutral bar. To what is the neutral bar connected?
   - A. The third lug in a main disconnect panel
   - B. The load side of the hot bus bar on the disconnect panel
   - C. The earth or ground
   - D. The return hot leg on the transformer at the utility pole

8. Using American Wire Gauge (AWG) measurements, what copper wire size is used for a branch-circuit conductor for lighting and small appliances with 15-A circuits?
   A. 12
   B. 14
   C. 15
   D. 20
9. What is the minimum voltage difference across a gap that must be present for an arc to jump spontaneously?
   A. 120
   B. 240
   C. 350
   D. 1200
10. How do electrical conductors appear when exposed to arcing through char?
   A. Long lengths of the conductor melt and appear as a brassy color.
   B. The insulation on the conductor melts off the entire length.
   C. The sporadic arcing localizes melting and produces melting spots.
   D. Impossible to determine, as fire destroys this evidence.
11. What can an investigator assume if, in conducting an investigation, he or she finds a 14 AWG conductor with a 20-A circuit protection?
   A. There is a large safety factor, and this does not necessarily indicate a fire cause.
   B. A fire would occur in a short period of time.
   C. A 20-A circuit breaker cannot be physically placed in a 15-A circuit breaker connection.
   D. Although it would take an extended period of time, this would eventually cause a fire.
12. What is caused when electricity flows through a loose connection?
   A. Resistance heating through an oxide interface
   B. Short circuit to the hot leg
   C. Short circuit to ground
   D. Static electricity discharge
13. What is lightning?
   A. Static electricity flowing from cloud to ground or cloud to cloud
   B. 60-Hz current flowing from cloud to ground
   C. Current flow of 600 V going from cloud to cloud or cloud to ground
   D. An energy plasma arc between water droplets
14. What is the cause of standard copper conductors found fused together lengthwise?
   A. Alloying of materials such as aluminum and zinc with the copper
   B. Melting caused by fire
   C. Arcing through the char along the length of the conductor
   D. Electrical arcing along the length of the conductor

15. What is the result of a conductor-to-conductor short circuit on a home wiring branch circuit?

A. The short circuit can often ignite the insulation on the conductors and allow fire to propagate.

B. Circuit protection will usually operate very quickly, possibly with a short parting arc.

C. The conductors will heat along the entire length, causing the insulation to melt.

D. Retro arcing will continue until the circuit protection opens, often allowing nearby combustibles to ignite.

16. In ordinary residential construction, what might be ignited by the electrical arc of a short circuit?

A. Ordinary combustibles near the arc

B. The insulation on the conductors, which will then ignite the frame materials of the structure

C. Fuels with high surface-to-mass ratios such as gases

D. Combustible materials, caused by the heating of other metal surfaces nearby

Building Fuel Gas Systems

OBJECTIVES

Upon completion of Chapter 7, the user will be able to

- Discuss how a building's fuel gas systems influence the way a building burns
- Define the different types of fuel gases and the appliances and equipment that they utilize
- Reference NFPA 54, *National Fuel Gas Code,* 2002 edition, and NFPA 58, *Liquefied Petroleum Gas Code,* 2004 edition

This chapter is based on Chapter 9 of NFPA 921. It outlines how a building's gas supply is distributed through the building and to the appliances. It provides a basic understanding of the influence these systems may have on fire cause and spread.

Read the introduction and basic descriptions of fuel gases in Section 9.2 of NFPA 921.

FUEL GASES

A building's fuel gas systems are used to control the indoor climate, to heat water, to cook, and to provide energy for manufacturing processes.

Impact of Fuel Gases on Fire and Explosion Investigations

Fuel gas supplied to a building or to appliances can influence a fire both to ignite and accelerate the fire's spread and growth. When the fuel delivery system fails and the fuel comes into contact with an ignition source, the fuel gases can act as the source for the original ignition sequence by acting as the first fuel. The fuel gases then can act to influence the spread and growth of the fire. If fuel gases become involved in a fire, they are usually a factor in the fire's spread and growth.

Fuel Sources

Fuel involvement most frequently results from compromised fuel delivery or containment systems. The fuel usually escapes from the containers that hold the bulk fuel supply, the piping from the bulk supply to the appliance, or systems in the appliance itself. Fuel that has escaped is often referred to as *fugitive gas*. Fugitive gas readily mixes with the air and ignites easily, even by ignition sources that have very little energy. The ignition temperature of most fuel gases ranges from 723°F to 1170°F (384°C to 632°C). The minimum ignition energy for these gases can be as low as 0.2 mJ.

Ignition Sources

Pilot lights and open flames from the appliances served by fugitive gases often function as ignition sources. Ignition sources can also be from static arcs from the appliances, electrical arcs, arcs from switches or contacts in appliances, or other electrical equipment. Often it is difficult to isolate the source of ignition for fugitive fuel gases because there are so many possibilities.

Fugitive fuel gases may develop around gas appliances or equipment that is served by the fuel gas. Once these appliances become involved, the fire will ignite combustibles in the vicinity of these appliances.

Both Fuel and Ignition Sources

Fuel gases can serve as both the first fuel (source of the ignition) and the fuel for the fire. Such incidents can occur when the source of ignition is, for example, the pilot light on an appliance. The fugitive fuel then provides additional fuel for the fire to grow. The amount of energy provided by the fugitive fuel gases depends on the size of the leak or the amount of fuel that is escaping. Fugitive fuel gases can greatly accelerate the spread and growth of a fire, especially if the leak continues during the fire growth and there are combustibles in the area that will allow for the fire growth. Fuel gases that are simply burning into the air with no combustibles nearby will not accelerate fire growth.

Additional Fire Spread

Fire can spread to other areas of the building due to this fugitive fuel gas. They may appear as separate fires, but in reality, they are related to the release of the gas. This phenomenon often occurs when the gas is released for a period of time prior to ignition and is allowed to pool in other areas. It can also occur when the gas is released under pressure, spreading the fire to other areas of the building. Table 7.1 describes some of the characteristics of natural gas and propane (liquefied petroleum, or LP-Gas), the fuels that are most common in residential use.

Odorization

Odorization is often added to fuel gas. There are cases in which natural gas does have an odor directly from a well head, but LP-Gas and natural gas frequently have no odor themselves. The supply company adds it as a safety feature so that people can detect the presence of the gas. The odor must be detectable when the fuel gas is at a concentration of not less than $1/5$ of its LEL (lower explosive limit). Butyl mercaptan is most often used in natural gas. Ethyl mercaptan or thiophane is most often used in LP-Gas.

TABLE 7.1 Characteristics of Natural Gas and Propane

	Natural Gas	*Propane*
Composition	Hydrocarbon gas Primarily methane	95% propane and propylene 5% other gases
Density	Lighter than air Vapor density of 0.59 to 0.72	Heavier than air Vapor density of 1.5 to 2.0
LEL	3.9 to 4.5%	2.15%
UEL	14.5 to 15%	9.6%
Ignition temperature	900°F to 1170°F (483°C to 632°C)	920°F to 1120°F (493°C to 604°C)
Delivery	Delivered via a distribution system directly to the customer	Stored in tanks on the customer's property and delivered via tank trucks

If the investigation reveals that fuel gas may be involved, the investigator should verify the presence of an odorant. Often the investigator can smell the odorant during the investigation. However, it is often advisable to get someone who is experienced in this field to take samples to test the gas for the proper odorant.

Review Section 9.3 in NFPA 921 on natural gas systems. ❮❮❮

NATURAL GAS SYSTEMS

The systems that transport various fuel gases are similar in that they all pipe the gas directly into consumers' buildings and appliances. One common difference, however, is that natural gas is supplied from a central location via underground supply service lines. LP-Gas or other types of storage gas are contained in a bulk supply, often at the consumer's location. Natural gas is supplied through a transmission pipeline, then through distribution pipelines, and then through service mains or service laterals before it is connected to the meter at the consumer's location. Table 7.2 provides a description of the natural gas supply system and typical pressures found.

LP-Gas Systems

LP-Gas distribution systems are similar in operation to natural gas systems except that the LP-Gas storage supply is often located at the consumer's site. LP-Gas storage containers can also be housed in bulk storage locations and piped underground, similar to natural gas systems.

Read Section 9.4 in NFPA 921, which discusses LP-Gas systems. ❮❮❮

TABLE 7.2 Natural Gas Supply System from Supplier to Consumer

Type of System	*Typical Use*	*Pressure*
Transmission pipeline	Used to transport the natural gas from storage/production facilities to the local utility	1200 psi (8275 kPa)
Distribution pipelines (mains)	Used to distribute the gas in a centralized grid system	Varies; seldom exceeds 150 psi (1035 kPa); typically 60 psi (414 kPa) or less
Service lines (service laterals)	Used to connect customers to the distribution pipelines	Ranges from 4- to 10-inch water column (1.0 to 2.5 kPA)
Metering	Measures the volume of gas passing through a pipe	Depends on the consumer's needs, from 4-inch water column (2.74 kPa) to several psi

LP-Gas Storage Containers

LP-Gas is stored in containers before being transported or used by the consumer. The differences between tanks, cylinders, and containers are discussed in NFPA 58, *Liquefied Petroleum Gas Code,* 2004 edition. The following discussion provides an introduction to those systems.

Tanks. A tank is defined as a storage container with greater than 1000-lb (454 kg) water capacity or 800-lb (363 kg) LP-Gas capacity. The design and construction of LP storage tanks are governed by regulations of the American Society of Mechanical Engineers (ASME). The typical working pressure in the storage tanks varies depending on the temperature of the LP, but can range as high as 200 to 250 psi (1379 to 1724 kPa).

Cylinders. Cylinders are considered upright containers. They have a water capacity of 1000 lb (454 kg) or less. The design and construction of cylinders are governed by regulations of the U.S. Department of Transportation (DOT). The pressure in the cylinders can be the same as that in tanks or containers, ranging up to 200 to 250 psi (1379 to 1724 kPa). Cylinders are most frequently used in rural homes and businesses, mobile homes, recreational vehicles, and for outdoor barbecue grills and motor fuel.

ACTIVITY 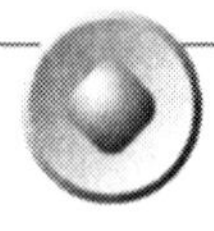

Read NFPA 921, Paragraphs 9.4.2 through 9.4.2.5, on container appurtenances. ◀◀◀

Container Appurtenances

The devices that are connected to the openings in tanks and other containers—such as pressure relief devices, control valves, and gauges—are called *container appurtenances.* Following is a discussion of these devices.

Pressure Relief Devices. Cylinders are governed by DOT regulations, and DOT cylinders are required to have pressure relief valves or fusible plugs. These pressure relief devices are designed to open at a specific pressure, usually around 250 psi (1724 kPa). The pressure relief device is generally placed in the container where it releases the vapor, although there are some exceptions. The pressure in a container is directly related to the temperature of the propane. As the temperature rises, the pressure continues to rise. NFPA 58 provides methods to determine the pressure in a container if the temperature of the propane is known.

Connections for Flow Control. Some container appurtenances control the flow of propane from the container. The most common is a shutoff valve. In some installations, there are excess-flow check valves that monitor the flow from the container; these valves activate if the flow exceeds a set amount. There are also valves called backflow valves that prevent the propane from backing up into the container.

Liquid Level Gauging Devices. Often, there are appurtenances on containers that indicate the liquid level of the propane. The types of liquid level gauging devices include fixed level gauges, which are fixed at a maximum level, and variable gauges. Types of variable liquid level gauges include float, magnetic, rotary, and slip tube.

Pressure Gauges. Pressure gauges are another type of container appurtenance, in this case depicting the internal pressure of the tank. The gauges are connected directly to the tank or sometimes through the valve. Pressure gauges do not indicate the quantity of liquid propane in the tank.

Fusible Plugs. Cylinders regulated by the DOT are required to have pressure relief valves or fusible plugs. Fusible plugs are devices that relieve at a set amount of pressure, and once activated, they cannot be reused. Aboveground storage tanks with less than 1200 lbs (544 kg) of water capacity may have fusible plugs. The melting temperature of these plugs is between 208°F (98°C) and 220°F (104°C).

Pressure Regulators. Pressure regulators are attached through the gas supply system to reduce the pressure so that it can be utilized by the appliance or utilization equipment. In some propane systems there are two regulators that reduce the propane pressure in two stages. Often, when there are two-stage regulators, one of the regulators will be located at the tank, and the other will be located near the appliances or when the fuel delivery system enters the structure. It is common to see the LP pressure reduced for use in the appliances to approximately 11-inch water column to 14-inch water column (2.74 kPa to 3.47 kPa).

The propane pressure should be reduced from tank pressure to where it can be utilized by the equipment. The tank pressure will vary depending on the temperature of the liquid. Table 7.3 compares temperature of the propane to the pressure in the tank.

TABLE 7.3 Typical Pressure Inside Propane Tank as It Relates to the Temperature of the Propane

Temperature	*Pressure*
0°F (–18°C)	28 psi (193 kPa)
70°F (21°C)	127 psi (876 kPa)
130°F (54°C)	286 psi (1972 kPa)

Vaporizers. Equipment is sometimes used to assist in vaporizing the liquid propane. Vaporizers are frequently used when there is a demand for large quantities of propane, such as for industrial uses or in cold weather environments. Vaporizers heat the liquid propane, converting it to a gas.

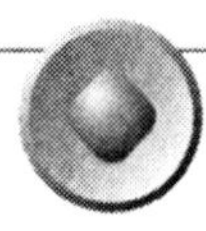

ACTIVITY

Read NFPA 921, Section 9.5, "Common Fuel Gas System Components."

COMMON FUEL GAS SYSTEM COMPONENTS

This section describes the various components of fuel gas delivery systems.

Pressure Regulation (Reduction)

The most common types of pressure regulators are the diaphragm type, which has a spring that is set to control the pressure, and the lever type. The vents on the regulators must be clear for the device to operate properly. If the vent becomes plugged or obstructed, the pressure regulator device might not function properly. In colder environments, it is important not to place regulators where water or ice accumulations can obstruct the vent openings. Table 7.4 gives a description of the general utilization pressures for appliances.

Service Piping Systems

Service piping provides for delivery of the fuel gas from the distribution lines to the user. The materials for these gas distribution mains and services are most often constructed of wrought iron, copper, brass, aluminum alloy, or plastic. Service lines are often buried underground and are regulated by NFPA 58. Underground piping should be buried deep enough to prevent physical damage and should be protected from corrosion. If the piping runs under a building or areas that are utilized by the public, it should be encased in an approved conduit.

Valves

Valves will be installed on the piping system to control the flow of the gas. These valves are often installed where the gas piping exits to ground and prior to the meter or regulator. Valves are also located prior to each appliance and in appliances. There are several types of valves used in a common fuel gas system. Some of the valves are automatic and are controlled by the appliance to be either closed, open, or partially open. Some of these automatic valves are configured to shut off automatically in an emergency condition such as fire or excess flow. Some of the

TABLE 7.4 Utilization Pressures for Appliances

Use	*Pressure*
Nonindustrial natural gas applications	4 to 10 in. w.c. (2.74 to 3.47 kPa)
Nonindustrial propane applications	11 to 14 in. w.c. (2.74 to 3.73 kPa)

common valves in appliances are individual main burner valves, main burner control valves, and manual reset valves.

Gas Burners

Most gas utilization equipment uses some type of gas burner. The gas burners and/or orifices on appliances may not be interchangeable between natural gas and propane. Ignition of the gas from a burner can occur by several methods, including manual ignition, pilot lights, and pilotless igniters. When a burner is ignited by manual ignition, a person provides the spark or flame. Gas burners may also be ignited from pilot lights, small flames in the appliance near the burners. The pilot light must be of sufficient size and close enough to ignite the gas as it escapes from the burner. In some designs, the pilot light burns constantly, and in other designs, the burner is ignited electronically. In electronically lighted piloted appliances, electric arcs ignite the pilots, which in turn ignite the burners. Pilotless burners do not have pilots but are ignited by electronic arc or resistance heating elements. Many systems stop the flow of gas in the event that the burner does not ignite.

Exhibit 7.1 shows burners and the gas control valve for a typical furnace. Exhibit 7.2 shows the pilot line and gas control valve on a water heater.

Read NFPA 921, Section 9.6, which discusses common piping in buildings. ◀◀◀

COMMON PIPING IN BUILDINGS

This section provides discussion of the common piping components in buildings and the utilization equipment for fuel gas piping systems. The size of the piping is determined by the maximum flow required by the equipment that is connected to that pipe. Common piping in the building is often similar to that described earlier in this chapter in the section on service lines. Piping material can include wrought

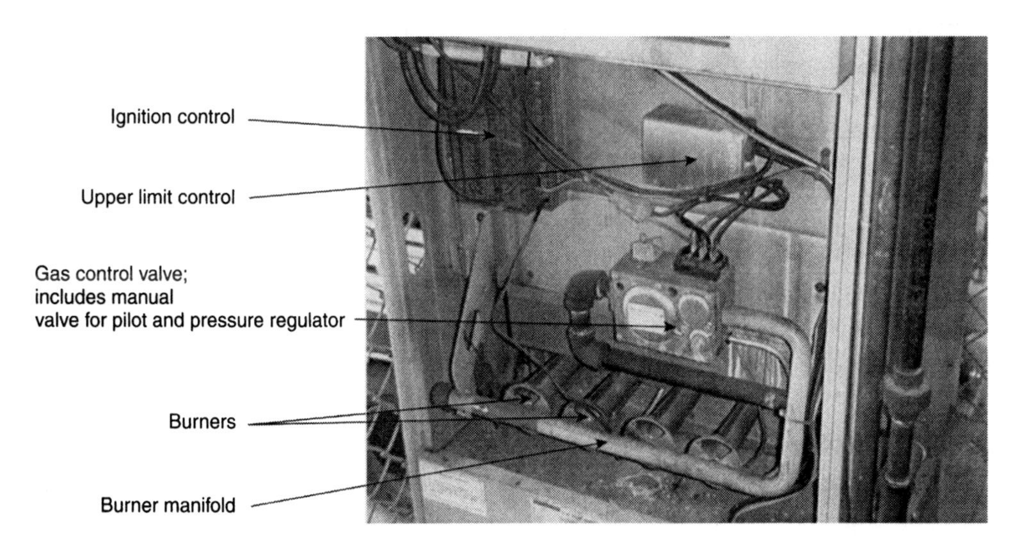

EXHIBIT 7.1 Furnace Burners and Gas Control Valve

EXHIBIT 7.2 Water Heater Pilot Line and Gas Control Valve

iron (black pipe), copper, and brass. Aluminum alloy may be used but is not suitable for underground applications. Plastic may not be used indoors, although it may be used for exterior or underground applications.

Joints and Fittings

Joints and fittings may be screwed, flanged, or welded. Screwed fittings are fittings with threads. Flanged fittings are conducted with the use of a special device that flanges the end of the pipe and requires a fitting to complete the connection, or fittings may be welded. Under some circumstances, compression fittings—which include a device that compresses the fitting to the pipe for a seal—may be used. Plastic piping, which may be used outdoors only, may not be threaded.

Piping Installation

Piping installation is regulated by NFPA 54, *National Fuel Gas Code,* 2002 edition, and NFPA 58. Piping installations should not weaken the building structure. The piping must be supported with the proper devices and not by other pipes or appliances. The piping should be equipped with drip legs where appropriate, usually near appliances to capture foreign material. All unused outlets or openings in piping should be capped, not simply controlled by a valve, to prevent gas from escaping.

Main Shutoff Valves

Main shutoff valves should be located on the exterior of the structure, although in older installations they may be located inside the building. The main shutoff valve controls the flow of gas throughout the entire building.

Prohibited Locations

As required by NFPA 54 and other recognized gas installation codes, natural gas piping cannot be run through air ducts, clothes chutes, chimneys, gas vents, ventilating ducts, dumbwaiters, or elevator shafts. Furthermore, gas piping should never be installed in areas that are subject to damage or exposed to corrosion.

Read NFPA 921, Section 9.7, which discusses appliance and equipment requirements. ◀◀◀

COMMON APPLIANCE AND EQUIPMENT REQUIREMENTS

This section discusses the common requirements for the basic installation of appliances where there is relatively low gas pressure. The installation should be conducted in accordance with NFPA 54. The appliances and accessories should also be approved as acceptable by the authority having jurisdiction.

The appliance must be compatible with the type of gas it is using. When appliances are designed for natural gas and are used with propane or when appliances are designed for propane and are used with natural gas, the results can be catastrophic, including fire or explosion.

Appliances should be placed where they will not be subject to damage. They should also not be placed where flammable vapors may accumulate because the appliance may ignite the gas vapors. In certain conditions, gas appliances can be installed in locations such as an automobile garage if they are located above floor level.

Appliance Installation

Where the system pressure is greater than the appliance is designed to handle, an additional regulator should be installed. (This situation was described earlier in this chapter in the discussion of multiple regulators.)

A gas appliance should be located where there is easy access for service or shutdown. The location should also have sufficient clearance between the appliance and combustibles, including building construction, to ensure that the heat of the appliance does not ignite the combustible. The minimum required clearance is often displayed on the appliance itself.

The electrical connections to the appliances should be in accordance with NFPA 70, *National Electrical Code*®, 2005 edition.

Venting and Air Supply

Exhaust venting is required to prevent buildup of products of combustion inside the building. Codes require that fuel-burning appliances be vented. Some equipment, however, does not require venting—including ranges, ovens, and small space heaters—because if they are used properly, they will not produce products of combustion. However, if these appliances are used improperly, they can allow products of combustion to build up enough to cause harm.

venting
The removal of combustion products as well as process fumes (e.g., flue gases) to the outer air.

Fresh air supply or combustion air supply venting from the exterior of the building is often necessary for proper operation of gas-fired appliances. If the volume of the room is sufficient, sometimes the installation of combustion air is not required. Many new appliances are designed so that the combustion air directly enters the appliance.

 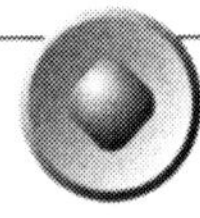

ACTIVITY Read Section 9.8 in NFPA 921, which discusses fuel gas utilization equipment.

COMMON FUEL GAS UTILIZATION EQUIPMENT

NFPA 921 provides a list of common fuel gas utilization equipment and a description of that equipment. The investigator may encounter other types of gas utilization equipment, but all have similar features in their requirement to regulate the gas, control the flow of the gas, and burn the gas. The most common forms of gas utilization equipment include air heating, water heating, cooking, engines, illumination, and incinerators.

Several different types of appliances use fuel gas in their operation. These appliances include water heaters, furnaces, clothes dryers, and ranges. (See Exhibits 7.3 through 7.5.) The investigator should be familiar with the design features and operation of these four common fuel gas appliances. (Also see Chapter 22, "Appliances.")

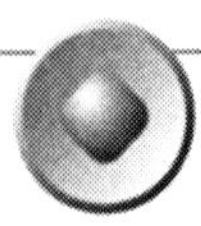

ACTIVITY Read Section 9.9 in NFPA 921, "Investigating Fuel Gas Systems."

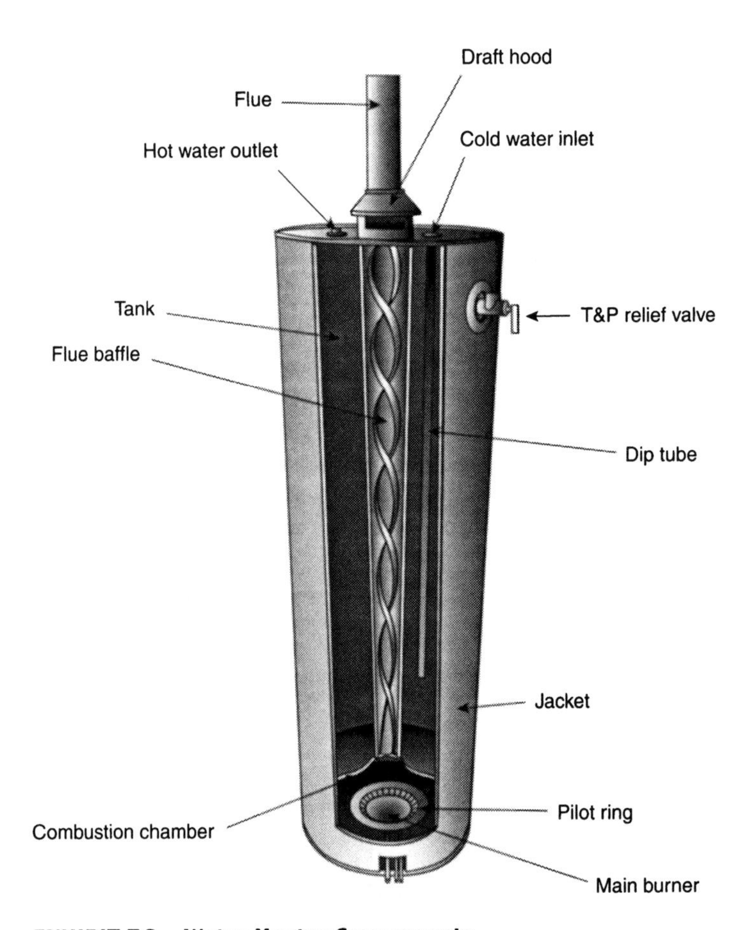

EXHIBIT 7.3 Water Heater Components

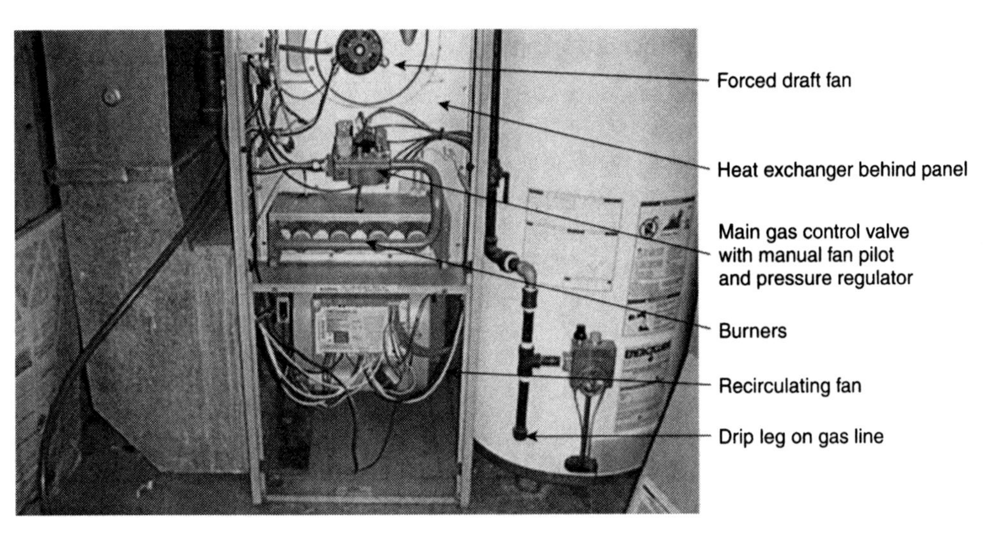

EXHIBIT 7.4 Furnace Components

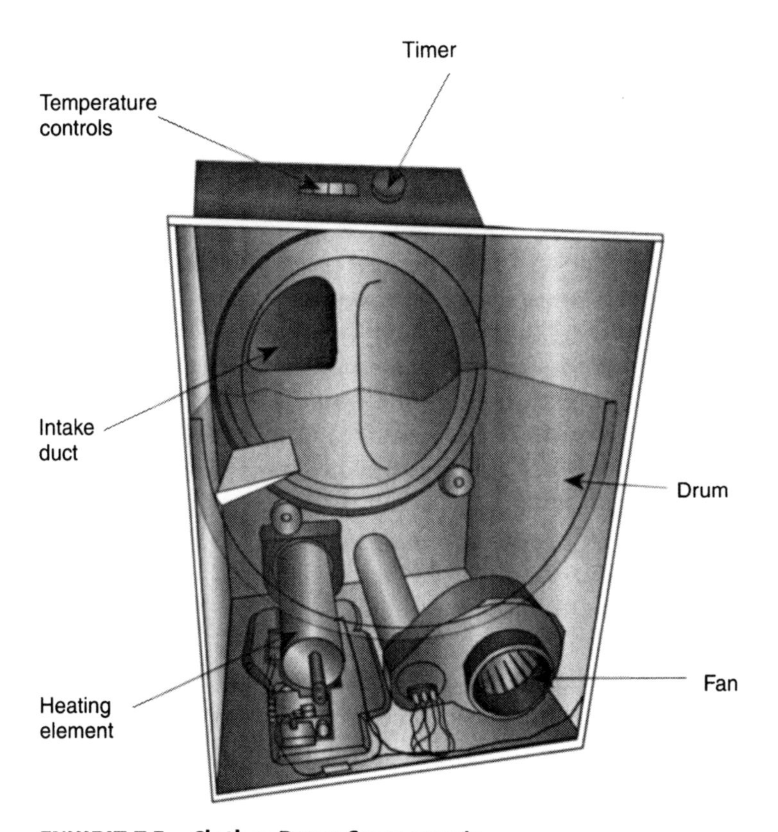

EXHIBIT 7.5 Clothes Dryer Components

INVESTIGATING FUEL GAS SYSTEMS

This section provides a description of how to conduct an analysis of a building's fuel gas system. Analysis of the fuel gas system can provide information if it was involved in the origin or cause of the explosion or fire. As with all parts of a fire investigation, the analysis of the fuel gas system should be done in a systematic manner to ensure thoroughness. NFPA 921 provides guidance in conducting the investigation.

Compliance with Codes and Standards

The investigator should evaluate the design and installation of the equipment and system to determine compliance with accepted codes and standards, calling on additional expertise or resources if necessary. There are differences in installation requirements for the various types of gas systems. The design and construction of the appliance itself are often regulated by an approval agency such as the American Gas Association. The installation manual for the appliance often provides information as to its installation requirements.

Leakage Causes

The main causes of gas-fueled fires and explosions involve leakage in the fuel gas delivery system or in the appliance itself. The most common leakage occurs at pipe junctions, unlit pilot lights or burners, uncapped pipes, malfunctioning appliances and controls, areas of corrosion in pipes, and points of physical pipe damage.

- *Pipe junctions:* Leaks can occur in pipe junctions if the threading is inadequate for a good connection, if the junction between the pipe and coupling is improperly joined or threaded, or if there is an improper use of pipe joint compound.
- *Unlit pilot lights or burners:* Most modern gas utilization equipment stops the flow of the gas to the burners if the pilot light is not lit or the burner is not lit. It is possible to have gas leakage in these systems if the sensing system malfunctions or if the automatic shutoff valve does not operate. The amount of gas that flows from an unlit pilot light is generally not sufficient to cause an explosion or fuel fires unless the gas is somehow confined to a space. Burners that have been turned on but not lit can provide sufficient gas leakage to fuel an explosion or fire. Some equipment contains a safety device that monitors the flame from the burner, and if there is no flame, it shuts off the flow of gas. This equipment can allow the flow of gas from the unlit burner, however, if there is a failure of the sensing device or a gas valve fails. It is common to see such burners without this sensing equipment, especially in cooking appliances.
- *Uncapped gas supply pipes:* Appliances may have been disconnected from the fuel gas line without a cap on the end of the gas pipe. If the shutoff valve is open, gas escapes from the uncapped pipe.
- *Malfunctioning appliances and controls:* Fugitive gas may leak around the controls, valves, fittings, and pipe junctions within the appliance control. Additional expertise or resources may be needed to analyze the equipment.
- *Gas pressure regulators:* A failure can occur within the internal diaphragm, which would allow a leak from the regulator and potentially allow excess pressure to the system. Failures can also occur on the rubberlike seals that control the input of the gas, which would allow a leak around the regulator and potentially allow excess pressure into the system. The vents to the regulator can become plugged or fail, allowing the system to become overpressurized.
- *Corrosion:* Statistics indicate that corrosion may cause as many as 30 percent of all known gas leaks. Corrosion can take place above or below ground. Generally, this type of failure takes a long time to develop. Corrosion can typically be caused by rust, electrolysis, or microbiological organisms.
- *Physical damage to the system:* The investigator should keep in mind that the point of the leak may be remote from the actual damage. The strain may

appear at pipe junctions or unions. Damage often occurs at the threaded portions of the pipe. Hidden or underground pipes are often damaged by construction.

Testing for Leakage

If the investigator suspects that fuel gas may be a factor in the fire investigation, he or she should test for leaks of the gas supply system prior to its disassembly. Expertise in conducting such evaluations is needed.

- *Pressurization of the gas supply system:* It is recommended that the gas supply system be pressurized, with the pressure not exceeding that which was contained within that system. Sometimes the pressure may be as little as 1- to 4-in. w.c. (248.8 to 996.4 kPa). Damaged sections of the piping should be isolated so that the investigator can test undamaged sections. Care should be taken not to unscrew or disassemble components before testing, as evidence can be inadvertently destroyed at the point where the leak occurred. The investigator should attempt to isolate the piping for testing in an area where the leak did not occur, and where the proper fittings can be attached.
- *Gas meter testing:* Testing can also be done with a gas meter, but only if it is safe to reintroduce gas into the system. The meter can detect the gas escaping from the system.
- *Pressure drop method:* To check gas piping, the system is pressurized with air or an inert gas. This type of test is appropriate for operating systems of 0.2 psi (3.4 kPa) or less. The system is pressurized, the air supply capped, and the pressure is monitored by a pressure gauge.
- *Soap bubble solution:* A simple method to locate leaks is the application of soap to the piping in question. If bubbling is observed, then there is a leak at that location.
- *Combustible gas indicator:* This device is used in gas detector surveys and may also detect the presence of other gases. When using this method on the exterior of the building, the investigator should test for fugitive gases at every possible ground opening, including all drains or pipes that exit the ground. In gas detector surveys performed on the interior, the investigator should test junction boxes and unions of gas piping.

combustible gas indicator

An instrument that samples air and indicates whether combustible vapors are present. Some units may indicate the percentage of the lower explosive limit of the air/gas mixture. (Source: NFPA 921, 2004 edition, Paragraphs 3.3.29 and A.3.3.29)

If the investigator suspects an underground gas leak, an appropriate procedure is the use of bar holes. After a series of underground holes is drilled in the vicinity of the underground piping, readings are taken at each hole and graphed or charted. This type of survey can potentially lead to the area of the leak.

Exterior surveys also include vegetation surveys. Vegetation situated in the vicinity of a long-term gas leak will die, turn brown, or be stunted.

Underground Migration of Fuel Gases

It is possible for gases to migrate great distances underground before entering a structure or exiting the ground. The gas can enter underground gas lines, underground sewer lines, underground electrical and telephone conduits, or underground drain tiles. Gas can enter structures by migrating through cracks or holes in foundations. Weather changes (such as frozen ground) can cause the gas to be routed in a different direction or to migrate underground for greater distances.

When gas passes through a filter medium such as soil, the medium can "scrub" the odorant out of the gas. The gas then becomes odorless and might not be detected.

ACTIVITY 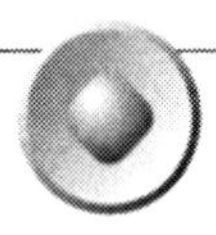

NFPA 54 requires that gas meters be installed at least 3 feet (0.9 meters) from sources of ignition and protected from damage. Visually examine the meter at your home or the home of a friend. Does the installation meet or exceed the intent of NFPA 54?

ACTIVITY 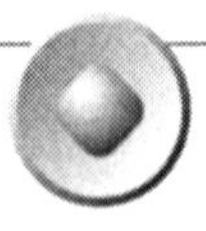

Accompany a trained gas expert or investigator during the examination of a fuel gas system and observe how the expert checks the system for leaks. A local fire marshal may provide this assistance, or your local gas utility company may know of recourses to assist you with this task.

▸▸▸ QUESTIONS FOR CHAPTER 7

1. List the ranges of the following properties of natural gas and commercial propane.

   Natural Gas

   Relative weight to air: ______________________
   Vapor density: ______________________
   Lower explosive limit (LEL): ______________________
   Upper explosive limit (UEL): ______________________
   Ignition temperature: ______________________

   Commercial Propane

   Relative weight to air: ______________________
   Vapor density: ______________________
   Lower explosive limit (LEL): ______________________
   Upper explosive limit (UEL): ______________________
   Ignition temperature: ______________________

2. What is the percentage of odorant to be added to a combustible gas as mandated by law?

   A. 0.20
   B. 0.25
   C. 0.33
   D. 0.35

3. What two agencies govern the design and fabrication of portable tanks and cylinders that carry LP-Gas?
4. List the common requirements of fuel gas appliances.
5. What is recommended by NFPA 54 if no leak is detected with the gas meter test?
6. How does a vegetation survey provide information on possible gas leaks?

Fire-Related Human Behavior

OBJECTIVES

Upon completion of Chapter 8, the user will be able to

- Analyze fire-related human behavior
- Integrate human behavior into the total investigation
- Reference standards, guidelines, and regulations dealing with safety warnings and product design

This chapter is based on Chapter 10 of NFPA 921. A fire's origin, development, and consequences are all related—either directly or indirectly—to the actions and/or omissions of human beings. Understanding the behavior of witnesses or occupants is integral to the fire scene examination. By understanding why the occupants or witnesses behaved in a particular manner, the investigator can better evaluate the effects of a particular fire.

Read NFPA 921, Chapter 10, "Fire-Related Human Behavior." Read an additional reference listed in Chapter 2 or Annex B of NFPA 921. Alternatively, study the NFPA publication entitled *NFPA Ready Reference: Human Behavior in Fire Emergencies,* 2003 edition.

GENERAL CONSIDERATIONS OF HUMAN RESPONSE

Research conducted within the past several decades indicates that an individual's or group's behavior before, during, and after a fire can be predicted. Factors that affect this behavior include characteristics of the individual, characteristics of the group or population to which that individual belongs, characteristics of the physical setting where the fire occurs, and characteristics of the fire itself.

Characteristics of the Individual

An individual's actions are shaped by psychological factors including his or her physical limitations, limitations of cognitive comprehension, and knowledge of the physical setting. These factors all affect how the individual recognizes a threat and reacts to it.

Physical Limitations. An individual's age, physical disabilities, cognitive incapacitation due to mental or physical conditions, and/or chemical impairment can adversely affect the appropriateness of that person's actions before and during a fire. The very young and very old are most susceptible to these limitations.

Cognitive Comprehension Limitations. Along with physical limitations, cognitive/mental limitations can affect an individual's ability to react appropriately to a fire or explosion. Some factors that can limit a person's cognitive ability are age (very young or very old), level of rest, alcohol use, drug use (legal or illegal), developmental disability, mental illness, and inhalation of smoke and toxic gases. Cognitive comprehension limitations can cause a significant delay in an individual's response time, thereby placing that person at increased risk for injury or death.

Familiarity with Physical Setting. An individual's knowledge of the setting can make escape more likely, although cognitive impairments can result in a person's becoming lost in his or her own home. In larger, unfamiliar structures, individuals tend to leave by the same route they took to enter—perhaps because of their lack of knowledge of their surroundings or perhaps because during extreme situations the brain fixates on the most recent method of entrance.

Characteristics of the Group or Population

An individual's response to a threat is affected not only by his or her own circumstances. Other significant factors are the size and structure of the group that the individual is with, the permanence of the group, and the group's roles and norms.

Group Size. An individual's response to a threat or purported threat is tempered by the size of the group with which the individual is associated. Smaller groups aid in the response time of the individuals associated with it. Larger groups have a tendency not to respond within appropriate time frames. This fact can be attributed to the individual's desire not to be first or to a disinclination to disrupt the workings of the group.

Group Structure. When a group has a formalized structure, individuals who perceive themselves as members but not in leadership or authoritarian positions delay their response until those who are in the leadership or authoritarian positions respond to the situation. Examples of this type of group are school populations, hospital populations, and religious facility populations.

Group Permanence. Research has indicated that the degree of familiarity among the individuals in a group also affects response times. If the group is established and its members know one another well, the individuals react and notify each other in a more timely manner than they do if the group is newly formed or its members are unfamiliar with each other. Examples of groups with a high degree of permanence are families, sports teams, choirs, and clubs.

Roles and Norms. A group's roles and norms in terms of gender, social class, education, and so forth can also affect its response to threats. Traditional gender roles are good examples. Males are more likely to engage in activities to counter the threat, while females will engage in reporting and escaping activities.

Characteristics of the Physical Setting

The characteristics of a burning structure will affect the response of the individuals within it. Some factors that affect response are the location of exits, the number of exits, the structure's height, and the operation of its fire warning systems and fire suppression systems.

Location of Exits. It is important for fire exits to be clearly marked. If occupants are unfamiliar with the building, they often react with panic and increased levels of stress, which in turn can result in unpredictable behavior. Furthermore, as was mentioned previously, occupants who are unfamiliar with a building try to exit by the door that they used to enter, even if that behavior moves them in the direction of the threat.

Number of Exits. The number of exits available for escape drastically affects the occupants' behavior. If there are too few exit routes or if the exits are blocked or restricted, occupants often panic. Such an unpredictable situation can expose them to additional danger.

Height of the Structure. Most people believe that they are less safe in a tall building during a fire. This misconception can result in unpredictable behavior, such as jumping out of windows, during an emergency.

Fire Alarm Systems. Fire alarm systems can help occupants to recognize a threat. Unfortunately, if there have been a number of false alarms, people tend to delay their response until the actual fire emergency is confirmed. This delay can result in the occupants being trapped and unable to escape.

Fire Suppression Systems. The presence of fire suppression systems can positively and negatively affect the response of the occupants. With fire suppression systems in place, occupants traditionally have more time to react to threat. Unfortunately, the amount of time can become exaggerated in the minds of the occupants to the extent that they remain in the structure far too long.

Characteristics of the Fire

An individual's or group's response to a threat is tempered by their senses. Visual and olfactory indicators shape responses. Characteristics of a fire that can influence people's behavior include the presence of flames, the presence of smoke, and the effects of toxic gases and oxygen depletion.

Presence of Flames. Most individuals do not understand the threat presented by flames. They have limited exposure to fire and incorrectly believe that the presence of small flames is not a hazard. They do not understand that the fire can grow exponentially and can release toxic gases and products into the air. Because they do not understand these principles, individuals tend to disregard small flames as a source of danger.

Presence of Smoke. The presence of smoke in a structure is also misunderstood. There is a general lack of knowledge about the toxic content of smoke and its incapacitating effects on humans. There is also a misperception that lighter-colored smoke is not as dangerous as darker-colored smoke, when in reality both have the potential to be fatal.

Effects of Toxic Gases and Oxygen Depletion. Many people do not understand the fire tetrahedron. (See Chapter 3, Exhibit 3.1.) As a result, they do not understand that fire consumes oxygen during the burning process. In an enclosed space, this fact equates to diminishment of the available oxygen. Depletion of oxygen results in impairment of motor and mental skills and can be fatal to occupants. In addition, the combustion process releases toxic by-products that, when inhaled, adversely affect the occupant and can also result in death.

FACTORS RELATED TO FIRE INITIATION

Fire and explosion incidents frequently occur as a result of an act or omission by one or more individuals. Three main areas relating to fire initiation are improper maintenance and operation of equipment or appliances; careless housekeeping; and failure to follow product labels, instructions, and warnings. Additional areas include recalls, violations of fire safety codes and standards, and other considerations.

Improper Maintenance and Operation

During the lifetime of most equipment, there is a prescribed maintenance and cleaning schedule that should be followed to prevent malfunction. When the equipment is capable of explosion or starting a fire, the required maintenance becomes critically important. Likewise, the operating procedures for equipment or appliances are designed to ensure safety. If either of these two areas is neglected or improperly performed, the lapse can lead to a fire or explosion. The investigator should examine all maintenance records and operating instructions carefully. If a fire can be traced to improper maintenance or operation, the person or persons responsible might not provide these records voluntarily.

Housekeeping

A majority of household equipment—specifically, equipment capable of initiating an explosion or fire—has instructions delineating standoff distances for combustibles. For example, paper products or liquid accelerants should not be stored adjacent to the pilot light of a water heater. Carelessly discarded smoking materials, such as cigarettes or matches, can ignite a fire. Grease buildup in cooking areas or improperly stored cleaning solutions are other hazards. The investigator should note any housekeeping irregularities that may have contributed to the fire or explosion.

Product Labels, Instructions, and Warnings

Labels, instructions, and warnings are placed on products to prevent their misuse or abuse. Manufacturer's labels inform the user of the product's capabilities. Instructions provide the user with all pertinent information about how the product was intended to be used. Warnings alert the user about the catastrophic events that can occur if the product is not used as intended. A proper warning contains four key elements: an alert word to signal danger, a statement of the danger, a statement of how to avoid the danger, and an explanation of the consequences of the danger.

Recalls

Recall notices are a method of notifying consumers of a product defect that was identified after the product was released for consumption. For the most part, recalls are the result of identifying a dangerous situation, even when the product is used in its intended manner. If an individual disregards a recall notice and continues to use the product, the result can be a fire, explosion, or other catastrophic event.

Violations of Fire Safety Codes and Standards

The investigator must be aware that noncompliance with fire safety codes and standards can result in a fire or explosion. Noncompliance may be deliberate or unintentional. When an investigator conducts an origin and cause determination, it is frequently difficult to determine whether there was deliberate misuse or abuse of the product, carelessness, or some other factor that contributed to the event. The investigator should examine training records, maintenance records, and other documentation to determine whether there is a pattern that will point toward the potential cause of the event.

CHILDREN AND FIRE

Children are drawn to setting fires for several reasons, such as curiosity, frustration, anger, revenge, or attention. The investigator should be aware of this fact during the origin and cause determination. There are three recognized age categories of juvenile firesetters: child firesetters (ages 2 to 6), juvenile firesetters (ages 7 to 13), and adolescent firesetters (ages 14 to 16).

Child Firesetters

Fires set by children in the 2 to 6 age group are traditionally curiosity fires set in hidden locations, out of sight of adults. Children at this age are not usually able to form the intent of causing damage with fire.

Juvenile Firesetters

Fires set by children aged 7 to 13 are typically a symptom or indicator of a psychological or emotional problem. These fires are often set in and around the home or in an educational setting.

Adolescent Firesetters

Fires set by teens aged 14 to 16 are usually symptomatic of stress, anxiety, anger, or another psychological or emotional problem. The most usual targets of these fires are schools, churches, vacant buildings, fields, and vacant lots. The fires are frequently associated with disruptive behavior, a broken home life, or a poor social environment. (See Chapter 20, "Incendiary Fires," for a discussion of the motives associated with such fires.)

RECOGNITION AND RESPONSE TO FIRES

In a fire situation, the occupant's ability to recognize the danger is critical to his or her survival. The occupant must be able to respond appropriately to the perceived and actual dangers associated with the event. Sensory perception is affected by mental state and chemical inebriations (alcohol, drugs).

The occupant's actions are based on the following four factors:

1. *Sight:* Direct view of flames, smoke and visual alarms, or flicker
2. *Sound:* Crackling of flames, failure of a window, audible alarms, a dog barking, children crying, voices, or shouts
3. *Feel:* Temperature rise or structural failure
4. *Smell:* Smoke odor

When the occupant identifies a threat (whether fire or explosion), the person must make a decision about how he or she is going to react. The occupant has several choices, including ignoring the problem, investigating, fighting the fire, signaling an alarm, rescuing or giving aid to others, reentering the structure (after successfully escaping), fleeing the fire, or remaining in place. The final decision will be affected by the occupant's state of mind.

The ability to escape is affected by the perceptibility of escape routes, distance to the escape routes, fire conditions (such as smoke, heat, or flames), the presence of dead-end corridors, the presence of obstacles or people blocking the escape path, and the individual's physical disabilities or impairments.

Interviews with event survivors can provide the investigator with information that will be beneficial in determining how people actually behaved before and during the fire or explosion. Personal interviews can help to establish the following:

- Prefire conditions
- Fire and smoke development
- Fuel packages and their location and orientation
- Victims' activities before, during, and after discovery of the fire or explosion
- Actions taken by individuals that resulted in their survival (e.g., escaping or taking refuge)
- Decisions made by survivors and reasons for those decisions
- Critical fire events such as flashover, structural failure, window breakage, alarm sounding, first observation of smoke, first observation of flame, fire department arrival, and contact with others in the building

ACTIVITY 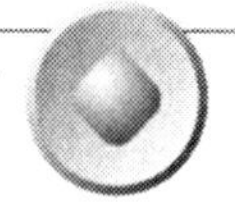

Obtain and read the NIST report on the Pulaski Building fire by Harold E. Nelson (http://fire.nist.gov/bfrlpubs/). Review the activities of the occupants on the floor of fire origin. What conclusions can be drawn about the fire-related behavior of the occupants?

▸▸▸ QUESTIONS FOR CHAPTER 8

1. What are five physical limitations that affect an individual's response to fire?
2. What are five cognitive comprehension limitations that affect an individual's response to fire?
3. Describe the human behavior that corresponds to the circumstances listed below. What fire-related reactions are predicted by the circumstance described?
   - **A.** Unfamiliarity with the physical setting
   - **B.** Being part of a formal, structured group, as in a church or office setting
   - **C.** Being part of a transient group of random individuals, as in a mall setting
   - **D.** Being a woman
   - **E.** Being a man
   - **F.** Having inadequate identification of exits or number of exits
   - **G.** Being in a tall building
   - **H.** Hearing verbal alarm systems
   - **I.** Hearing alarm bells and sounds
   - **J.** Having fire suppression systems (such as sprinklers) activate
   - **K.** Seeing small flames
   - **L.** Seeing light gray smoke
   - **M.** Inhaling toxic gases and experiencing oxygen depletion
4. According to federal regulations, what are the key elements of a proper warning on a label or product?
5. Outline the three recognized age groups of children with regard to setting fires, including the motive and location of fire for each group.
6. List three main conditions in which visual and olfactory indicators shape people's response to a fire.
7. How does the size of a group affect an individual's response to a fire threat?

Legal Considerations

OBJECTIVES

Upon completion of Chapter 9, the user will be able to

- Recognize the legal issues and considerations that may arise in an investigation
- Identify the appropriate response to those issues and considerations
- Implement the appropriate response to ensure compliance with applicable legal standards

This chapter is intended to provide information on the legal considerations that may arise during and after a fire investigation. The investigator must confront legal issues from the time of first arrival at the fire scene through the presentation of testimony and evidence at trial. Without an understanding of the legal issues, the entire investigation can be jeopardized. The results and findings can be inadmissible at trial if the investigator has failed to follow proper legal procedures.

The information in this chapter is necessarily generalized, since laws and legal procedures vary from jurisdiction to jurisdiction. Even within a particular jurisdiction, legal requirements vary depending on whether the case is a civil or criminal proceeding and whether the trial takes place in state or federal court. The investigator must be aware of those differences to meet the specific requirements in any particular case.

In every jurisdiction, the law is constantly evolving. New case decisions and revisions to existing laws and ordinances can substantially change the legal requirements that the investigator must meet. Therefore, the investigator must stay current with new developments in the law and seek legal advice whenever necessary.

The information in this chapter is based on current U.S. law. Investigators conducting fire scene investigations in other countries should always consult with a qualified local attorney.

In the United States, investigators should be aware of safeguards found in amendments to the U.S. Constitution that apply to criminal investigations and prosecutions. This topic is covered later in this chapter in the section entitled "Constitutional Issues in Witness Interviews."

ACTIVITY Review NFPA 921, Section 11.2, "Constitutional Considerations."

PRELIMINARY LEGAL CONSIDERATIONS

Authority to Conduct Investigations

The legal authority to conduct a fire scene investigation can be either statutory or contractual. For law enforcement and public sector investigations, the authority is conferred by statute, ordinance, or agency rule. The scope and limitations on the power of the investigator are contained in those laws and the court decisions related to construing those laws. There can also be constitutional limitations on the investigator's authority under applicable state or federal law.

In the private sector, the investigator's authority is a contractual right that is derived from the person or institution authorizing the investigation. In cases in which the property owner is a third party to the investigation (as in insurance claims), additional legal considerations can be involved. Authority to investigate insurance claims is granted under the terms of the insurance policy, which generally requires that a claimant allow the insurance company to conduct a fire scene investigation before the claimant is eligible to receive payment for the loss.

Right of Entry

The authority to investigate a fire does not necessarily grant the investigator the right of entry at the fire scene. Public sector/law enforcement investigators are bound by constitutional limitations under search and seizure law. A fire scene investigation is considered a search and seizure, just as with any other criminal investigation. Under the Fourth Amendment of the U.S. Constitution and parallel provisions under state constitutional law, every entry onto a fire scene must be justified. There are four general circumstances that justify a search and seizure to conduct a fire scene investigation: consent, exigent circumstances, administrative search warrant, and criminal search warrant.

Consent. A consent search is perhaps the best method of entry to ensure compliance with the Fourth Amendment. Whenever there has been a delay in starting the fire scene investigation, consent to search should be obtained to remove all legal concerns about the timing and duration of the investigation. However, there are strict requirements for obtaining lawful consent to search. First and foremost, consent must be obtained from the proper party. Although the owner of the property is often the proper person to give consent, this is not always the case. The test to be applied is "common authority and control" over the premises. Property that has been leased or rented to an individual usually requires consent from the tenant or occupant rather than the owner, even though the owner or landlord may have the right to enter and inspect the premises under the terms of a lease agreement.

The person who grants consent can authorize the search of only those areas under his or her common authority and control. For instance, although a child might be authorized to grant consent to search common areas such as the living room or kitchen, the child cannot authorize a search of a parent's bedroom or home office. If the child is under age 16, the consent to search generally extends only to those areas under the child's direct control, such as the child's bedroom. In the case of roommates, one roommate cannot authorize a search of the other's private areas, such as a separate bedroom or personal closet. In all consent searches, only those areas within the common authority and control of the person giving consent may be searched.

For a fire investigator in the private sector, there are no constitutional issues to confront. The private investigator utilizes consent entry almost exclusively—either express consent from the owner or occupant or implied consent under the provisions of an insurance policy. A private investigator may not gain entry to the fire scene if the owner or occupant refuses to allow access. However, if an insured claimant refuses to allow access to the fire scene, the insurance claim may be denied.

Of all the methods of entry onto a fire scene, consent is always preferable when it can be properly obtained. Even though one of the other methods might be appropriate to the situation, consent entry should always be requested whenever possible. The investigator should bear in mind, however, that when a fire scene investigation is conducted pursuant to consent, the person who gives the consent may limit the scope of the search and may even revoke consent at any time. Consent to search a fire scene should always be obtained in writing and be witnessed.

Exigent Circumstance. Public sector–law enforcement investigators have the authority to conduct a search and seizure whenever an exigent circumstance (or "emergency") exists, requiring an immediate response in the interest of public safety. The U.S. Supreme Court decisions in *Michigan* v. *Tyler* (436 U.S. 499) and *Michigan* v. *Clifford* (464 U.S. 287) outline the scope of a fire investigator's right of entry under this rule. Every fire is considered a public safety emergency that requires a response by fire service. When a fire is reported, the exigent circumstance allows entry onto property to suppress the fire. As an extension of this concept, the determination of a fire's cause is part of the public safety response. Thus, the investigation of a fire scene is authorized as part of the exigent circumstance surrounding every fire.

However, there is no open-ended authority to enter a fire scene and conduct an investigation; the investigation must take place within a "reasonable time" following suppression and extinguishment of the fire. Whether a fire scene investigation has been conducted within a reasonable time depends on a variety of facts and circumstances. Generally, the investigation must commence at the first available opportunity following suppression and extinguishment of the fire. Any delay in beginning the investigation must be justified for the investigation to be considered part of the exigent circumstances surrounding the fire. A delay to wait for daylight is generally considered acceptable. However, waiting until later in the day when the investigation could easily have begun earlier may result in an illegal search and seizure. The investigator should always respond immediately when called to investigate a fire.

Administrative Search Warrant. An administrative search warrant is a third method of entry for the public sector–law enforcement investigator, expressly authorized by the U.S. Supreme Court in the *Michigan* v. *Tyler* case. It is seldom used by investigators and even less often recognized by courts, however. An administrative search warrant is intended for situations in which the reasonable time after a fire has already passed and consent to search is refused or is otherwise not available. It requires a sworn affidavit presented to a judge in a court of competent jurisdiction for the limited purpose of gaining entry to a fire scene to determine origin and cause. The applicant must show in the affidavit (1) that he or she has the administrative authority to conduct fire scene investigations in the jurisdiction, (2) that a fire has occurred but the investigator cannot lawfully enter the premises at this time, and (3) that the court's authorization is needed to fulfill the administrative responsibilities of investigating the fire to determine its origin and cause. An administrative search warrant cannot be employed once there is evidence of arson or some other crime, as outlined in *Michigan* v. *Clifford.* When a fire scene investigation pursuant to an administrative search warrant uncovers evidence of a crime, any further search of the premises requires a criminal search warrant.

Criminal Search Warrant. A criminal search warrant is the fourth alternative for gaining entry to a fire scene for the purpose of conducting an investigation. As with the administrative search warrant, this method is a judicial authorization based on the filing of a sworn affidavit by the investigator. However, a criminal search warrant is only issued upon a showing of probable cause, establishing that a crime has occurred and that evidence of the crime is believed to be located in the premises to be searched. The affidavit must demonstrate reliable information and specific facts known to the applicant who is seeking the warrant. The premises and the specific areas within the premises where the search will be conducted must be precisely identified. It is strongly recommended that a prosecutor or legal advisor assist in the preparation of the affidavit and application for warrant.

ACTIVITY 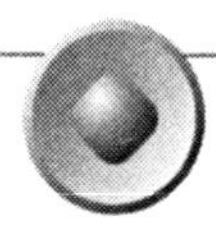

Read Section 11.3, "Legal Considerations During the Investigation," in NFPA 921. ◀◀◀

CONSTITUTIONAL ISSUES IN WITNESS INTERVIEWS

ACTIVITY 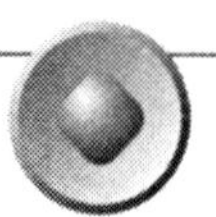

Read Paragraph 11.3.4, "The Questioning of Suspects," in NFPA 921 for legal background on witness interviews. ◀◀◀

There are special constitutional issues that an investigator must address when conducting witness interviews in order for the statement of a witness to be legally admissible at trial. These issues involve the Fifth Amendment protection against self-incrimination and the Sixth Amendment right to counsel, that is, to have an attorney present during questioning. These issues are the foundation of the interrogation guidelines of the *Miranda Rule* (*Miranda* v. *Arizona,* 384 U.S. 436). Under the Fifth Amendment, a person cannot be compelled to testify against himself or herself, nor to provide incriminating testimony that can be used later in a prosecution. The Sixth Amendment contains the constitutional "right to counsel," which is also part of the protections afforded under the Miranda Rule. The Miranda Rule requires that a witness being interrogated or interviewed in a custodial setting receive certain warnings before the statement can legally be taken and used as evidence later. The Miranda Rule has four components:

1. The right to remain silent and refuse to give a statement
2. The warning that any statement made can later be used as evidence against the witness
3. The right to have an attorney present to represent the witness during questioning

4. The right to have an attorney provided free of charge, before any questioning begins, if the person cannot afford to hire an attorney

Whenever a witness is interviewed under custodial circumstances, the investigator must ensure compliance with the full set of Miranda warnings before the witness's statement can be held admissible at trial. Mere questioning of a witness does not in itself invoke the Miranda Rule. The interview or interrogation must take place in a custodial setting. In determining whether the interview or interrogation took place in a custodial setting, a court will consider various factors, including these:

- *Apparent authority of the interviewer to detain or arrest the witness:* A private citizen or a public official who lacks arrest powers will generally not be required to comply with the Miranda Rule.
- *Location of the interview:* For the Miranda Rule to apply, the setting must be custodial, meaning that the witness does not have complete freedom of movement, including the ability to walk away from the interview.
- *Length and context of the interview itself:* The longer the interview lasts and the more intense it becomes, the more likely that it will be considered a custodial interrogation—even if it began as a simple witness interview.
- *Participants (and observers) at the interview:* A large number of interviewers and spectators, especially law enforcement officials, can be considered a circumstance intimidating enough to change a mere witness interview into a custodial interrogation.

The basic witness interview statement should be taken under circumstances appropriate to the nature of the interview. When a suspect in a criminal investigation is being interrogated, however, the interview should be treated differently, with full Miranda warnings given to the suspect before questioning ever begins. When there is any doubt about the circumstances of the interview or the individual being questioned, the best practice is to advise the individual of all Miranda rights. When an individual has been given a Miranda warning and agrees to make a statement, a written waiver of rights should be signed and witnessed. The interview itself should be tape-recorded or videotaped whenever possible.

SPOLIATION OF EVIDENCE

ACTIVITY

Read NFPA 921, Paragraph 11.3.5, as well as Paragraphs 11.3.5.1 through 11.3.5.6, which cover spoliation of evidence. ◀◀◀

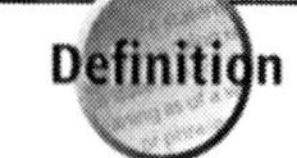

Definition

spoliation

The loss, destruction, or material alteration of an object or a document that is evidence or potential evidence in a legal proceeding by one who has the responsibility for its preservation.

Spoliation of evidence as a legal issue has existed for hundreds of years but has taken on great importance only in recent years. It is now an issue that arises in almost every fire scene investigation. While spoliation was traditionally an issue reserved for civil litigants, it can also appear in criminal matters.

Spoliation of evidence is an inevitable by-product of any fire scene investigation. In a very real sense, an entire fire scene is critical evidence in a case. When an investigator has conducted a scene examination by moving fire debris and other evidence at the scene, the scene itself is necessarily altered. When a thorough fire scene examination has been conducted with the movement of debris using a methodical layering process, the result can be a scene that is devoid of virtually all evidence, leaving only a bare slab. When evidence is collected and removed from the scene, further change may occur. The examination of equipment and appliances found at a fire scene may involve destructive testing, which constitutes another form of change, alteration, or destruction. In each of these situations, the investigator can face claims of spoliation when other parties later find that they are unable to independently investigate the evidence from the fire scene because that evidence is no longer available as it existed immediately after the fire.

When spoliation has been found to occur, a court may take any number of actions against the responsible party as sanctions for the conduct that caused the spoliation. Those sanctions may include prohibiting any testimony about the spoliated evidence by the responsible party, prohibiting the responsible party from presenting any testimony at all about the fire scene investigation, instructing the members of a jury that they may infer that the spoliated evidence would have been damaging to the responsible party's case, or even striking the responsible party's case and entering judgment in favor of the party victimized by the spoliation. Moreover, courts are now recognizing the tort of spoliation as a separate legal cause of action against a spoliator.

The concept of spoliation is a matter of fundamental fairness. In cases in which one party has lost, destroyed, or altered physical evidence that prejudices another party, the courts will use the doctrine of spoliation to balance the scales. The sanctions that a court may impose are designed to correct the prejudice that the injured party has suffered by being unable to independently examine and test the evidence in order to defend itself against claims relating to the evidence or to use the evidence to assert its own claims against other parties.

The law in the area of spoliation is rapidly evolving and the claims that can be asserted under a theory of spoliation are expanding, especially in the field of fire investigation. Most jurisdictions require some element of deliberate intent on the part of the spoliating party for sanctions to be imposed. In some jurisdictions, the standard is negligence, even though the offending party might not have actually intended to prejudice another party. For public sector investigators, there is still limited exposure to claims of spoliation, although the doctrine of sovereign immunity might shield public investigators and their agencies. However, the intentional destruction of evidence to purposely prejudice another party is likely to be actionable against even a public sector investigator and should always be avoided. For all investigators, claims of spoliation can best be avoided by using a commonsense approach. Evidence should never be needlessly discarded, destroyed, or lost. Whenever evidence is uncovered that could be important to the interests of another party (such as in product liability cases), the evidence should be immediately preserved, and notification should be sent to all parties that have a potential interest in the evidence to allow them the opportunity to examine it.

The inevitable alteration of evidence as fire debris is examined and moved during a fire scene examination generally does not constitute actionable spoliation, especially if the scene has been properly documented and photographed before, during, and after movement of the evidence. Thus proper fire scene documentation—which should already be employed in every investigation—is the first line of defense against claims of spoliation. When specific evidence is uncovered (such as equipment or an appliance) that is believed to have been a factor in the fire, those items should always be carefully preserved, and notification should be immediately sent to all interested parties. With this as an established investigative protocol, the threat of spoliation is substantially eliminated. NFPA 921, Paragraph 11.3.5 and Paragraphs 11.3.5.1 through 11.3.5.6, provide a detailed discussion of this issue.

EVIDENCE

The collection, preservation, and use of evidence at trial are governed by constitutional law, statutory law, and the applicable rules of evidence in the jurisdiction. Evidence that has not been properly collected, including evidence that is gathered during an unauthorized search and seizure, will be inadmissible at trial. Evidence that has not been properly preserved, including a record of the chain of custody for certain physical evidence, will be inadmissible as well. Likewise, evidence that has not been properly documented, authenticated, and presented at trial will be inadmissible.

Federal Rules of Evidence

The federal rules of evidence are the model for the discussions on evidence in NFPA 921. Those rules apply in all federal court proceedings throughout the United States. For state court proceedings, the applicable state rules of evidence will control. Although there are some variations in the rules of evidence in different states, most of them are based on the federal rules and are similar in their provisions.

Demonstrative Evidence/Physical Evidence

Demonstrative evidence is an important form of evidence used at trial. This is any type of tangible evidence, such as a photograph, diagram, or chart, that is used to demonstrate an issue relevant to the case. Demonstrative evidence includes items that were gathered from the fire scene, as well as other tangible evidence items (as opposed to witness testimony). The collection of physical evidence at the fire scene is governed by the constitutional limitations on search and seizure. The right to collection and possession of physical evidence requires compliance with those constitutional standards.

The actual method of collection of physical evidence requires compliance with technical standards for collecting and handling evidence as outlined in NFPA 921. To be admissible at trial, evidence must be properly transported and stored. Photographs, videos, slides, or digital media must be properly authenticated by showing that they are a "fair and accurate depiction" of the subject. Such evidence is almost always found to be admissible when it has been properly authenticated in this way. It should be noted that those forms of demonstrative evidence (film or other media) generally do not require authentication by the person who takes the photographs (although it is preferable to do so), nor is there a requirement for proving the chain of custody for such evidence. Evidence that has been altered, destroyed, or lost is termed spoliated, and not only is it inadmissible at trial, but its existence may result in other evidence being deemed inadmissible as well. In some cases, the investigator who is responsible for the spoliation may be prohibited from testifying. In extreme cases, spoliated evidence can result in outright dismissal of the case.

The most common forms of demonstrative evidence used at trial are diagrams, charts, timelines, models, and other such evidence that is specially created for the trial to illustrate or demonstrate a point to the jury. The basic legal requirements for such evidence are relevance to the issues at trial and usefulness to the jury in understanding those issues.

Documentary Evidence

The second form of evidence used at trial is *documentary evidence.* This category includes any type of written record or document that is relevant to the case. Business records, incident reports, telephone records, bank records, insurance policies, claim documents, correspondence, written statements of witnesses, investigative notes, and transcripts of recorded interviews are all examples of documentary evidence.

Once again, there are a number of legal issues to consider. First, the evidence must have been properly obtained. The investigator must consider Fourth Amendment issues of search and seizure law, as well as Fifth Amendment issues of self-incrimination (if, for example, the subject of a witness interview is a criminal suspect). All documentary evidence must be properly authenticated at trial under the standards of relevance and reliability. In the case of business records, bank records, telephone records, and other such documentary evidence, authentication will usually require the testimony of a records custodian to confirm the authenticity and accuracy of the documents. Documentary evidence may be subject to a legal challenge as hearsay if the information was obtained from an out-of-court source. However, there are many exceptions to the rule against hearsay evidence that may allow the documentary evidence to be used once a proper legal foundation has been laid. To avoid hearsay challenges, investigators should work with the prosecutor or attorney handling the trial to ensure that documentary evidence is properly presented.

Testimonial Evidence

The third form of evidence at trial is *testimonial evidence,* which is verbal testimony of a witness given under oath or affirmation and subject to cross-examination by the opposing party. Although relevance and reliability are the baseline requirements for admitting testimonial evidence at trial, it is subject to the full range of requirements under the rules of evidence and may be prohibited for a variety of reasons. For instance, a witness might not be competent to testify at all, a witness might not be competent to testify about a specific issue, the testimony might be irrelevant or immaterial to the case, the testimony might include improper hearsay statements from other persons, or the testimony might be speculative or unfounded; the testimony may contain the opinion of a witness who is not qualified to offer opinion testimony (usually, only an expert witness is qualified to offer opinion testimony). In the case of expert testimony, especially the testimony of an origin and cause expert, recent developments in the law have imposed new requirements for qualifying an expert and presenting expert testimony. Those requirements are discussed in the next section.

ACTIVITY 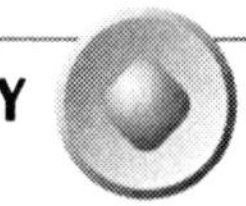

Read NFPA 921, Paragraph 11.5.2.3, "Testimonial Evidence," and Paragraphs 11.5.2.3.1 and 11.5.2.3.2. ◀◀◀

EXPERT TESTIMONY

Expert witnesses assume a special role in the courtroom. A fact witness may only testify to observations and statements of fact that are known to the witness. A fact witness may not express an opinion about issues in a case—the only exception being the common observations and opinions that any person would be qualified to express, such as commenting that it was a "hot" day, that the vehicle was traveling "fast," or that the structure was "fully involved" in flames. For the fire investigator who is testifying as an expert witness, the opinion as to the fire's origin and cause is the ultimate objective of his or her testimony. In a select few jurisdictions, even an expert witness is not allowed to express this type of opinion. However, in virtually all other jurisdictions, opinion from an expert witness is permitted and is always the focal point of the trial.

As with all other forms of evidence, testimonial evidence from an expert must show relevance and reliability. However, the measure of reliability for an expert witness can be a complex analysis. In federal court cases, and recently in a number of state court jurisdictions, expert testimony is now governed by the *Daubert Rule* (*Daubert* v. *Merrell Dow Pharmaceuticals,* 509 U.S. 579). Under this new standard, the trial judge takes a proactive role in screening the testimony of prospective expert witnesses. The judge must assess the reliability component of an expert's testimony by challenging the expert as to both the conclusions (opinions) to be presented and the techniques and methodologies used to reach those conclusions. The judge may consider various criteria to establish the reliability of the expert's testimony, including the following:

1. Can the theory or technique be tested?
2. Has the theory or technique been subjected to peer review and publication?
3. Is there a known or potential rate of error for the theory or technique?
4. Has the theory or technique gained general acceptance in the expert's field?

In addition to those general criteria, a trial judge may ask any other questions that are appropriate to the particular facts of the case to ensure that the testimony is reliable enough to be heard by a jury.

In jurisdictions where the Daubert Rule applies, expert witnesses can expect a particularly strong challenge to their testimony and findings by both the opposing party and the trial judge. In responding to a Daubert challenge, the investigator must be able to demonstrate that both the methodologies used in the investigation and the conclusions reached by the investigator can be scientifically validated as reliable by reference to objective and documented sources. Perhaps the best source of such validation is NFPA 921 itself. Familiarity with the entire document and the ability to readily cite the appropriate sections addressing any issue in question will enable an investigator to respond effectively to a Daubert challenge.

Successful testimony by an expert witness at trial requires the use of demonstrative evidence as discussed in the preceding sections. The proper presentation of expert testimony requires a knowledge and understanding of these issues, as well as the most important consideration of all: preparation. Whether testifying at a deposition or at trial, the investigator should always be fully prepared and organized with a full command of the investigative file. A conference with counsel to review the anticipated testimony and the expected challenges to the testimony should be a part of the investigator's preparation in all cases.

CRIMINAL PROSECUTION

ACTIVITY 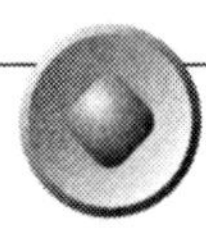 Read NFPA 921, Paragraph 11.5.6, "Criminal Prosecution."

The investigator must be familiar with the criminal laws and statutes that pertain to arson and other fire-related crimes. When new statute books are published each year, they should be reviewed for any new amendments or revisions. The prosecuting authorities in a jurisdiction should keep investigators informed of any new case decisions on those laws. Professional journals and publications are other sources of information on changes in the laws and their interpretation by the courts.

Arson

Arson is one of the oldest criminal offenses under the law. It dates back to the early common law of England, when it was one of only three capital offenses (along with treason and murder). The crime of arson under the common law of England was originally defined as the malicious burning of another's home at night. For several centuries, this was the only form of intentional fire setting that was considered to be a crime. Today, the crime of arson can be committed in many forms. According to *Black's Law Dictionary*, *arson* is generally defined as causing unlawful and intentional damage, by fire or explosion, to designated types of property, including residences, buildings, sheds and other structures, vehicles, aircraft, watercraft, and personal property. There are significant differences in the arson statutes from state to state, and an investigator must be closely familiar with the arson statute in his or her jurisdiction.

Most arson offenses are felony crimes punishable by imprisonment and fines. Arson to an occupied structure is generally considered to be first-degree arson, punishable by substantial prison time ranging up to a life sentence in some jurisdictions. Acts of second-degree arson or third-degree arson are generally punishable as felony offenses as well. The intentional burning of other types of property that are not designated under the arson statutes may be considered vandalism or criminal mischief, punishable as a misdemeanor, depending on the type and value of the property burned.

NFPA 921, Paragraph 11.5.6.3, includes a list of factors to be considered when an investigator is determining whether the crime of arson has occurred. However, it must be remembered that the arson statutes of each jurisdiction contain requirements specific to that jurisdiction.

Other Fire-Related Criminal Acts

In addition to arson, other criminal offenses may be related to a fire scene investigation. These fire-related crimes include the following:

- Burning to defraud
- Insurance fraud
- Unauthorized burning
- Manufacture or possession of fire bombs and incendiary devices
- Wildfire arson
- Domestic violence

- Child endangerment/child abuse
- Homicide and attempted homicide
- Endangerment/injury to fire fighters
- Obstruction/interference with fire fighters
- Disabling of fire suppression/fire alarm systems
- Reckless endangerment
- Failure to report a fire
- Vandalism or malicious mischief
- Burglary/trespass

These are only some of the potential offenses that may come to light in a fire investigation. In addition to these offenses, virtually any other type of criminal activity may be uncovered in the course of the investigation.

Arson Immunity Acts

Since the first immunity reporting legislation was enacted in Ohio nearly 30 years ago, every state has adopted some form of immunity reporting act that requires the release of information to public officials regarding fires that may have been the result of criminal acts. Most of the statutes are based on the model act created by the National Association of Insurance Commissioners. These acts are intended to facilitate the exchange of information between law enforcement authorities and insurance companies in the investigation of a suspected incendiary fire. The *arson immunity acts* require an insurance company to release all information in its possession concerning the investigation of a fire when requested to do so by a designated public agency. Additionally, many of the immunity acts require insurance company representatives to notify those public agencies whenever they have reason to believe that a fire claim under investigation might have been the result of arson, regardless of whether there has already been a request from public authorities. In exchange for providing this information, the insurance companies are granted statutory immunity from civil or criminal liability. If an insurance company engages in fraud or bad faith or if it acts out of malice in seeking protection under an immunity reporting act, the immunity may be lost. An insurance company that fails to respond to a request for information under the immunity reporting act may be subject to civil, administrative, or even criminal penalties.

In addition to the requirement for providing information to public agencies, many of the immunity reporting acts contain reciprocal provisions authorizing the exchange of information from those agencies to the insurance companies to assist in their civil investigations. Several of the acts also authorize the exchange of information directly between insurance companies.

Every investigation of a suspected insurance fraud fire by law enforcement agencies should include a request for information from the insurance company investigating the claim, under the appropriate immunity reporting act. Information in the insurance company claim file can be invaluable to the law enforcement investigator involved in the case. The claim file may contain original evidence needed for a criminal prosecution and may include such things as claim documents, payment records, origin and cause investigation reports, lab reports, forensic accounting analyses, bank records, business documents, photographs and videos (including pre-loss photographs), witness interviews, sworn statements or examinations under oath, and many other investigative materials that can greatly assist the public sector investigator.

There are specific requirements and conditions for invoking the release of information under an immunity reporting act. Investigators should be closely familiar with those laws in their jurisdiction to properly utilize the laws in their investigations.

CIVIL LITIGATION

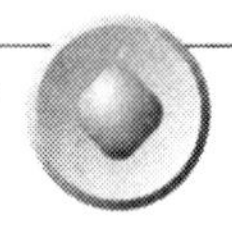

ACTIVITY Read NFPA 921, Paragraph 11.5.7, "Civil Litigation."

Civil litigation in fire investigation cases is far more prevalent than criminal prosecution. An investigator for a public agency who has conducted a fire scene examination may become involved in civil litigation related to that fire, regardless of the outcome of his or her own investigation and the results of any subsequent prosecution. For private sector investigators, the prospect of becoming involved in civil litigation at some point is a virtual certainty. Civil litigation may take the form of a civil arson case, product liability action, fire legal liability action, negligence claim, fire code violations claim, personal injury claim, or wrongful death claim. Every fire scene investigator must have an understanding of civil litigation principles in order to respond effectively when called to testify in a civil proceeding.

There are, of course, fundamental differences between the criminal and civil courts. First among these differences is the fact that the consequences of the proceeding are far more severe in a criminal case. An individual's life and liberty are at stake in a criminal case; a criminal defendant who is convicted at trial faces the prospect of imprisonment. Civil cases and their outcome are usually concerned only with the award of money damages. But although life and personal freedom may not be at stake in a civil case, the individual's financial freedom may be at risk. Fire litigation cases in civil court are usually high-stakes cases. For this reason, they are seriously and aggressively litigated by the parties and their counsel.

A fire litigation case in civil court can directly involve the actions of the public sector investigator and may ultimately rest on his or her findings. Frequently, one of the parties seeks to use a public investigator as an expert witness for its side of the case in order to have a presumably impartial witness support the case. Private sector investigators, on the other hand, are always challenged on the fact that they have been paid for their testimony and findings. When a public sector investigator can be used in a civil case, the stigma of having been hired to testify for that side of the case is not a factor.

The procedures in civil litigation are essentially the same as the procedures in criminal cases in most respects, yet there are significant differences. The discovery phase of a civil case is usually much longer and much more involved than that in a criminal case. Dozens of witnesses may be deposed, hundreds of records may be subpoenaed, and numerous procedural motions may be filed by the parties.

In federal court cases, witnesses who will be testifying as experts must make certain disclosures at the early stages of the litigation under Rule 26 of the Federal Rules of Civil Procedure. At the very outset of the litigation, all expert witnesses must be identified to the other side in a written disclosure. The substance of the expert's anticipated testimony must be disclosed, along with a copy of the expert's reports and analyses. An expert who has not yet prepared a report must do so to comply with Rule 26. The disclosure must include a detailed list of all information, records, and documents reviewed by the expert in reaching the opinions and conclusions in the case. Additionally, the expert must provide a current Curriculum Vitae or resume with complete information about his or her educational background, employment history, professional qualifications, fees charged, articles or publications written, and prior testimony as an expert. The information about prior testimony must be provided in detail, including the style (name) of each case, the court in which the case was tried, the attorneys who litigated the case, and the party who hired the witness. This information is required for every case in which the expert has testified as an expert witness within the preceding 4 years. For this reason, every investigator should maintain a complete history of all cases in which expert testimony has been given, including all of the specific information referenced above.

With increasing frequency, fire investigators are finding themselves more than mere spectators in civil litigation. Lawsuits have been filed against public sector investigators and private sector investigators alike, as well as their agencies and companies. Negligence claims and professional liability (malpractice) claims are being asserted against fire scene investigators for improper investigations, incorrect findings, erroneous conclusions, false arrest, and malicious prosecution in criminal cases, as well as for bad faith in insurance claims cases and the spoliation of evidence in both criminal and civil cases. An awareness of the theories of liability is essential to understanding both the processes of civil litigation and the best way to avoid becoming a part of civil litigation as a party defendant. Paragraph 11.5.7 of NFPA 921 discusses the various theories of legal liability.

Meet with a local prosecutor or an attorney working in the civil side and discuss the legal and constitutional issues outlined in this chapter as they relate to your investigations. If an attorney is not available, work with someone who is knowledgeable about these issues—possibly someone who has investigated losses in your area. ◖◖◖

▸▸▸ QUESTIONS FOR CHAPTER 9

1. Under which type of evidence would a fire investigator's report be classified?
   A. Demonstrative evidence
   B. Documentary evidence
   C. Circumstantial evidence
   D. Testimonial evidence
2. True or false: Questions regarding access to a fire scene are subject to constitutional protection under the Fifth and Sixth Amendments.
   A. True
   B. False
3. According to *Black's Law Dictionary*, which of the following when intentionally set on fire is not considered arson?
   A. Dwelling
   B. Automobile
   C. Person
   D. Business
4. What are the four ways in which a fire investigator can obtain right of entry to the fire scene? Give a brief description of each.
5. True or false: The most frequently used forms of demonstrative evidence at trial are photographs, sketches, maps, diagrams, and models.
   A. True
   B. False
6. True or false: A significant number of civil lawsuits involve issues regarding negligence, product liability, and fire code violations.
   A. True
   B. False
7. True or false: Samples of fire debris evidence do not require documentation and proof of chain of custody.
   A. True
   B. False
8. When referring to consent for the investigator to enter and remain on the property, which of the following statements is not true?
   A. Giving consent is a voluntary act on the part of the lawful property owner.
   B. Once consent is given, it cannot be taken back.
   C. If consent is given, it should be documented by the investigator.
   D. Written documentation is an effective method of documenting consent.
9. What is meant by a "reasonable period of time" as referenced by the Supreme Court in dealing with exigent circumstances authorizing entry onto a fire scene?
   A. 24 hours
   B. 36 hours
   C. As soon as practical
   D. As long as somebody remains present at the fire scene
10. True or false: Any removal of evidence from a fire scene that results in its deterioration or spoilage is known as spoliation.
    A. True
    B. False

Safety

OBJECTIVES

Upon completion of Chapter 10, the user will be able to

- Identify the various hazards at a fire scene that can cause harm to the health and safety of a fire investigator
- Identify the proper protective clothing, equipment, and practices needed to conduct a safe fire scene examination
- Perform a proper fire scene hazard assessment so that all hazards are identified and removed or mitigated prior to the initiation of the fire scene examination

Modern fire investigators are faced with a multitude of hazards. Fire investigators must be concerned not only with the physical hazards of a fire scene, but also with the chemical and other toxic substances present at the scene. Fire investigators must follow the safety principles that are utilized during fire-fighting operations. They must also be trained to recognize and protect themselves and others from the hazards of the fire scene.

Read NFPA 921, Chapter 12.

FIRE INVESTIGATION RISKS

The training that a fire investigator receives provides the necessary foundation for a long, safe, and healthy career. However, in today's rapidly changing society, the fire investigator must always update training and knowledge in the areas of safety. Safety should be one of the primary concerns at every fire scene. The fire investigator must also beware of the long-term effects of repeated exposures to toxins or even one-time exposure to a toxin at harmful levels. There are numerous cases of fire investigators who are fighting serious health problems as the result of investigations involving pressure-treated wood decks, plumbing supply warehouses, or automobile fires, to name a few examples. Groups of fire investigators, who would normally be expected to be healthy, are experiencing cancer at a higher rate than the general population. The fire investigator who dies 20 years after an incident because of exposure to a toxin at a fire scene is dead as a result of the response to that incident where the harmful exposure took place. All too often, fire fighters succumb to illnesses that were caused by their response to one bad fire. The fire departments in Fort Lauderdale (Florida), Newark (New Jersey), and New York City are among those that have experienced fires that caused responders to succumb to similar cancers years after the fire was forgotten. Therefore, the investigator must address the health hazards of fire investigation as well as the physical hazards at each fire scene to ensure the safety of all who operate at the fire scene during the investigation phase of the incident.

RESPONSIBILITY FOR SAFETY AT A FIRE SCENE

NFPA 921 states that fire scenes, by their nature, are dangerous places that present varied hazards that threaten the immediate and long-term health of the investigator. Therefore, investigators have a duty to protect themselves and others from hazards that are present at the fire scene. The investigator often must operate in hazardous areas without supervision or without benefit of the normal safety precautions that are taken during the fire-fighting operations. The investigative phase is not considered to be part of the emergency phase of the incident, and it is extremely rare for a fire investigation to be conducted under emergency conditions. Nevertheless, it is important to maintain the same safety standards that were in place during the initial phases of the incident. The fire investigator may be expected to operate in areas that contain toxic atmospheres, compromised structural systems, and sometimes even dangerous criminals, so caution must always be exercised.

Safeguarding Technical Personnel

The fire investigation often involves technical considerations that the investigator might not be qualified to address. This calls for additional experts to take part in the investigation. These experts, while they may be highly trained in their respective fields, are often untrained in how to operate safely at the fire scene. It is the responsibility of the investigator to ensure that all of the personnel who assist with the investigation are made aware of the hazards and of the precautions that have been taken to make the work area safe.

Safeguarding Bystanders

The fire scene is a scene of destruction that is foreign to most people. For this reason, the fire scene—especially a large-loss fire scene—often attracts crowds of people who come to observe the devastation. Some fire scenes take on almost a circus atmosphere, with numerous people loitering in the area. These individuals often attempt to get as close as they can to the incident, exposing themselves to the hazards of the fire scene. The fire investigator is responsible for the scene, and this responsibility includes the safety and protection of anyone who may wander into a hazardous area. (See Exhibit 10.1.)

All of these factors mean that the fire investigator must wear many hats during the course of the average origin and cause investigation. The fire investigator must always be cognizant of the hazards present at each scene and must stay in control of the scene for safety as well as scene preservation concerns.

EXHIBIT 10.1 Example of a Hazardous Area

See the section entitled "Occupational Safety and Health Administration" later in this chapter for information related to site safety plans.

INVESTIGATING A FIRE SCENE

The fire investigator must also be able to operate at the fire scene in conjunction with the fire fighters. The investigation must be conducted within the incident command system (ICS), through the incident commander.

Liaison with Fire Fighters: Using the Incident Command System

The fire investigator has a responsibility to contact the incident commander when he or she arrives at the scene. The incident commander can relay any pertinent information about the fire that could be valuable to the investigation. During this transfer of information, the fire investigator should also determine whether the fire fighters have discovered any hazards. The incident commander can relay information about structural stability, chemicals, or any other hazards present at that scene.

While investigating large losses in large buildings or in areas where it is difficult to keep track of the investigation team—especially if suppression activities are still in progress—an ICS, including an accountability system, must be utilized. The investigation team must report to the accountability officer and verify the areas where the investigators will be working and the specific jobs in which they will be involved while operating in that area. When they are finished in that location, the investigators should inform the accountability officer of the completion of that portion of the investigation and then verify the new location where they will be working. This information allows the accountability officer and the incident commander to monitor all personnel operating at the fire scene.

When the fire suppression units clear the scene, there must be a face-to-face transfer of command from the incident commander to the fire investigator in charge of the investigation. The transfer of command should identify any areas of concern that can affect the safety of any personnel remaining at the scene. The investigator at that point becomes responsible for the control of the scene and for the safety of everyone at the scene. Any personnel who are brought to the scene must be briefed on all of the known hazards, protective equipment requirements, evacuation signals, and emergency medical procedures. The lead investigator is also responsible for the security of the scene. It is essential to prevent the destruction or removal of evidence and to prevent bystanders from taking a "sight-seeing trip" through the fire scene and becoming injured or removing evidence.

Entering the Fire Scene

There is rarely a good reason for a fire investigator to enter the burning building before the fire has been brought under control. In the initial phases of the fire incident, the fire investigator will be busy conducting interviews and documenting the scene as well as searching the surrounding areas for potential evidence. If an occasion arises in which the investigator feels that there is a need to enter the building before the fire has been extinguished, the investigator must be equipped with structural fire fighting gear and a self-contained breathing apparatus (SCBA). This protective clothing and equipment must meet or exceed the requirements of NFPA. The fire investigator must also coordinate all investigation activities with the suppression crews and abide by the direction of the sector commander or the incident commander. The urge to "freelance" should always be resisted. All operations must be carried out under the direction or knowledge of the incident commander. During this phase of the incident, the priority is to extinguish the fire and protect lives, not to conduct an investigation, although the two can be done simultaneously.

The fire investigation community teaches fire fighters not to destroy the fire scene. Classes on the fire fighter's role in the origin and cause investigation stress the need to keep overhaul to a minimum until the scene has been documented and examined by an investigator. Because of this training, overhaul is commonly kept to a minimum, and fire-fighting operations are suspended when the forward progress of the fire is stopped.

Exiting the Fire Scene

Whenever the investigator enters a fire-damaged building, he or she must identify two separate means of exit from the area of operation. One means of exit could be a ladder being placed to a window or to the roof. If this type of exit is necessary, the ladder must remain in place at all times when the area is occupied by the investigator. The separate means of exit should be monitored to ensure that they remain free of obstructions during the investigation.

The fire investigator must beware of a "rekindle" occurring in the building. (See Exhibit 10.2.) The rekindled fire could grow to the point at which the fire investigator's escape is cut off. For this reason, the means of exit should be at opposite ends of the work area, and hoselines should remain in place during the initial phase of the investigation. When the investigation is complete, over-

EXHIBIT 10.2 Example of Fire Rekindling

haul operations must be conducted to ensure that the fire has been completely extinguished.

FIRE SCENE HAZARDS

The investigator should always guard against complacency, which can lead to injury and/or exposure to the toxins present at every fire scene. The fire investigator cannot afford to take a chance or risk. These risks and exposures over the expanse of a career can accumulate and lead to life-threatening health problems. Materials that are used in today's society are causing every fire scene to be volatile and hazardous. Toxins such as benzene, toluene, formaldehyde, and cyanide, among others, can be found at every fire scene. Blood tests show that nearly all fatal fire victims have toxic levels of cyanide in their systems—levels so high that some medical examiners are questioning whether carbon monoxide (generally the most prevalent toxin at a fire scene) or cyanide is killing fire victims. The investigator must also remember that toxins remain in the air after the fire has been extinguished. During the investigation phase of the fire incident, the toxic gases have cooled and are settling or mixing with the air throughout the room, where they are affecting the fire investigator through inhalation or absorption through the skin.

Other hazards that potentially exist at fire scenes occur because fire consumes oxygen and can create an oxygen-deficient atmosphere in confined areas. Chemicals that are present at the fire scene may have been released due to the failure of their containers. Carbon monoxide, a common by-product of fire, can cause chemical asphyxiation. Carbon monoxide joins with hemoglobin 250 times more readily than oxygen does, thereby preventing the body's cells from receiving oxygen. This effect can be cumulative and will build after several exposures, causing a collapse in less than toxic atmospheres. Cigarette smoking is an example of the cumulative effect of carbon monoxide, which can raise the carbon monoxide level in an individual to 10 to 15 percent even without the individual's being exposed to a fire.

Personal Health and Safety

The physical hazards that cause trip, fall, or crushing injuries are well known to everyone in the field. These injuries occur because fire scenes are cluttered and dangerous places. The fire investigator must constantly be aware of his or her footing and also take note of what is standing nearby or hanging above the work area. As dangerous as these conditions are, there are also toxins and hazardous materials present at every fire scene. These toxic materials pose a problem when they get to the susceptible target organ at levels above the toxicity limits or in a form that the body cannot overcome. The toxic reaction depends on the dose and the duration of the exposure. The dose can be acute, whereby a single incident exposes the individual to harmful levels of the toxin. The dose can also be chronic, whereby the individual is exposed at a lower concentration, but the toxin is absorbed over an extended period of time until a harmful accumulation is present in the victim's system.

People are injured by toxins via the four routes of exposure: inhalation, absorption, ingestion, and injection.

Inhalation. The most obvious of toxin injuries is the inhalation injury, caused when the investigator inhales a toxin that is present in the air. The lungs are coated with a thin membrane that allows oxygen to enter the bloodstream, but this membrane also allows toxins to enter the bloodstream rapidly. The toxin can be a gas such as carbon monoxide or a solid such as asbestos. In any event, the toxin gets into the lungs and either prevents the flow of oxygen (chemical asphyxiation, as occurs with carbon monoxide) or displaces the oxygen available.

Absorption. Absorption, the next route of exposure, is prevented by the body's largest organ, the skin. Intact skin is a good defense for many of the harmful materials that people come into contact with every day. However, some toxic materials, such as benzene, can be absorbed through the skin. Harmful materials can also enter the body through breaks in the skin. Intact skin protects a person from cutaneous anthrax. (That is, when the skin has a cut or an abrasion, the *Bacillus anthracis* spores can enter the body and become active, causing subcutaneous anthrax.) Because many of the chemicals and toxic materials present at a fire scene are skin-absorbable, the investigator must limit the exposure of bare skin to the by-products of the fire. Toxins can also be absorbed through the eyes (periocular). This route of exposure also allows the toxin to rapidly enter the bloodstream.

Ingestion. Although fire investigators would never ever eat anything they find at a fire scene, it is still possible to ingest harmful materials through poor hygiene practices. The investigator who steps out of a fire scene and eats, drinks, or smokes a cigarette can ingest the harmful materials that have adhered to the hands. Once ingested, these materials can be absorbed into the body through the stomach and intestinal walls. The mere act of washing hands reduces the risk of exposure greatly. All fire investigators should have some means to wash their hands at the fire scene. Hand wipes or other cleaners can be utilized until the investigator has access to a sink with soap and water, but they should not replace the soap and water.

Injection. Injection may not be an obvious form of exposure at the fire scene—injection normally occurs in a doctor's office when the physician or nurse administers an inoculation. However, the investigator must be alert to the possibility of this type of injury. Metal objects cool faster than other materials at the fire scene, and the cooling process can cause toxins to condense on the surface of a metal object. Then, if the fire investigator is cut by metal at the fire scene, the toxins adhering to the metal are injected into the bloodstream.

Physical, Biological, Chemical, and Radiological Hazards

The hazards present at a fire scene may be physical, biological, chemical, or radiological. Depending on the type of occupancy to which the fire investigator responds, any combination of these hazards may be present. The physical hazards are discussed later in the chapter. The investigator must also be aware of the other hazards present at the scene. For instance, the investigator who responds to a medical or construction facility can be exposed to radiological materials that are stored at these facilities. In addition, every fatal fire scene contains biological hazards. Rarely does the fire destroy all of the bodily fluids in a fire victim. The fire investigator must be able to assess the fire scene and choose the proper protection by blocking the routes of exposure from the hazards present at each specific fire scene. A fire investigation is not a life-or-death situation. Therefore, the investigator should not risk health and safety to determine the origin and cause of any fire.

The fire investigator must utilize all protective equipment (Exhibit 10.3) or engineering controls available to protect the investigative team from harmful materials by blocking the above named routes of exposure. This is sometimes easier said than done. But with the proper vigilance and by resisting complacency, the fire investigator can be protected from the known harmful materials present at every fire scene.

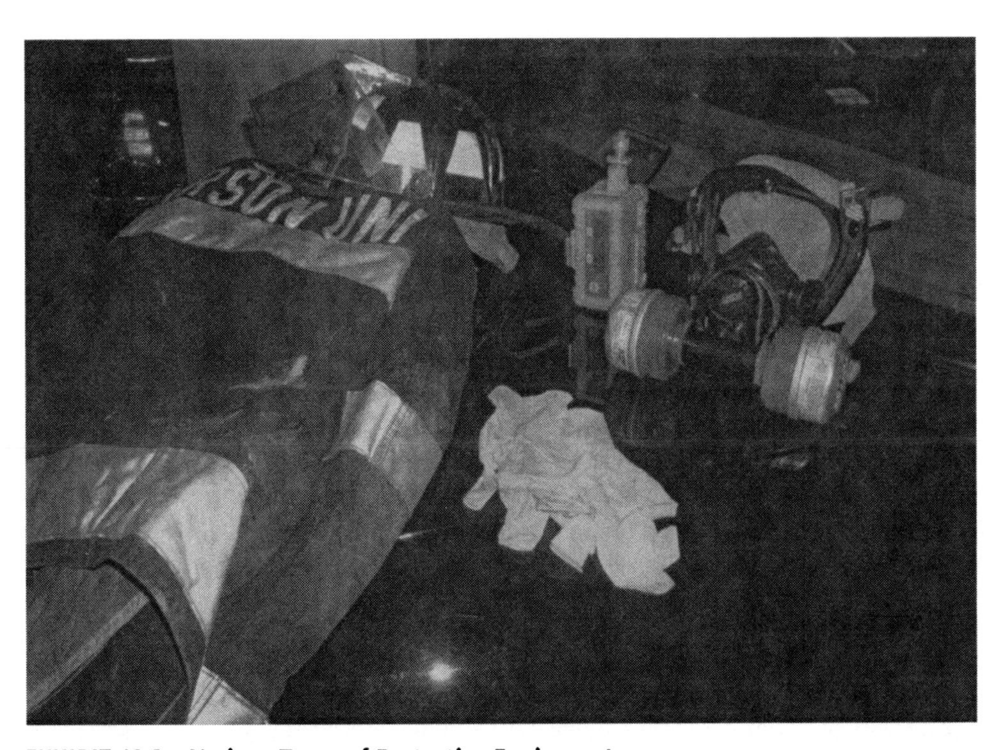

EXHIBIT 10.3 Various Types of Protective Equipment

STRUCTURAL STABILITY HAZARD IDENTIFICATION

Before entering any fire scene, the investigator must assess the stability of the structure and eliminate any physical hazards by either the use of shoring materials or demolition. Modern building construction methods utilize lightweight materials to carry heavier loads, and this lightweight construction allows the construction of larger buildings at economical costs.

Truss Roofs

During the 1970s, fire fighters were taught to beware of the bowstring truss roof, which killed far too many fire fighters. Unfortunately, almost every building under construction today has a truss in one form or another, and these trusses are just as dangerous as the bowstring truss. When any portion of a truss fails, the truss can no longer carry its intended load, and the surrounding trusses—which are also being exposed to the fire—then have to carry that additional load. This strain was not intended by the building's designers. Trusses, without the readily identifiable bow in the roof, are still killing fire fighters today. The fire investigator must be able to assess the condition of the building before entering and continuously examine the structural conditions as the investigation progresses.

A common building method in wood frame construction is the use of engineered trusses that are constructed at a truss company and transported to the construction site. These trusses are held together with gusset plates, which are metal plates with portions pushed out to form small spikes. These gussets are then pressed into the wood trusses. Unfortunately, the metal plates conduct heat rapidly. This heat then heats the wood surrounding each metal tab of the gusset plate, weakening the connection. At times, if a truss is damaged during transport, the construction crew may repair the truss using a wood gusset plate. The failure of this wood plate would cause the entire truss to fail.

Collapse Zones

Collapse zones that are set during the fire-fighting operations must be maintained during the investigation until the condition of the structure is examined. This examination should take into account not only the condition of the structure, but also the expected weather. An increase in wind or the presence of snow may affect the structural stability. The collapse zone should also be evaluated to determine whether it is sufficient for the hazard presented by the compromised structure. For instance, it is not uncommon for the spray of bricks from a collapsing wall to travel a distance equal to twice the height of the wall. When the investigator assesses the conditions, the type of materials used to construct the wall should be taken into consideration. When they hit the ground, brick or brick veneer walls will send bricks a greater distance than wood frame walls send the wooden studs or plywood siding. The type of ground also determines the distance of the spray of materials during a collapse. A cement walk or other paved surface allows the collapsing material to bounce and travel farther than soft ground does.

Structural Stability

The fire investigator must not take for granted that any portion of a building that remains standing after the fire is secure and safe. It is important to know that a collapsing wall normally does not make any sound until it hits the ground or whatever is standing between the falling material and the ground.

Freestanding chimneys pose a substantial risk to both fire fighters and fire investigators. These structures depend on the building structural members for their stability. When the building is compromised, so is the chimney. Also it is common for fire fighters who spend extended periods of time manning master streams to have "target practice" on a chimney or other freestanding object. The fire-fighting hose streams can place a torsional (twisting) load on the chimney that it is not intended to endure. The streams may also wash away the mortar from an older chimney, further weakening it.

The portions of the building that are not supported by trusses are also vulnerable to collapse. Very few buildings are constructed by using extra building materials to ensure structural stability. Most buildings are built with the minimum amount of materials required by engineering principles and local building codes. The removal of any portion of a building places unintended loads on the remainder of the building, which in turn puts the building in danger of collapse. When the roof collapses, the walls are now largely free standing. If a wall collapses, the roof and other walls are now in danger of succumbing to the forces of gravity. All components of the building's construction are interdependent. The removal of any component alters the system that was put into place to form the building.

Building Load

A building has two different types of building loads: the dead load and the live load. See Chapter 5, "Building Systems," for more information on building loads.

When the structure is designed, the designer takes into account its intended use and the expected live load. Often, due to changing occupancies, buildings are used in ways that the designer did not intend. Walls are moved when occupancies change from residential to commercial to industrial. These changes normally do not take into account the design of the building, thereby placing the building in stress. Poor maintenance, poor repairs, and aging of the building materials induce further stress on the structure. A fire can be the final straw or the means to set in motion the cause of a later collapse. When wood studs or joists are compromised or destroyed by fire, the surrounding studs or joists must carry the load. The studs may be able to accomplish this for only a short period of time.

Impact Load

Another consideration related to structural stability is the *impact load,* which is a sudden added load that can be caused by a fire fighter jumping from a ladder onto a roof or a fire investigator jumping down onto a floor after peering into the ceiling or over an appliance, for example. This sudden load may cause the collapse of already weakened or overburdened structural members. A partial building collapse may also provide the impact load that brings down the remainder of the building.

The most stable type of structure is the pyramid. The pyramid utilizes a decreasing pattern of masonry building materials. This "pile" will resist collapse. If we take these same building materials and stack them on top of each other, we cause instability. A freestanding wall, whether it is a parapet wall (the portion of a wall that extends above the level of a roof) or any wall that is no longer supported by other structural members, should never to be trusted without shoring. The safest way to handle a freestanding wall is to allow gravity to accomplish its goal and knock the wall down in a manner and at a time that the fire investigator determines.

Structural Members

Fire-fighting operations themselves can add to the stress on a building's structural members. Water weighs 8.35 lbs per gal (3.8 kg per l). A 2½-in. (6.35-cm) or even a 1¾-in. (4.45-cm) hoseline putting out up to 250 gal (946.4 l) per minute will place an additional live load of 2087.5 lbs (946.9 kg) into the building every minute that that hoseline is in operation. This additional load must be drained before the investigation can begin. During cold weather investigations, the removal of this water can be difficult or even impossible. Fire fighters often cut holes in floors to remove water from the structure. The fire investigator must be aware of this possibility and identify any holes in the floor. Often in urban areas, the fire fighters use doors from the building to cover these holes and any holes caused by the fire itself. These doors must not be relied on to hold the weight of the fire investigator.

An investigation in which the stability of the structure is under question is a strain on any fire investigation team. The team should make the work area safer by mitigating the hazardous condition. If the stability of the building is a concern, the investigator should utilize a controlled demolition and stabilization of the structure. Demolition is the safest because the hazard is removed. The demolition team should work under the direct supervision of a trained fire investigator who has intimate knowledge of the goals of the investigation. This investigator can then direct the equipment operator to remove the portions of the structure that are unstable without destroying the areas of interest to the investigation team. A good heavy equipment operator working in conjunction with an experienced fire investigator can remove the hazardous portions of a fire building without destroying evidence or the investigation. This controlled demolition will make the fire scene safer and allow for a successful completion of the origin and cause investigation.

INVESTIGATING THE SCENE ALONE

The fire scene contains many hazards that can affect the health and safety of the fire investigator. Some of these hazards have long-term effects, while others present an immediate risk to the investigator's health. Whenever an investigation is conducted where a hazardous condition exists, the investigator should not work alone. The presence of at least two fire investigators provides for another set of eyes and hands to assess the dangers and handle any problems that might be encountered. If an investigator becomes injured or trapped, the second investigator can provide assistance or call for help. Ideally, when the investigation is conducted under hazardous circumstances, a rapid intervention team should stand by to provide rescue assistance if needed. This team should be staffed by experienced fire fighters who have been trained in urban search-and-rescue techniques.

Unfortunately, due to monetary or staffing restrictions, fire investigators are often required to operate alone at fire scenes that do not present an immediate hazard. When this is the situation, the investigator should notify a responsible person of the location, the expected work area at that scene, and the expected time required to complete the investigation. The investigator should always carry a two-way radio, cellular telephone, or other means of communication. If there is an unexpected delay, the investigator should notify the responsible person of the delay and adjust the estimated completion time and area of operation. To prevent complacency, the investigator should be diligent in communicating with the responsible person. If the emergency contact becomes accustomed to an investigator who forgets to advise when the investigation is completed, that investigator might not receive help if an emergency occurs.

CHEMICAL HAZARDS

When the hazard is of a chemical nature, the investigative team must utilize protective equipment appropriate to the specific hazard. This equipment, however, requires specialized training and a staffing complement that may be prohibitive to smaller investigation units. The safer action might be to mitigate the material that is causing the hazardous conditions. Even this action would require the investigator to work closely with the hazardous materials team to protect the scene and preserve any evidence of the fire's origin and cause. After the area has been made safe, the investigative team can operate without the use of restrictive personal protective equipment.

The fire investigator should have the training and experience to assess any chemical hazards that may be present at the scene and properly address the hazardous condition without compromising the investigation. The investigator should also be able to determine the size and makeup of the investigative team that is required for a specific incident.

SAFETY CLOTHING AND EQUIPMENT

As was mentioned earlier in this chapter, fire investigators must protect themselves from physical, biological, chemical, and radiological hazards while conducting an origin and cause investigation. Toxic gases, hazardous chemicals, asbestos, flammable gases, and oxygen-deficient atmospheres are a few of the dangers at almost every fire scene. All of these hazards have the potential to cause serious harm or even death to anyone who enters a fire scene unprotected and unaware.

Fire investigators should be aware of the protective clothing and equipment that are available—but how many investigators actually wear and utilize these items while investigating a fire scene? Just as police officers are saved by their bulletproof vests and fire fighters are saved by their turnout gear and SCBA, fire investigators have protective clothing and equipment available to them that should always be worn and used at all fire scenes. This protective equipment should be of a quality that is capable of providing the desired level of protection. The equipment should also meet the standards of an impartial organization such as NFPA or ASTM.

The Helmet

The protective ensemble should start with a helmet, which should be worn during all fire investigations. It is important to choose a helmet that is sturdy enough to provide the desired level of protection. The helmet must also be comfortable so that the investigator will wear it during extended operations. This helmet must protect the wearer from impact and from electrical hazards. An improperly constructed helmet or a helmet with unapproved additions could deliver electric current directly to the wearer's head. There should also be a suspension system to absorb all but the most severe impact. The helmet should have a chinstrap to hold it in place after an impact. The chinstrap should be used whenever the helmet is worn in potentially hazardous areas, because without it, the helmet may fall off, and additional falling debris could cause severe head injuries. The helmet should also have some form of built-in eye protection that is easily deployed by the wearer. This eye protection should be constructed of a material that resists scratches.

ACTIVITY 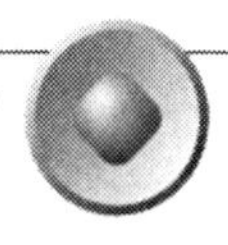

Identify the protective clothing and safety equipment that would be needed to safely conduct an origin and cause investigation. ◀◀◀

Protective Clothing

The investigator should wear protective clothing to prevent contamination of street clothes. This clothing could be fire-fighting turnout gear or coveralls. The fire investigator is exposed to toxins at every fire scene. These toxins become embedded in the investigator's clothing and in anything the investigator contacts after the exposure—including the investigator's vehicle, office furniture, and home—if proper precautions are not taken. This cross-contamination exposes the investigator, coworkers, and their families to the toxins from the fire scene. The coveralls or turnout gear should be washed regularly in accordance with the manufacturer's instructions. This procedure often calls for machine washing with a mild soap and water—but not in a washing machine that is normally used to clean other clothing. A separate washing machine or a company that specializes in cleaning fire-fighting equipment should be used. Certain toxins that are present at every fire scene (such as benzene) are absorbable through the skin. The improper cleaning of protective clothing can cause repeated exposure to this carcinogen as well as others.

Respiratory Protection

Today's fire investigator also needs to address respiratory protection at every fire scene. This respiratory protection could come in the form of SCBA, cartridge filter masks, or engineering controls such as mechanical ventilation. No matter which form of respiratory protection is chosen, the work area should be monitored at all times during the investigation. The ideal protection can be obtained with the use of engineering controls to remove the hazard from the work area. Positive pressure ventilation will flood the work area with fresh air efficiently and effectively. There may be times when positive pressure ventilation is not practical. During these investigations, the fire investigator can use a combination of positive pressure ventilation and negative pressure ventilation to help remove the toxic gases from the area or reduce their concentrations. The fire investigator should have a good grasp of the theory behind mechanical ventilation to be able to ventilate any area where the fire investigation may take place.

When respiratory protection is utilized, the employer must have a written respiratory protection program in place. All employees who will be utilizing respiratory protection should be trained in the selection, use, and maintenance of the proper equipment for the hazards encountered. The training should also address emergency procedures for any equipment failures. Medical surveillance is also required to determine the physical abilities and limitations of the user. Periodic medical examinations may be necessary to determine whether any employee has been exposed to a toxin that has the potential of causing a medical problem.

Self-Contained Breathing Apparatus. There may be times when ventilation alone cannot bring the levels of the toxins down to a safe level. During these investiga-

EXHIBIT 10.4 Self-Contained Breathing Apparatus

tions respiratory protection must be utilized. SCBA provides the highest level of respiratory protection. (See Exhibit 10.4.) Unfortunately, SCBA limits the amount of time that the user can be protected, due to the finite amount of fresh air in the compressed air cylinder. Usually, a compressed air cylinder is rated to provide 30 to 60 minutes of breathing air, although this time varies depending on the physical condition of the wearer and the level of physical and mental strain the user is experiencing. The SCBA face piece also restricts the visual abilities of the user, thereby limiting the investigator's ability to observe the scene. An alternative, the supplied air breathing apparatus (SABA), provides breathing air for longer periods, but the hose is limited to 300 feet. The investigator who wears any breathing apparatus should work in 30-minute periods to properly allow for rehydration, rest, and monitoring of physical condition due to the stress and strain of wearing a breathing apparatus. If the investigator is utilizing SABA, it is tempting to extend the 30-minute work period, overexposing the investigator to dangerous strains and exposures to heat stress, which should be avoided.

Air-Purifying Respirator. The air-purifying respirator (APR) can be utilized at certain fire scenes. However, this type of respiratory protection has severe limitations. Air monitoring is a must when using an APR, because the user is not protected from oxygen-deficient atmospheres or from areas where toxins exist at a level that is deemed "immediately dangerous to life and health" (IDLH). The APR is toxin-specific, so the user must know all of the hazards that are present at the scene to be able to select the most effective filter. Unfortunately, there is no possible way to identify all of the hazards present at any fire scene. Whenever unknown quantities of two or more molecular compounds break down due to their exposure to an unidentified amount of heat, as in a fire, it is impossible to determine

EXHIBIT 10.5 APR with Eye Protection

EXHIBIT 10.6 APR Without Eye Protection

exactly what material is ultimately created and what its level of toxicity might be. Investigators can determine general hazards, but the unknown is always present at any fire scene.

Exhibit 10.5 shows an APR with eye protection. Exhibit 10.6 shows a form of APR without built-in eye protection.

Eye Protection

The fire investigator should wear eye protection that provides the maximum level of safety. Safety glasses are more effective than the face shield on a helmet when fire debris splashes the user. To fully protect the eyes, safety glasses should be used along with the eye protection provided by the helmet. The investigator should remember that beyond obvious eye injuries, any toxins that enter the eyes will quickly enter the bloodstream. Eye protection should also shield the sun's ultraviolet rays.

Foot Protection

Proper foot protection can prevent major and minor health-threatening injuries. A safety shoe protects the foot from the hazards associated with stepping on nails or sharp objects, as well as from heavy objects falling on the toes. A properly waterproofed shoe protects against exposure to toxins that are present at fire scenes. Rubber boots provide a high level of protection, as do the traditional thigh-high fire-fighting boots when they are pulled up over the knees and used in association with hard kneepads.

Gloves

The fire investigator should wear gloves at every fire scene. Traditional fire-fighting gloves provide limited protection from heat and physical injuries, but they do not provide any chemical protection. Therefore, the fire investigator should consider double-gloving during the investigation, wearing an inner glove that provides a limited level of chemical protection. The inner glove should be made of a material that has been chosen after conducting an assessment of the hazards at the fire scene. For instance, latex gloves provide protection from biological agents but are not effective during extended periods of wear under wet conditions. For the average fire scene, nitrile gloves with some puncture resistance are commonly used and provide a higher level of protection. When the fire investigator arrives at a scene, the inner glove should be worn from the beginning of the investigation. To prevent cross-contamination, this inner glove can be changed after every sample is collected.

Safety Equipment

Each fire scene is different, and at times, additional equipment may be needed to provide a safe work environment. All work areas should be well lit to prevent trip-and-fall injuries and to enable the investigator to effectively assess the hazards and complete the investigation. Portable lighting is required at almost every fire scene, but at no time should a generator be brought into an enclosed work area; the buildup of carbon monoxide would quickly become a serious health issue (the action level for carbon monoxide is 25 PPM). Furthermore, when an internal combustion engine is brought into the building, the fire investigator can no longer state that the investigation team did not bring an ignitible liquid into the scene.

Some investigations require lifelines and fall protection. This equipment should be maintained and stored in accordance with the manufacturer's specifications. The investigator should not depend on any ladders that are present at the fire scene other than those brought by the fire department. Any ladders that were in the fire building may have been exposed to the effects of the fire, or the owner of the building may have abused or improperly maintained the ladders before the fire. If a ladder is required, the investigator should be comfortable with its condition and should ensure that it is the proper height and style for the intended use.

Some safety equipment requires specialized training and experience. Even if the investigator is capable of operating specialized equipment, an operator who routinely works with the equipment is more skillful and should be utilized if available. For instance, shoring equipment or other urban search-and-rescue techniques may be required to protect the investigator at a fire scene. All of this equipment and the techniques that are used require training beyond the levels of basic fire investigation courses and should be part of the fire investigator's continuing education.

HAZARD ASSESSMENT

NFPA 1033, *Standard for Professional Qualifications for Fire Investigator,* 2003 edition, states that a fire investigator must meet the requirements of NFPA 472, *Standard for Professional Competence of Responders to Hazardous Materials Incidents,* 2002 edition, Paragraphs 4.2.1 through 4.2.3, "Competencies for the First Responder–Awareness Level." (See Annex B.) Individuals who are trained to the first responder–awareness level must be able to identify a chemical hazard and initiate the next level of response to a hazardous materials incident; the responder is not expected to come into contact with the hazardous material or to mitigate the hazard. Under normal circumstances, this awareness level is sufficient for the fire investigator. However, there are many times when the fire investigator will be responding to an incident during which chemicals have been released. Any scene that has experienced the uncontrolled release of a hazardous chemical can be considered a hazardous waste site. If the scene is officially declared a hazardous waste site, an investigator who is trained only to the first responder–awareness level can be denied entry. In some cases, the fire investigator may be in command of a scene that requires the mitigation of a hazardous material. Personnel performing these actions must be trained to the hazardous materials technician level according to 29 CFR 1910.120, HAZWOPER ("Hazardous Waste Operations and Emergency Response"). The fire investigator who is trained to this level would be able to be in complete control of any fire scene.

The investigator must take into account all occupancies and the hazards that are present prior to starting the investigation. (See Exhibit 10.7.) This hazard assessment ensures a safe investigation.

Training

As was mentioned previously, according to the HAZWOPER standard, any employee who is expected to encounter a hazardous material and be involved with its mitigation should be trained to the hazardous materials technician level. This training involves the hazard assessment process, monitoring of hazards, and the selection and use of protective equipment and clothing. The training should also include methods for containing and mitigating a hazard and monitoring the scene with gas meters or other detection equipment. All fire investigators can be expected to encounter hazardous materials during the course of an investigation, and it may be necessary to mitigate these hazards. At large-loss investigations, representatives of OSHA may already be present at the scene, and the investigator may not be permitted access to the site if he or she does not possess the Hazardous Materials Technician certification. This level of training is becoming necessary even for the average fire investigator, due to the scenes that the investigator is called to investigate.

EXHIBIT 10.7 Occupancies and Hazard Assessment

Air Monitoring

To safely operate at any hazardous fire scene, the investigator should also require air monitoring. This monitoring must be conducted by a Hazardous Materials Technician who has been trained in the operation of the monitoring equipment used to determine the levels of a multitude of hazardous or toxic materials that are present at nearly every fire scene. In monitoring for an unknown hazard, the sequence should be to test first for presence of radiation, then for oxygen levels, then for flammability, and finally for toxicity.

Radiation. A radiological survey meter is used to determine the presence of any radiation (alpha, beta, gamma, and x-rays). This meter should be able to detect all forms of radiation, as alpha, beta, gamma, and x-rays exist naturally in many locations. (See Exhibit 10.8.) Furthermore, during any incident involving an explosive, the fire investigator should monitor for the presence of radiation to rule out the possibility that the explosive device contained radioactive materials placed by a bomber to cause more damage and panic.

Oxygen. The next test that is conducted should determine the percentage of oxygen present in the workspace. Normally, there should be 20.9 percent of oxygen present in the air. Any variation from this norm should be addressed and considered a serious problem. An oxygen-deficient atmosphere could be caused by a fire in a poorly ventilated space, but other hazards can also cause this condition. Most flammable gases are heavier than air. These gases, among others, will displace oxygen if they are released into an enclosed or poorly ventilated space. Any time the percentage of oxygen falls below 18 percent, there is a serious threat to the

EXHIBIT 10.8 Radiological Survey Meter

investigator's health. On the other hand, an oxygen level above normal is also an indication of a problem. The fire investigator must determine why oxygen exists at an above-normal level in this space. It should be considered a hazardous condition—an indication that a chemical reaction might be taking place or that there has been a release of pure oxygen. The added oxygen will affect the flammability ranges for any combustible gases in the area and also make ordinary combustible material easily ignitible.

Flammability. Flammability can be measured at the same time as oxygen in most combustible gas detectors. It is important that the combustible gas detector always be utilized in conjunction with the oxygen meter. All flammable gases have an explosive range within which they can burn. (It is important to remember that all gases ignite explosively.) A gas that is present below its lower explosive limit (LEL) will not have enough fuel to support the fire, and a flammable gas that is present at a concentration above its upper explosive limit (UEL) will have too much fuel but not enough of an oxidizing agent to burn. The combustible gas detector determines the percentage of the LEL to which the investigator is exposed, and then the user determines whether the area is safe to work in. For the purpose of a fire investigation, any presence of an ignitible gas should be cause to evacuate the area and ventilate and control the leak before the investigation can continue.

Exhibit 10.9 shows a gas detector that can be used to determine whether a harmful gas is present. Such detectors can be purchased with sensors for the most common hazards in a specific response area.

Toxicity. Finally, air-monitoring procedures test for toxicity. Toxicity can be measured by various meters or colorimetric tubes, such as the one shown in Exhibit 10.10. These items are product specific, and the investigator must have an idea of the hazards present to effectively measure for toxicity of a suspected material. All of these meters require specialized training. The user should be familiar with the detector or the meter to be able to effectively understand what the instrument is reading.

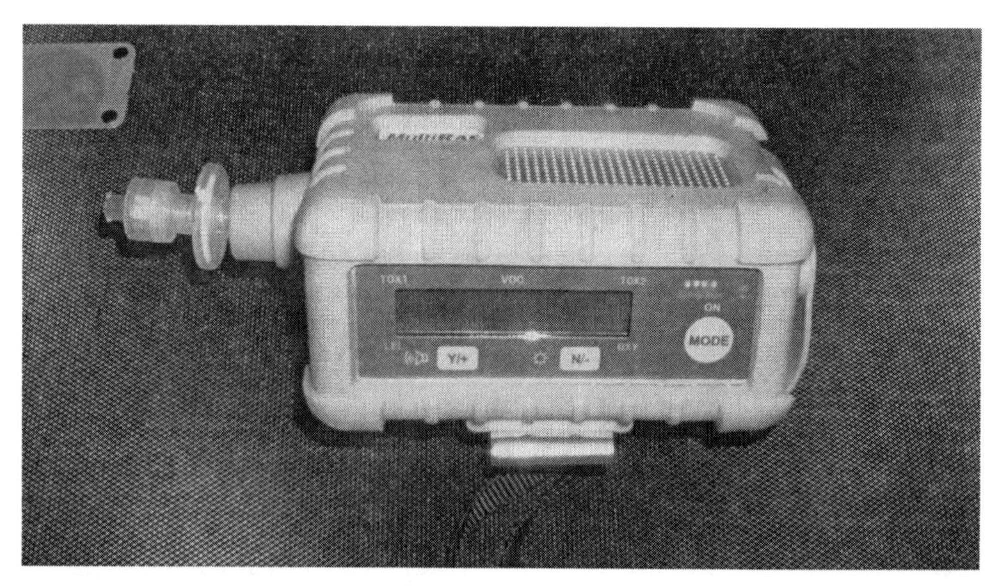

EXHIBIT 10.9 Gas Detector

EXHIBIT 10.10 Colorimetric Tubes

Each meter or detector that is used is different and must be tested before each entry. The meter sounding an alarm might not be telling the user what he or she thinks it is. Meters are calibrated to specific gases, and the user must employ conversion equations to determine the specific levels of a noncalibrated gas. Unless the investigator has received specialized training in the use of these devices, he or she should not depend on them for health and safety.

OCCUPATIONAL SAFETY AND HEALTH ADMINISTRATION (OSHA)

The following OSHA standards most directly affect the fire investigator.

29 CFR 1910.120, HAZWOPER

OSHA regulations in 29 CFR 1910.120, HAZWOPER standard deal with operations at a hazardous waste site and the emergency response to hazardous materials incidents. By their nature, all fire scenes are potential hazardous waste sites. The

HAZWOPER standard, as referenced in 40 CFR, "Protection of Environment," Part 311, states that the provisions in 29 CFR 1910.120 apply to state and local employees who are involved in hazardous waste operations. This protects all employees, even public employees operating in states that do not have a state OSHA plan. All fire investigators should be familiar with the following NFPA documents, which can clarify the HAZWOPER standard:

- NFPA 471, *Recommended Practice for Responding to Hazardous Materials Incidents*
- NFPA 472, *Standard for Professional Competence of Responders to Hazardous Materials Incidents*
- NFPA 473, *Standard for Competencies for EMS Personnel Responding to Hazardous Materials Incidents*

As stated in these documents, the investigator is expected to understand the states of matter and recognize hazards by occupancy, shape of containers, markings, and odors reported to the responder.

29 CFR 1910.146, Permit Required Confined Spaces

The "Permit Required Confined Spaces" standard deals with the safety aspects of entering a confined space. Before this standard was enacted, 60 percent of all confined space incident deaths were incurred by would-be rescuers. The fire investigator should be able to identify a confined space, and at no time should a confined space be entered without the proper precautions in place. A *Permit Required Confined Space,* as defined by OSHA, includes the following criteria:

1. Any space that an employee can bodily enter and perform assigned duties
2. With a limited or restricted means of entrance or exit
3. That is not designed for continuous employee occupancy

Entry into a confined space requires special training and further requires the continuous monitoring of the atmosphere, fall protection, ventilating the space, and wearing a supplied air breathing apparatus.

29 CFR 1910.147, Lockout/Tagout

Whenever an investigative team works around equipment or wiring where an *unexpected* energization, start-up of machines, or release of stored energy could result in the injury of the investigators, the team must consider OSHA's "Control of Hazardous Energy (Lockout/Tagout)" standard. This standard utilizes a lockout/ tagout device that disables the equipment. Every investigator who works in the area places a padlock on the device, and the only key for the locks remains with the investigator. Then, on completing the investigation, each investigator leaving a hazardous area must unlock the device before power can be restored or energy released. This way the lockout device cannot be removed until every investigator has left the hazardous area and has been accounted for. This standard should be followed whenever the investigator is working in or around heavy equipment or electrical equipment or where a valve can be operated remotely, releasing a material that can engulf the investigator.

BYSTANDER SAFETY

All bystanders, including the occupants of the fire-damaged structure and nonessential responders, should be kept at a safe distance from the fire scene. The fire investigator should establish or adjust and enforce collapse zones. The collapse zone must be identified by markers or specialized scene tape. All responders should be made aware of the collapse zones verbally, since most police officers, fire fighters, and contract personnel are not normally restricted by the use of "scene tape." These zones are set to prevent injury as well as prevent evidence contamination. Bystanders who wander too close to the fire ground could be injured by any one of the myriad of hazards present at the fire scene. The fire investigator should also identify a safe, secure place to conduct interviews, away from the distraction and hazards of the fire scene.

ADDITIONAL SAFETY CONCERNS

One thing that many fire fighters forget when they become involved in fire investigation is that the public no longer holds the image of them as public servants who are there to help. They have become part of the law enforcement establishment. Like police officers, they must realize that the criminal is not the only threat to their safety. Private sector investigators often work alone and must be cognizant of additional hazards. The investigator must realize that a fire may have been set intentionally inside of the structure. The person who commits arson does not consider the safety of others before his or her own interests. When an investigator is on the street, he or she should wear a bulletproof vest and utilize the street survival tactics that were learned in training.

The fire investigation field can be a rewarding career. The fire investigator must realize that there are hazards that go along with this new career. The origin and cause investigation is not a life-threatening investigation, and no lives should be risked to conduct this investigation. Any risks must always be weighed against the gain. If no lives are at risk, the investigator cannot justify risking his or her own life or health to conduct a fire investigation.

This activity may be done alone or with several colleagues. Utilizing fire scene photographs, identify safety hazards present at fire scenes. Identify safety practices that would remove or mitigate those hazards. ◖◖◖

▶▶▶ QUESTIONS FOR CHAPTER 10

1. True or false: Fire scene examinations should not be undertaken alone.
   - A. True
   - B. False
2. What are the four main hazard categories of which an investigator must be aware when investigating a fire scene?
3. Who grants permission for a fire investigator to enter a scene before the fire is completely extinguished?
   - A. State fire marshal
   - B. Investigator's immediate supervisor
   - C. Incident scene commander
   - D. Fire suppression personnel
4. What should an investigator know about the structure's utilities before entering the fire scene?
5. What steps should be taken when an unauthorized individual is found within the fire investigation scene area?
   - A. Identified, noted, and required to leave
   - B. Arrested on suspicion of arson
   - C. Identified, noted, and held for questioning
   - D. Arrested for loitering

Sources of Information

OBJECTIVES

Upon completion of Chapter 11, the user will be able to

- Expand a fire investigation beyond the immediate fire scene through a variety of sources of information
- Differentiate between public and protected sources of information
- Determine potentially useful types of information

A thorough fire investigation starts with examination of the fire scene and evaluation of the documentation of the scene. Additional information must be researched and analyzed by the fire investigator. These combined efforts provide the investigator with the opportunity to establish the origin, cause, and responsibility for a particular fire.

Read Chapter 13 of NFPA 921 for an overview of the variety of information that is available.

ACCESS TO INFORMATION

The availability of information to the fire investigator is governed by legal considerations. The Freedom of Information Act provides for public access to information held by the federal government. Most states have laws that give similar access to state records. Privileged and confidential communications, however, are protected by law from forced disclosure. These include husband–wife, attorney–client, and doctor–patient communications.

FORMS OF INFORMATION

Important information is available to the fire investigator in interviews, written records and electronic data, and visual and scientific documentation. These three categories summarize the formats that are maintained by governmental agencies and private industry.

Interview Sources

Recorded verbal accounts can provide useful and accurate information, even if the interviewee was not an eyewitness to the event. Credible witnesses may not include those with a specific interest in the outcome of a fire investigation, and these individuals should be approached with caution. Effective interviews are the result of good planning, developed by an investigator who has a thorough understanding of all facets of the inquiry. Timeliness and the setting of the interview are also important considerations. Complete documentation of the interviewee's identity is necessary to ensure the reliable evidentiary value of the interview.

Written Records and Electronic Data

Government records provide a wealth of information that is accessible to the investigator. Street maps, building permits, blueprints, and property ownership records contain some of the vital data that investigators need for a fire scene investigation. Municipal, county, state, and federal governments all maintain records that should be accessed as appropriate. A government agency that maintains extensive databases of fire incidents is the U.S. Fire Administration, which oversees the National Fire Incident Reporting System (NFIRS) and the Arson Information Management System.

Private sources of valuable information include the following organizations:

- American Society for Testing and Materials (ASTM)
- International Association of Arson Investigators (IAAI)
- National Association of Fire Investigators (NAFI)
- National Fire Protection Association (NFPA)
- Society of Fire Protection Engineers (SFPE)

The real estate and insurance industries maintain valuable records concerning structures and their owners, which can aid in the detection of fraud and arson. Utility companies and trade organizations are two examples of the diverse private sources of useful reference materials.

Visual and Scientific Sources

Still photography and video footage provide uniquely informative documentation of a fire incident. In addition to these visual sources, lightning and weather data can play an important part in causal analysis of a fire scene. Local television stations, lightning detection networks, and the U.S. National Oceanic and Atmospheric Administration (NOAA) are some of the alternative resources that should be considered.

◗◗◗ QUESTIONS FOR CHAPTER 11

1. Which of the following may NOT be a component of a thorough fire investigation?
   - **A.** Photos of the fire scene
   - **B.** Interviews with witnesses
   - **C.** Visiting the fire scene
   - **D.** Opinions of owners who have pending insurance claims
2. Which of the following is generally NOT accessible under the Freedom of Information Act and related state laws?
   - **A.** Birth certificates
   - **B.** Real estate transactions
   - **C.** Attorney–client communications
   - **D.** Legal descriptions of property
3. The U.S. Fire Administration
   - **A.** Maintains an extensive database of fire incidents
   - **B.** Oversees the Freedom of Information Act
   - **C.** Prepares lightning reports
   - **D.** Writes codes and standards for fire investigations
4. When investigating possible arson, which of the following would NOT be relevant?
   - **A.** Information from NFIRS
   - **B.** SFPE research
   - **C.** IAAI resources
   - **D.** FBI records
5. Which of the following would NOT be a source for visual or scientific data?
   - **A.** NOAA
   - **B.** Television studios
   - **C.** Interviews
   - **D.** Lightning detection systems

Planning the Investigation

OBJECTIVES

Upon completion of Chapter 12, the user will be able to

- Identify the types of information that affect the organization of a fire investigation
- Organize an investigation team

For any fire investigation, planning is an important first step. Considerations should include size and complexity of the fire scene, level of safety at the scene, potential for loss of life, number of investigators, staffing, and budget.

It is important to utilize the team concept whenever possible. A fire scene investigation includes photography, sketching, evidence collection, witness interviews, and other varied tasks that require diverse skills. By using the team concept, the investigator in charge can delegate these functions to the individuals best qualified to perform them and thus ensure a thorough and professional investigation.

Use of outside contractors, such as heavy equipment operators, should be considered as well. To save valuable time when the investigation actually begins, the investigator should develop these resources before they are needed.

BASIC INCIDENT INFORMATION

Although all fires differ from each other in many ways, certain basic information must be gathered at all fire scenes. The investigator must remember that the accuracy of the information that is gathered at this point is paramount to the investigation, because it determines many of the procedures that follow, and a mistake at this point could jeopardize the results. Basic data should include the location of the incident, the date and time of the incident, the weather conditions at the time of the incident, the size and complexity of the incident, the type and use of the structure involved in the incident, the nature and extent of the damage, the security of the scene, and the purpose of the investigation.

INVESTIGATION ORGANIZATION

The person who is responsible for planning the investigation should identify functions and responsibilities and then organize the team members according to their skills. It is important to keep the team concept in mind at this stage of the planning process. There are basic functions that are commonly performed in each investigation. These assignments or duties include leadership, coordinating, safety assessment, safety officer, photography, note taking, mapping, diagramming, interviewing witnesses, searching the scene, evidence collection, and preservation.

PREINVESTIGATION TEAM MEETING

Under the team concept, the preinvestigation meeting is the ideal forum for the team leader or lead investigator to explain the goals of the team, introduce team members, and hand out assignments. The team leader should address issues such as jurisdictional boundaries, legal authority to conduct the investigation, assignment of specific responsibilities, and collection of basic information.

The team leader should also ensure that each member of the investigation team is aware of the conditions at the scene and any safety considerations. The team leader should discuss the safety clothing and equipment that might be necessary and should verify that team members have that equipment when they get to the site.

The team leader should remind team members that plans made in this preinvestigation meeting do not preclude revised plans later. Once the investigation begins, there will be daily briefings for all team members to relay pertinent information that was gathered during the day's activity. Team members will get an updated overview and will have the opportunity to request that additional information be sought by interviewers, for instance, or to request the assistance of specialized personnel.

ACTIVITY

Read NFPA 1561, *Standard on Emergency Services Incident Management System,* 2005 edition. ◀◀◀

SPECIALIZED PERSONNEL AND TECHNICAL CONSULTANTS

Planning should always allow for the possibility that at some time during the investigation, it will be necessary to bring in additional personnel or consultants—people with specialized expertise to assist in analyzing the incident or specific aspects of the investigation. Experts may be available from colleges or universities, government agencies, professional societies, trade groups, or consulting firms.

It is advantageous to identify these resources in advance of the incident. However, the investigator must ensure that there is no conflict of interest with any of the outside experts. Types of specialists to consider should include a materials engineer or scientist, a mechanical engineer, an electrical engineer, a chemical engineer/chemist, a fire protection engineer, a fire engineering technician, an industry expert, an attorney, an insurance agent/adjuster, and an accelerant detection canine team.

This activity may be done alone or with several of your colleagues. Develop several fire investigation scenarios, from small scale to large scale. Work through the planning process with the goal of developing a resource list and an investigation worksheet. ◀◀◀

QUESTIONS FOR CHAPTER 12

1. What should a fire scene investigator try to avoid when bringing in outside specialized or technical experts?
2. True or false: The need to preplan investigations is largely dependent on the number of people involved and size of the fire or explosion.
   A. True
   B. False
3. With what should each person be equipped when investigating a fire scene?
4. If the investigator has established a team, what should the team members do before conducting an on-scene investigation?
5. List five of the specialized or technical people who could be enlisted at a fire scene.

CHAPTER 13

Documentation of the Investigation

OBJECTIVES

Upon completion of Chapter 13, the user will be able to

- Document conditions at a fire scene using various media
- Utilize various types of photographic equipment
- Describe the importance of note taking and the use of sketches and diagrams to analyze a fire scene

In recording any fire or explosion scene, the investigator's goal is to record the scene through a medium that will allow the investigator to recall his or her observations at a later date to document the conditions at the scene. Common methods include the use of photographs, videotapes, diagrams, maps, overlays, tape recordings, and notes.

PHOTOGRAPHY

The primary method of recording the scene is through the use of visual media. Photography—whether digital, still, or video—provides the investigator with pictures of the scene that can be used as points of reference when writing the report. These pictures present the investigator and other examiners with the most concise depiction of the condition of the scene, thus improving the process of pattern identification. Furthermore, because the report may be written some time after the actual site investigation, the visual recordings are an effective way of reminding the investigator of the condition of the fire scene at the time it was investigated.

Recording the scene through visual representation should be accomplished as early as possible in the investigation. This helps to identify when debris, furnishings, or other contents might have been moved during the course of suppression, overhaul, and reconstruction, or whether the structure or its contents were altered or destroyed. In addition, by documenting the scene early, the investigator can ensure accurate representation of the condition of the scene in case of subsequent building collapse or other structural or mechanical defects that can render the scene unsafe.

Photographs are acceptable for presentation in court. Typically, courts accept photographs that are objective and do not inflame or exaggerate. The presenter of the photographs to the court must be able to affirm the accuracy of the depictions as compared to what he or she saw at the scene.

Before the visual recording process begins, however, the investigator should have a fundamental understanding of photography, including familiarity with the equipment and accessories, lighting and movement, and film and film speeds. These issues are important because the predominant color encountered in a fire or explosion scene is black. In interior photography of a structure that has been subjected to the effects of thermal and mechanical insult, various shades of black must be realistically depicted on film. Proper lighting and film speed allow the investigator's recordings to accurately reflect what he or she observes.

Cameras

General Considerations. A multitude of cameras are available at a wide range of prices. The investigator is limited in camera choice primarily by financial resources and his or her personal skill level. The commercially available cameras can be categorized according to price and by manual versus automatic operation. Automatic cameras are the easiest to operate. They determine the primary scope of the photograph and then focus on the most obvious image in the viewfinder. These cameras can provide a sense of comfort to some investigators because they automatically determine film speed, light requirements, and focusing, thereby removing many potential pitfalls for the inexperienced photographer. Some investigators, however, prefer manual cameras because they allow the user to adjust focus and other settings specifically to suit the immediate circumstances. In addition, manual cameras allow bracketing, taking a series of photographs with sequentially adjusted exposures, which ensures at least one good picture when circumstances make it hard to determine the correct exposure setting.

Film Speed. Film speed is an important consideration that can determine the ultimate quality of photographs taken at the fire scene. Film speed can vary from 100 to 1600 ASA for color photography and up to 6400 ASA for black-and-white photography. The higher the ASA, the better the photograph is able to depict objects in darker situations. Unfortunately, with higher ASA numbers, the quality of photo enlargements decrease—they appear grainy, and detail is lost. Most investigators use film that has a rating of 100 to 400 ASA.

Digital Photography

Digital cameras record photographs onto one of three devices: floppy disk, compact disc (CD), or memory card/stick. Once the digital image has been captured, the image should be preserved on a nonalterable medium, such as a CD-ROM. The investigating agency should have a written policy regarding this practice. Once the image is preserved on a suitable medium, it can then be manipulated to enhance views to be printed to a high-quality paper. This manipulation does not change the original photo preserved on the CD-ROM.

Lenses

Variety of Lenses. Camera lenses can accentuate the quality of the photograph. The investigator should understand that various camera lenses produce different results. Telephoto lenses allow the photographer to see detail from a distance or to accentuate minute detail. In addition, particular types of camera lenses can distort the appearance of detail. For example, a fish-eye lens can distort the peripheries of the photograph, exaggerating curves in objects. The focal length of a lens refers to what the camera sees through a given lens. Lenses range in size from 50-mm, which gives a view similar to that of a human eye to wide-angle lenses and telephoto lenses.

The area of clear definition, or depth of field, is the distance between the farthest and nearest objects that will be in focus at any given time. The depth of field will determine the quality of the photograph and is affected by the focal length of the camera lens. For a given focal length lens, detail is depicted in the size of the aperture opening. The smaller the aperture opening, the larger is the depth of field. Conversely, the larger the aperture opening, the smaller is the depth of field.

The investigator should seek the aid of a person knowledgeable in the use of digital cameras when buying or using digital cameras with interchangeable lenses. The focal lengths for lenses used on traditional 35mm cameras are vastly different from the focal lengths used on digital camera lenses.

Lens Filters. Color filters are an accessory that is available to the investigator, although they are not recommended unless the investigator has a thorough understanding of their capabilities. Color filters can alter the photograph dramatically by changing the color of the subject of the photograph.

Lighting

Lighting plays an integral part in photography. The easiest light source available to the investigator is the sun, although the circumstances of the examination may dictate recording the scene at a time other than during daylight. In those circumstances, an alternative light source should be used. The most popular is the flash attachment, which may be permanently mounted on the camera, temporarily mounted, or separate from the camera. Other alternative light sources include portable lights such as floodlights and flashlights. It is recommended that the investigator utilize flash units that can be separated from the camera system, allowing the investigator to angulate the light source as needed under the circumstances without adverse reflection. Separate flash assemblies can also be used to "paint" light across a detailed photograph taken at a time of limited visibility. For example, by locking the shutter in an open position and using the flash remotely from the camera system, the investigator can accentuate a particular view and light it with several flash exposures from different directions. The resulting photograph depicts the view with the appearance of light from multiple directions.

Composition and Technique

Photographs are an integral part of the examination and should reflect the condition of the scene as seen by the investigator. The photographs should be taken in a predetermined manner and in accordance with accepted practices in the fire and explosion investigation field—for example, by implementing the philosophy of examination from areas of least damage to areas of most damage.

The photographic documentation of the scene should depict sequential observations. A photograph of a relatively small subject (e.g., a chair) is taken first from a distance (perhaps from the doorway to the room), followed by a sequence of photos taken at gradually closer distances, ending with an extreme close-up of the subject. Sequential photography allows the observer of the photograph to better understand the totality of the view and the relationship of the subject to the overall surroundings.

Another method for depicting the totality of the scene is to use a mosaic of photographs. Mosaic photographs are a series of photographs that encompass a large area by abutting the start of one photograph where the previous photograph ended. To create a mosaic, the investigator should identify the breadth of the photograph according to readily identifiable landmarks. Each ensuing photograph should encompass a portion of that landmark so that the final appearance is of one large photograph encompassing an overall view of a particular area.

Photo Diagram

When recording the scene, the investigator should annotate a diagram of the site, identifying the point from which each photograph was taken, the direction of the photograph, the placement of the item, and the photo number. There are particular occasions when the time that the photograph was taken is important as well and should be annotated on the diagram. The photographer should identify the diagram by affixing his or her signature, the date, location of the scene, and any other pertinent identifiers. A diagram like this is helpful to individuals who did not visit the fire scene. It gives them an overall picture of the condition of the source and what was observed there by the investigation team.

Video Photography

Video photography has become an acceptable medium for orienting viewers to the scene. When using video photography, the investigator must be objective and not exaggerate any one particular detail over another. In addition, video photography should not be the sole medium used to convey the conditions of the scene. Still photography is a good choice for alternative recording.

What to Document

It is important to document as much of the scene as possible. Some suggested activities to record are suppression, overhaul, observers, and origin and cause determination. The progression of the fire, its colors, its reaction to suppression activities, and the overhaul procedures employed are all important in helping the fire investigator to determine the origin and cause. Photographs can also document the extent of damage to the victims or structure.

Photographs of the crowd observing the fire scene activity can help the investigator to identify individuals who may have knowledge beneficial to the investigation. These photographs can also help the investigator to identify individuals who are seen at multiple fires or are known by the law enforcement or fire department community.

Photographs of the suppression activities can help the fire investigator to understand why the fire reacted in a particular manner. Also, when documenting suppression activities, the location of hydrants, engine companies, apparatus,

EXHIBIT 13.1
Street Signs

and hose placement can all help the fire investigator to understand the totality of the scene.

Exterior photographs are important and can be used to establish the location of the fire scene. Exterior photographs should include street signs (Exhibit 13.1) or other identifiable landmarks (Exhibit 13.2), surrounding locations of the fire scene, and all angular views of the exterior of the fire scene.

Structural photographs document the extent of damage to the structure. These photographs should be taken from multiple views to record heat and

EXHIBIT 13.2
Identifiable Landmarks

flame damage. The photographs that are taken should be useful to the investigator as he or she orients supervisors, insurance representatives, other individuals, or the court system—more fully characterizing the condition of the scene at the time of the origin and enabling more accurate cause determination.

Interior photographs will assist the investigator in documenting the scene and to more thoroughly describe the conditions within the structure. All significant points that were accessed or created by the fire should be photographed, along with significant smoke, heat, and burn patterns. The condition of rooms within the immediate area of the fire should also be photographed to document thermal and smoke insult. All heat-producing appliances or equipment should be photographed to document their condition at the time of the origin and cause determination. The condition of furniture and furnishings should also be documented in photographs depicting heat and smoke travel.

The condition of entrance and exit routes, especially doors and windows, should be documented. The photographs should depict their condition at the time of the fire (whether doors were locked or open, whether windows were open or closed, etc.). The condition of interior fire prediction devices should also be documented (whether sprinklers activated, whether extinguishers were employed, whether smoke detectors activated).

The utilities present at a structure should be thoroughly documented to include positioning of diaphragms, meters, transformers, and panels. The point at which the utilities enter the structure should be documented to include their location and condition at the time of the fire. The condition of fuses and/or circuit breakers is important in determining the timing of their activation/deactivation as a result of an event.

During the scene examination, photographs should be taken of all items of evidentiary value. For instance, the photo in Exhibit 13.3 documents the wick from a Molotov cocktail found at a fire scene. Photographs of potential evidence should allow the viewer to locate the item in the scene and its general appearance. More detailed photographs of evidentiary items can be taken after they are moved if the photographer is unable to obtain detailed photos due to scene constraints such as lighting, location, or hazards.

EXHIBIT 13.3
Wick

Photographs should also be taken of occupant or victim actions at the scene. These pictures should document any survivor's or victim's location at the time of the fire, any indicators of actions taken by them during the fire, and any end result such as serious injury or death. If the scene includes a fatality, the state of the victim should be thoroughly recorded. If the full condition of the victim cannot be documented at the scene because of lighting, scene hazards, or other obstacles, more extensive photographs may be taken later at an alternative location.

If, during the course of the scene examination, a witness or victim reports that he or she saw something that the investigator believes to be significant, the photographer should attempt to recreate that view as closely as possible. This can help the investigator to support or refute observation claims.

Aerial photographs can help the investigator to identify burn patterns that might not be readily visible from ground level.

Photographic Presentation

When displaying or presenting pictures of a fire scene examination, the investigator should choose the medium—video, slides, or print photographs—that presents the examination with the greatest clarity. The investigator should also determine what medium is most acceptable to the local court system, law enforcement agencies, fire departments, and insurance companies.

Video presentations are an effective means of conveying scene conditions and actions undertaken at the scene. The investigator should be aware of any audio that may have been recorded during the filming and ensure that it is not offensive or inappropriate. In addition, the investigator should check the video quality to ensure that it does not detract from the effectiveness of the presentation.

A slide presentation can be advantageous when showing large-scale items to an audience, but the investigator should remember that projectors are subject to breaking down and malfunctioning. When that occurs, photographs become a much easier method of displaying the visual information. Photographs do not require mechanical support for presentation and can easily be shared among several viewers.

Computer presentations are rapidly becoming an effective and comprehensive medium for exhibiting large quantities of visual material. They allow for integration of multiple formats into a single presentation. Unfortunately, computer presentations are susceptible to software incompatibilities, malfunctions, and so forth. Therefore, an alternate source should be secured prior to presentation.

NOTE TAKING

In addition to visual representation of the scene, the investigator should incorporate note taking to supplement those items that cannot be photographed or sketched. Note taking may include names and addresses, model or serial numbers, witness statements, photo logs, identification of items, or types of materials. The investigator should be careful not to rely solely on one medium to record the scene. Table 15.3.2 in NFPA 921 can be used to assist the investigator in data collection. See Appendix C of this *User's Manual* for some sample forms.

DRAWINGS

Various types of drawings—including sketches, diagrams, and plans—can be made or obtained to assist the investigator in documenting and analyzing the fire scene. Drawings are used to support the investigator's memory.

Types of Drawings

Several types of drawings can be used to document the scene. The differences among the types of drawings relate to the amount of detail that is incorporated, the type of construction of the structure, features of the structure, equipment, and other factors that are important as to the origin, cause, and spread of the fire.

When the fire investigator is determining the type of drawing to use, he or she should decide what needs to be shown on the drawing. For example, if a fire in the interior of a structure was substantially aided by the furnishings, the investigator would want to choose a drawing that showed the placement of those furnishings. Exhibit 13.4 is an example of a rough sketch showing location of furnishings in an interior fire. Exhibit 13.5 is the finished drawing prepared from the sketch in Exhibit 13.4. In the same vein, if a fire propagated from one structure to a second, the investigator would choose a drawing that showed the locations of the structures.

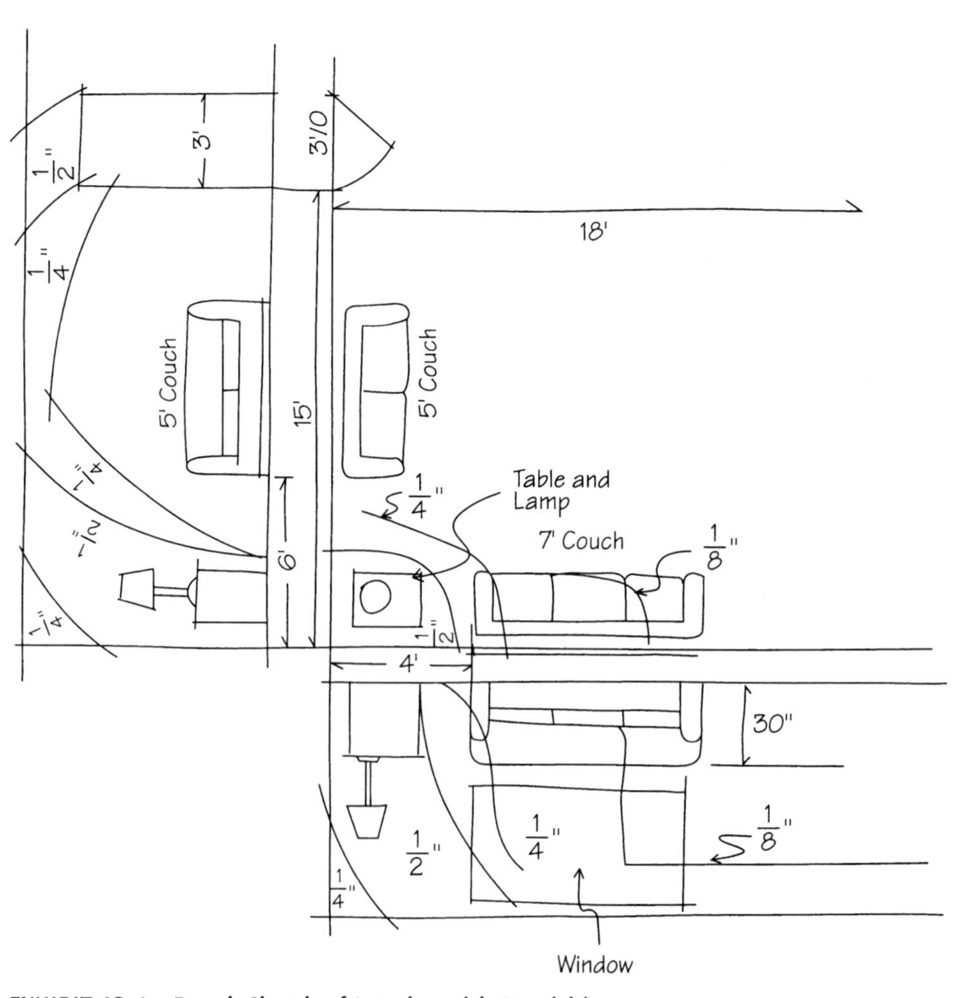

EXHIBIT 13.4 Rough Sketch of Interior with Furnishings

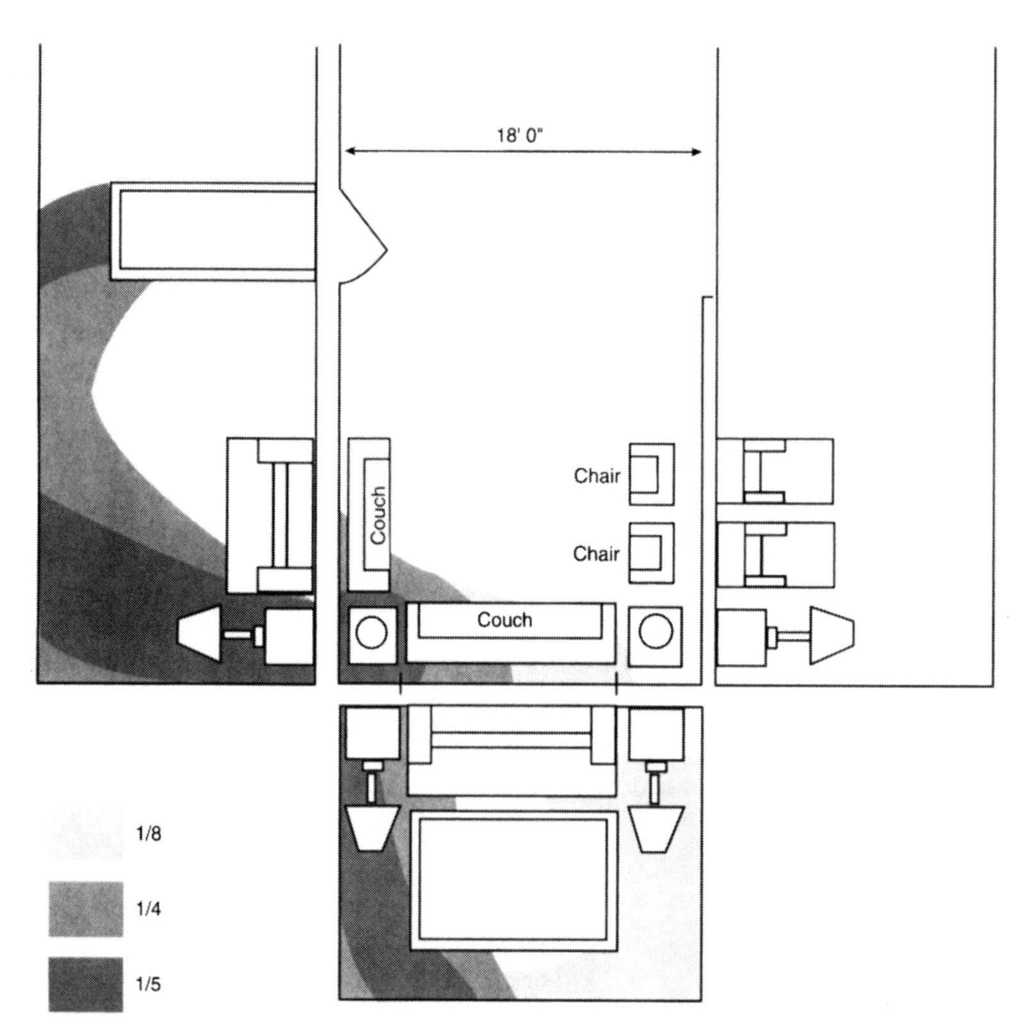

EXHIBIT 13.5 Finished Drawing of Interior with Furnishings

The fire investigator should create a legend for any drawing, indicating what the referenced symbols represent. The investigator must be consistent with the use of symbols, not using the same symbol for multiple purposes.

If a fire investigator is unable to create a detailed drawing or believes that such a drawing is unnecessary, he or she should, at the very least, create a rough sketch that annotates basic placement of the fire within the structure, furnishings, doors, and windows. Exhibit 13.6 is an example of a rough sketch that shows basic layout of the structure. Exhibit 13.7 is the finished diagram made from the sketch shown in Exhibit 13.6.

Several types of drawings can be used to illustrate fire scene conditions. Sometimes, a detailed diagram is necessary for proper documentation, but other times, a rough sketch will suffice. The type of drawing chosen depends on the degree of detail called for. Sketches are freehand drawings of concepts. Schematic design drawings are drafted drawings that show a preliminary design layout with little detail. Design development drawings are somewhat more complete, defining and detailing the elements shown in a schematic drawing. Construction drawings are drafted drawings with extensive detail, showing what was used by contractors to build the structure. As-built drawings are drafted drawings that also show field modifications made during construction, reflecting the finished structure.

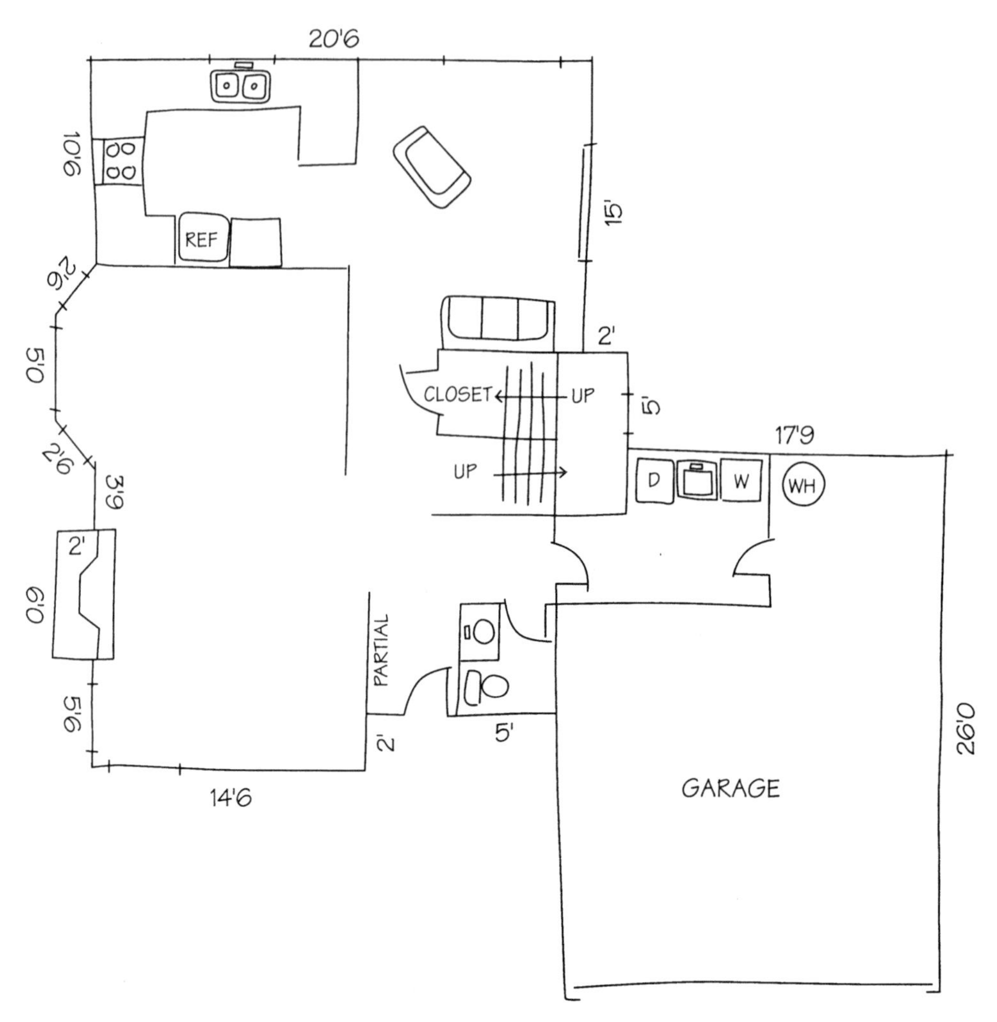

EXHIBIT 13.6 Rough Sketch, Showing Layout of Structure

Architectural or Engineering Schedules

In larger buildings, it often becomes necessary to detail the type of equipment used within the structure. A detailed list of this type is known as a *schedule*. Schedules may be broken down by equipment type and generally detail all of that particular type of equipment used within the structure. For example, a door schedule details the placement of all doors within the structure.

When an architect or engineer prepares a drawing, the types of material used in construction are listed on the specification sheet. The specification sheet matches materials to specific placement. For example, the specification sheet could annotate the use of R-17 insulation in the attic area, ⅝-inch plywood in a ground floor bedroom, and so forth. These drawings can usually be obtained from the building inspection department in the community in which the structure was built. These drawings also include the architect's elevation drawings. These are drawings of the outside of the building, which are helpful if the building is a total loss. If possible, as-built plans should also be obtained. These are architectural representations of how the structure was actually built and may vary from the original plans.

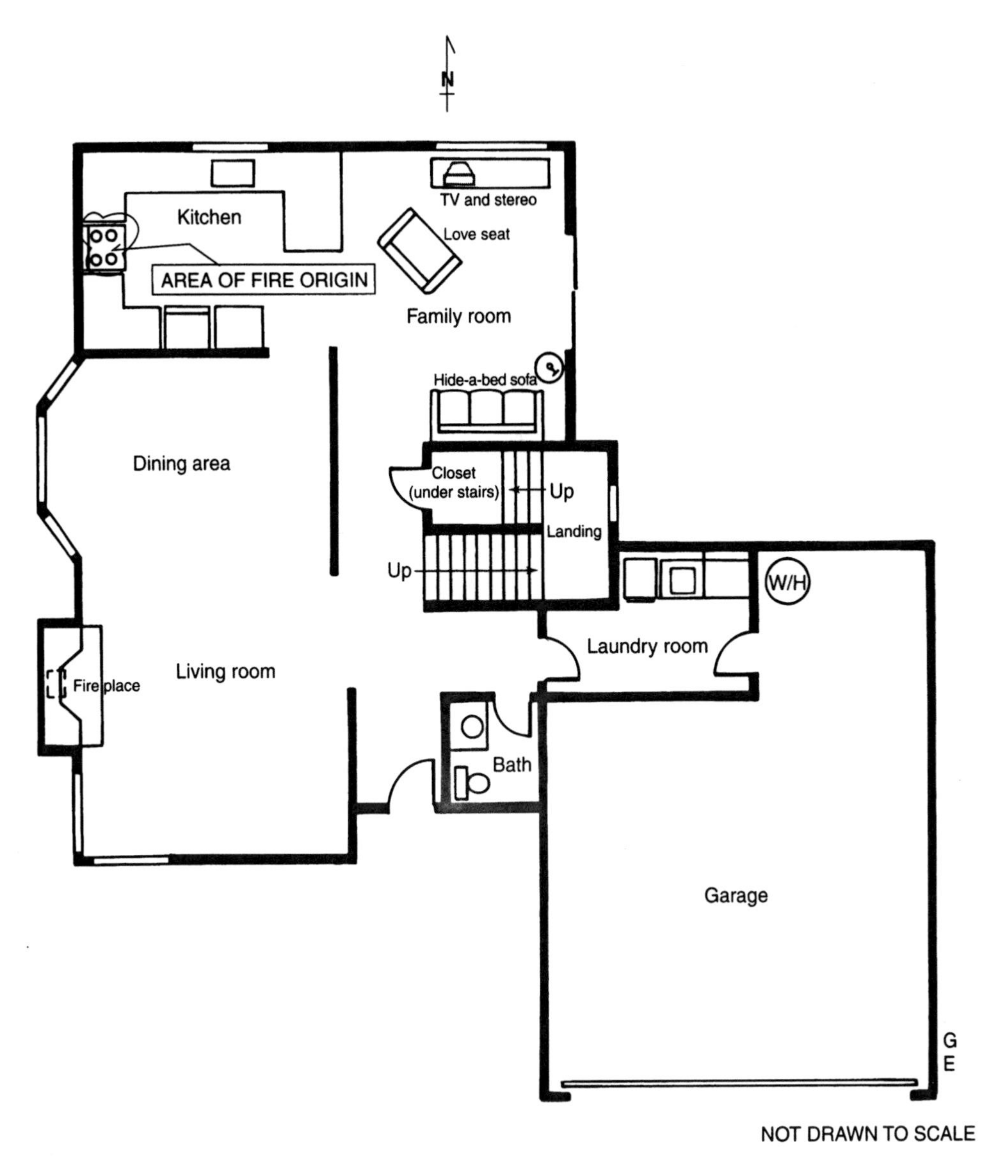

EXHIBIT 13.7 Finished Diagram of Structure

interFIRE TRAINING

Table 13.1 provides an interface between NFPA 921 and the interFIRE VR training program. It brings the user from the NFPA 921 section of interest to the corresponding area in interFIRE. The user can apply the interFIRE information or find more information on what NFPA 921 offers.

TABLE 13.1 NFPA 921/interFIRE VR Training

921 Section	Knowledge/Skill	interFIRE Tutorial Student Activity	interFIRE Scenario Student Activity	interFIRE Resource Section Student Activity	www.interFIRE.org Student Activity
Chapter 15—"Documentation of the Investigation"	*Demonstrate a working knowledge of fire scene photography and scene documentation methods*	Information on recording the fire scene is an integral part of the program. Review Tutorials with additional information on recording the scene to facilitate documentation and review. **Module: Fire Scene Examination** • Document the Scene • Handling Injuries and Fatalities (including Appendix on Additional Information) **Exercise:** Review Chapter 15 of NFPA 921. Write a summary/engage in a class discussion describing a method for systematic examination and documentation of electrical supply and utilization equipment in accordance with this chapter. Discuss the pros/cons of including a qualified electrical code inspector/electrical engineer on your fire investigation team.	Fire investigation students using interFIRE's Scenario can document evidence by the same techniques used in an actual fireground investigation. All work is retained in an investigation file, where it can be reviewed as the investigation progresses. **Witness interviews:** A written interview is produced with responses as you ask selected questions. **Photographs:** The investigator can take up to a total of 72 photographs to document the condition of the scene, recovery of physical evidence, and other uses. **Crime scene sketch:** When the investigator recovers potential evidence for laboratory examination, he or she must document it with a photograph and fix its location in a fire scene sketch. **Exercise:** Focus on documenting the interFIRE VR scene. Take photographs of: • All potential accidental causes that you eliminate • The condition of all doors and windows • All evidence that you collect • Any demonstrative burn patterns that help you demonstrate point of origin and fire spread Compare the photographs you take with the photographs taken by another class member. What photographs did you fail to take? Now present your photos to another member of the class. Treat that class member as if he or she was on a jury. Using the photographs as demonstrative evidence, walk your "jury member" through the fire and convince this person you have established the origin and cause. Focus on your scene diagram. Inspect every door and window at the house and note their condition on your diagram. Check with the first-in fire fighter to determine how the fire department made entry. Speak to the passerby rescuer (Dan Mezzi) to determine how he entered the location. What does this tell you about how entry may have been gained to the house if the fire was incendiary?	The Resource Section contains additional information on proper documentation methods of evidence commonly found at fire scenes. Beginning with the *browse* view (file drawer view) select the file titled "Fire Scene Examination." **We recommend a review of the following documents:** Documentation Overview • Arson Investigation Basics: Reports and Documentation • Photographic Processing and Analysis • Sample Photo Log • Written Notes • Crime Scene Search as a Process • NFPA 906 *Guide for Fire Incident Field Notes* • Documentation of the Fire Scene: A Legal Perspective	All of the reference and photographic information available in the CD-ROM Resource Section is available online at www.interfire.org. Additionally the following "Featured Articles" containing information and examples of Recording the Scene are available on the website: • *Full Scale House Experiment for interFIRE VR* • NIST *Fire Dynamics at Cherry Road* (May 2000) Check the website frequently for additional information on this topic.

▶▶▶ QUESTIONS FOR CHAPTER 13

1. List five ways to document a fire scene.
2. True or false: The fire scene should be photographed throughout the investigation process to document the sequence of events while a scene is being excavated.
   A. True
   B. False
3. What is a "must do" when digital photography is used to document the fire scene?
   A. Make sure the batteries to the camera and flash unit are charged.
   B. Transfer images from the memory stick/card to a nonalterable medium such as a CD-ROM.
   C. Be familiar with the use of the camera.
4. What ASA film rating do most investigators use when photographing fire scenes?
   A. 25 to 100
   B. 100 to 400
   C. 400 to 600
   D. 600 to 1200
5. What type of photograph is helpful in understanding the relationship of a small subject to its relative position in a larger area?
   A. Sequential photographs
   B. Photo mosaic
   C. Panoramic view
   D. Photo diagram
6. List five activities that should be documented at a fire scene by photography or videotape.
7. Taking what type of photograph is most challenging to the fire investigator?
   A. Exterior photographs
   B. Appliance photographs
   C. Evidence photographs
   D. Interior photographs

Physical Evidence

▶▶▶ OBJECTIVES

Upon completion of Chapter 14, the user will be able to

- Understand the importance of physical evidence in a fire investigation
- Describe nontraditional forms of physical evidence that may be found at fire scenes
- Decide what physical evidence to collect at a scene
- Describe acceptable procedures for protecting and preserving the scene
- Describe acceptable methods for collecting samples of various types of common physical evidence
- Process physical evidence using recommended and accepted methods

Physical evidence, also referred to as *real evidence,* is connected with the acts leading up to the actual commission of a crime or failure of a given source of ignition, utility, or appliance. This type of evidence can be produced in court or other proceedings if it is properly identified, documented, collected, preserved, and analyzed.

PHYSICAL EVIDENCE AND PRESERVATION OF THE FIRE SCENE

Defining Physical Evidence

Common physical evidence at a fire scene may include traces of ignitible liquid in flooring, a tool mark that is at a point of forcible entry or that indicates adjustment of a critical valve, a faulted electrical circuit, fingerprints, blood, or other physical item or mark that helps the investigator establish fact.

Physical evidence can be obvious, such as a melted accelerant container in a retail store lobby, or it can be latent (hidden), such as a fingerprint on a tool mark that might not even be visible until it is developed. It is the challenge of investigation to recognize potential physical evidence.

Consistently successful investigators are able to recognize objects that seem to be out of place. Examples might be a lid from a Sterno® can on a staircase leading from a basement where a fire occurred, a tool on a kitchen counter, and furniture that seems out of normal arrangement. The investigator should initially adopt a mindset of inclusiveness when it comes to physical evidence. If the object seems to be in the wrong place, the investigator can assume it is potential evidence until the investigation process proves otherwise.

Fire investigators can improve their results by getting trained in physical evidence recognition and collection and acquiring the needed equipment. Another, perhaps better, approach would be to establish a close working relationship with law enforcement crime scene evidence technicians who are already skilled in this process and possess all the needed equipment.

Usually, physical evidence requires laboratory examination and the testimony of an expert witness to establish its significance at a trial or proceeding.

evidence

Any species of proof or probative matter, legally presented at the trial of an issue, by the act of the parties and through the medium of witnesses, records, documents, concrete objects, etc., for the purpose of inducing belief in the minds of the court or jury as to their contention. (Source: Garner and Black, *Black's Law Dictionary,* 6th edition)

physical evidence

Any physical or tangible item that tends to prove or disprove a particular fact or issue. Physical evidence at the fire scene may be relevant to the issues of the origin, cause, spread, or the responsibility for the fire. (Source: Paragraph 16.2.1, NFPA 921, 2004 edition)

Physical evidence is something that can be observed by a judge or jury and differs from *direct evidence* (testimony of witnesses who observe acts or detect something through their five senses), *demonstrative evidence* (photographs, maps, x-rays, visible tests, and demonstrations), or *circumstantial evidence* (facts that usually attend other facts to be proven and are drawn by logical inference from them).

Preserving the Fire Scene

Physical evidence can take many forms and can exist in a variety of locations in a fire or explosion scene examination. The decision of what physical evidence to collect rests with the fire investigator. It is important to understand that although the most important physical evidence in a fire investigation is generally found within the area of origin, important evidence can also be found elsewhere within the fire scene and frequently outside the fire scene as well.

Skill in recognizing, documenting, collecting, and preserving physical evidence comes from specialized training and experience and is among the most important capabilities of any investigator. Experienced fire investigators understand the need to bring crime scene technicians or other evidence specialists onto the fireground to deal with identification and recovery of physical evidence that may exceed the investigator's own training or equipment capabilities.

Errors in identification, documentation, recovery, or preservation may preclude later analysis and use at trial. Inadvertent or intentional spoliation of physical evidence could potentially expose the investigator to legal sanctions.

Role and Responsibility of the Fire Service

The first stage of preservation of potential physical evidence on a fireground begins with the fire-fighting operation. Fire investigators in a jurisdiction can help themselves by offering training to fire crews and fire officers in how to recognize the importance of evidence preservation and basic ways to avoid inadvertent destruction of physical evidence. Fire investigation and law enforcement associations can influence basic fire fighter and fire officer training at the state or county level in the area of preservation of physical evidence at fires, explosions, traffic accidents, and other joint operations.

In initial fireground operations, the security of each door and window is an issue. During the size-up, well-trained fire fighters should always take note of existing conditions, including audible alarms, open or unlocked doors or windows, odors, and objects that do not seem to belong where they are located. Once life and safety issues are controlled, the incident commander (IC) should ensure that the fire scene is protected from any further destruction.

Management of fire-fighting operations and overhaul is crucial, especially within the area of fire origin. Walls and ceilings that reveal fire movement and intensity pattern evidence and damage patterns to furniture, appliances, and other equipment within the area of origin are critical evidence. Highly effective fire knockdown and overhaul techniques have been developed that can stop fire and minimize inadvertent destruction of evidence.

Old-fashioned fire-fighting operations, such as directing high-pressure, straight streams of water into potential areas of fire origin or overhauling all room contents out a window and pulling all walls and ceilings down, obviously destroy evidence and are avoided in well-managed fireground operations. Pulling salvage covers on top of the contents within the area of fire origin potentially introduces a source of cross-contamination and should be avoided.

Every effort must be made to leave items on the fireground, especially within the area of fire origin, in their prefire position.

Close coordination of fire suppression and fire investigation in a community can minimize the "time out of service" issue for fire suppression crews while maximizing the fire prevention, code enforcement, and law enforcement benefits of an effective fire investigation program. The solution to many of these traditional problems is to call experienced fire investigators to the scene as early in the fire-fighting process as possible. Inadvertent damage to physical evidence and loss of potential witnesses can be avoided by early intervention.

Protecting Physical Evidence on the Fireground

The investigation of fire causation differs from many other types of investigation, such as investigation of homicide or traffic accidents, because the actual cause of a fire might not be known until the full scene investigation has been completed. It is difficult to predict what the significant physical evidence will be at a given fire scene or where it is located. For this reason, the investigator should think of the entire scene as potential evidence until the process of investigation can narrow down the key areas.

The general rule is that the best physical evidence is usually located at the point of the most critical action of the person responsible for initiating the fire. In the large majority of incendiary and accidental fires caused by the negligence of a person, the area of fire origin is where one would normally expect to find the most important evidence.

In a majority of fire scenes, the fire service IC can recognize the area of fire origin simply by applying his or her experience and training. The IC may be able to obtain additional information from eyewitnesses who were present at the fire development or from the first public safety official to arrive at the scene.

Early suppression of fire within the area of origin should always be a priority to help preserve evidence there. In cases in which the area of fire origin is known and the fire has extended into other areas of a structure, it is desirable, whenever sufficient assets are present, to keep the area of origin "darkened down" by periodically directing suppression there. Fire attacks by indirect ceiling spray or fog streams are always best whenever possible because these methods are apt to have the least impact on potential evidence.

High-pressure, major appliance streams or positive pressure ventilation can create confusing fire damage patterns and can destroy physical evidence. They should be used only in conditions in which there is no alternative.

After a fire has been brought under control, interior crews may begin to discover potentially important physical evidence. There are a number of effective ways to protect physical evidence on a fireground. The best way is to post a fire fighter at the entrance to the area of fire origin or near a critical piece of evidence to restrict access, create a fire watch, and provide additional suppression as required. Trained investigators should be called to the fire scene as early as possible in this process.

Potential physical evidence, such as a gasoline container, a gun, a tool that may have been used in a burglary, an appliance that shows evidence of electrical faulting, or similar item, can often be "taped off" or protected by covering it with

EXHIBIT 14.1 Position of Knobs on a Range

a box or by placing a barrier to keep others away. The IC should always be notified when potential evidence is discovered.

Fire suppression personnel must avoid adjusting valves or knobs and switches on appliances or circuit breakers. (See Exhibit 14.1.) The original position of a switch or valve is often of crucial importance.

In general, physical evidence should be left where it is found until trained investigators can assume responsibility for the scene and guide documentation and recovery.

Before any evidence is removed from the area of fire origin, it needs to be photographically documented and fixed in a diagram that indicates its location and position before it is moved. Field note taking should document the condition of the evidence when it is discovered and should list the other people who are present. In Exhibit 14.2, the position of a Molotov cocktail is documented in a diagram. In Exhibit 14.3, a photograph is used to document the location and position of the evidence.

Skilled investigators, prosecutors, and medical examiners caution against moving physical evidence, such as containers of a suspected accelerant or dead bodies, from a fire scene before they can be carefully examined and their location documented.

Likewise, in cases in which there is evidence of forcible entry, pieces of an explosive device, or some other potential item of physical evidence, the general

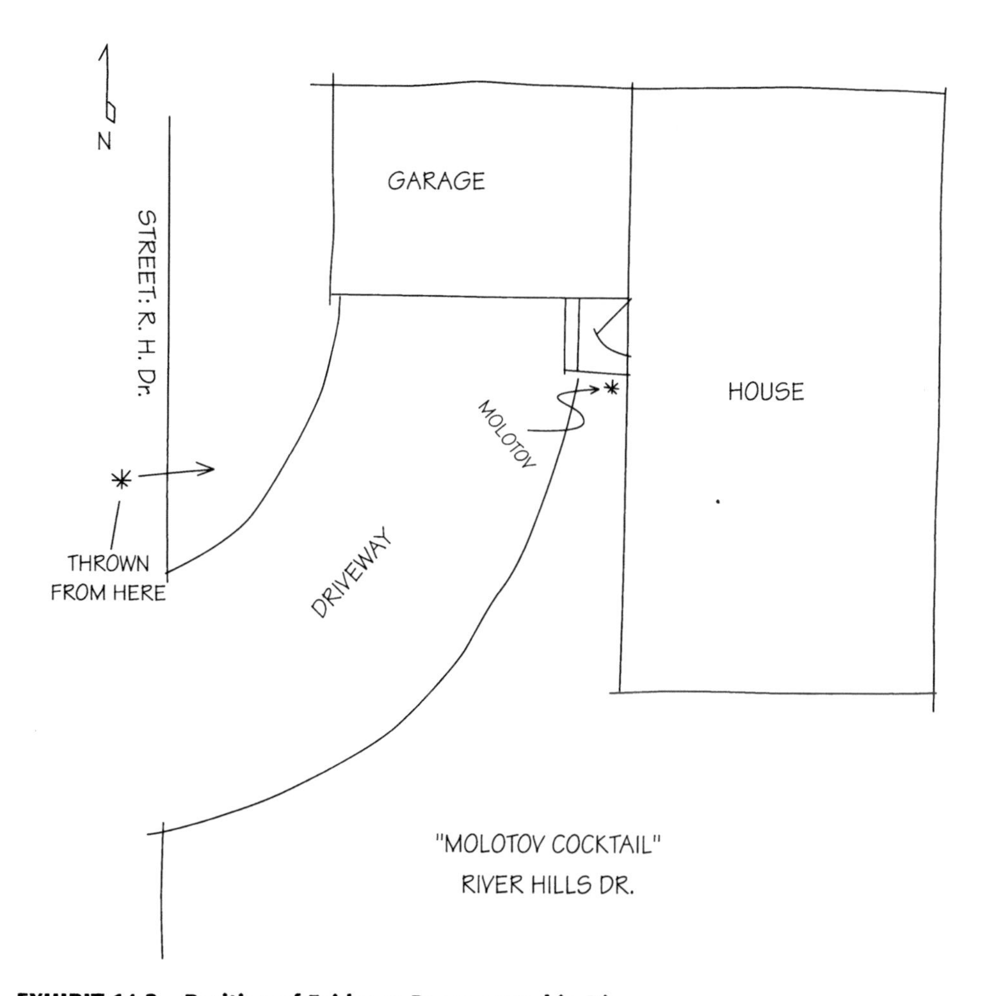

EXHIBIT 14.2 Position of Evidence Documented in Diagram

EXHIBIT 14.3 Position of Evidence Documented in Photograph

rule is that the evidence should be left undisturbed and fire-fighting operations, whenever possible, should be directed to a secondary position away from potential evidence sites.

It is safest for fire crews to interrupt electric or gas service at the street or to shut off the main valve or main breaker at the utility entrance. Individual gas appliances always have a shutoff valve behind the appliance, and this should be used to cut the fuel gas flow. To avoid destroying evidence, fire fighters should avoid practices such as tripping every single branch circuit breaker or readjusting valves on individual burner elements.

See the section on "Spoliation of Evidence" in Chapter 9, "Legal Considerations," of this *User's Manual.*

CROSS-CONTAMINATION OF PHYSICAL EVIDENCE

Avoiding contamination of physical evidence is of concern in any scene investigation. Opportunities to cross-contaminate evidence occur during fire fighting, overhaul and salvage operations, evidence handling, storage, and transportation.

Cross-contamination is the unintentional transfer of an ignitible liquid residue from one fire scene or location contaminated with ignitible liquid residue to an evidence collection site. There are four major potential sources of cross-contamination at a fire scene: tools, turnout gear, evidence cans, and emergency equipment. Fire fighters and fire investigators should comply with certain house-keeping procedures to help prevent cross-contamination.

Tools

Fire investigators should be equipped with a special tool kit to process fire scenes. These tools should be kept separate from other fire department equipment and must never be coated with any rust preventive material. After a fire scene examination has been completed, these tools should be rinsed clean with a strong stream of water.

It is recommended that fire investigators use steel blade tools (shovel, hoe, brick trowel, chisel, razor, etc.) and squeegees with hard rubber blades for excavation and ignitible liquid evidence sampling. Some types of common equipment (bristle brooms) and safety gear (fire fighter gloves) probably cannot be cleaned once they have been contaminated with ignitible liquid residue. Latex gloves should always be used when handling physical evidence.

It is necessary to clean each tool that was used for excavation or evidence sampling before taking it into the fire scene and, again, between evidence collection sites within the scene. Laboratory testing has found concentrated liquid dishwashing detergents that are effective in dissolving grease to be effective in dissolving ignitible liquid residue on steel tools when the tools are scrubbed with a clean scrub brush and flushed with clean water. The investigator should submit a sample of the liquid detergent to his or her forensic laboratory to ascertain its properties and ingredients.

If investigators have an accelerant detection canine or a sensitive hydrocarbon detector available, it can be used to double-check the tools after cleaning and before use.

Note that ignitible liquids derived from crude oil are generally not soluble in water alone.

Turnout Gear

It is important to clean boots before entering the area where samples are to be taken and during the scene examination process and to avoid walking through contaminated areas en route to the collection site. Investigators should not handle ignitible liquid residue samples with fire gloves on, but instead wear latex gloves to handle potential residue evidence. Several pairs of sterile latex gloves can be carried in an investigator's pocket or kit. Two latex surgical-type gloves will conveniently fit into an empty 35mm film container with a snap top. In cases in which there are two or more areas of fire origin in an incendiary fire, the investigator should always change gloves and clean collection tools between sites to minimize the chance of cross-contamination.

Evidence Cans

It is recommended that fire investigators carry a supply of one-pint, one-quart, and one-gallon paint-style evidence cans, or their equivalent, in which to store residue and comparison samples. It is a good practice to seal the cans immediately on receipt from the supplier and before putting them into a response vehicle. The reason for this is to prevent unintentional contamination of the interior of the evidence container by leaving cans open in the rear of a vehicle powered by a petroleum distillate and containing investigation gear.

Other types of physical evidence can be placed in cardboard boxes or, in some cases, plastic bags. It is important to be trained in the collection of physical evidence or to consult with an evidence technician at a fire scene to maximize the potential of collecting valuable evidence and minimize errors.

Emergency Equipment

It is common sense not to bring tools or generators powered by petroleum distillates into an area of fire origin. All of the housekeeping rules suggested here apply in any situation in which an investigator has handled anything that may contain a petroleum distillate before going into the area of fire origin where arson is suspected.

ACTIVITY 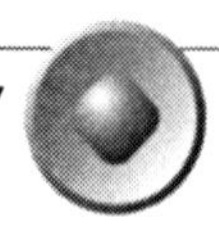

Review with fire suppression personnel how fire scenes can be contaminated by suppression actions and what steps can prevent this contamination.

ACTIVITY 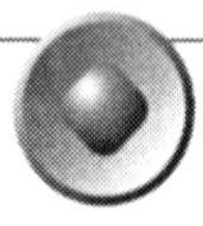

Review how fire investigators can contaminate the fire scene and what steps can prevent this contamination.

COLLECTION OF TRADITIONAL PHYSICAL EVIDENCE

Traditional physical evidence is defined as any trace, clue, impression or thing connected with the commission of the acts that resulted in the fire. Familiar types of traditional physical evidence include fingerprints, palm prints, footwear and tire track impressions, tool marks, fibers, and blood.

The collection of physical evidence is governed by the type, form, size, and condition of the evidence. An additional concern is whether handling and packaging a specific piece of evidence will inadvertently degrade fragile forms of evidence such as latent fingerprints or a toolmark. If there is any doubt in the mind of the fire investigator about how to properly handle and preserve a piece of evidence, the investigator should always obtain the assistance of a trained evidence technician.

Collection methods depend on the following factors:

- *Physical state:* Is it solid, liquid, or gas?
- *Physical characteristics:* What are its size, shape, and weight?
- *Fragility:* Will the evidence disintegrate or break?
- *Volatility:* Will the evidence evaporate?

In many jurisdictions, public safety fire investigators are principally concerned with fires of incendiary origin for which private fire investigators need to be concerned with evidence of arson and also evidence that establishes liability or negligence and that may be grounds for a subrogation recovery.

As a general rule, public fire investigators should not remove ordinary appliances, wiring, and the like that have malfunctioned and started a fire. When a noncriminal fire cause is established, public safety investigators should determine the insurance carrier from the property owner and make appropriate notification. In most cases, the property owner can provide the name of a local insurance agent or broker who can quickly make notification.

In cases in which the fire scene is structurally unstable or there is a possibility that someone may intentionally destroy or steal evidence, it is a good practice to thoroughly document the scene and remove the evidence to a secure storage facility.

After the physical evidence has been documented with field notes, photographs, and fixed in a diagram, it is ready to be collected and preserved.

Specific procedures for recovery of traditional physical evidence are entirely dependent on the type and condition of the evidence. Fire investigators who are trained and equipped to make recoveries of specific types of evidence can do so. They should bring in trained evidence technicians from local, state, and federal law enforcement agencies when the collection process exceeds their training or equipment capabilities.

Collection of Evidence for Accelerant Testing

An *accelerant* is any substance that is used to initiate or promote the spread of fire. The most commonly encountered arson accelerants are ignitible liquids such as gasoline or kerosene. These liquids share important physical properties that govern their behavior in a fire environment and especially as evidence that can be recovered by a skilled investigator using the proper tools.

The major skill of accelerant residue evidence collection is knowing what to collect and what not to collect. Ignitible liquids used as accelerants burn better than most of the surfaces onto which they are poured. The investigator can expect to find better, stronger samples in protected areas and inside absorbent materials within the pour pattern.

The following are the *most* desirable collection areas:

- Lowest areas and insulated areas within the fire damage pour pattern
- Samples taken from porous plastic or synthetic fibers
- Cloth, paper, and cardboard in direct contact with the pattern
- Inside seams, tears, and cracks
- Edges of burn patterns
- Floor drains and bases of load-bearing columns or walls

The following are the *least* desirable collection areas:

- Deeply charred wood
- Gray ash
- Edge of a hole burned through a floor
- Samples from absolutely nonporous surfaces
- The center of any burn pattern
- In general, areas that were exposed to the greatest heat or hose streams

Key Properties of Common Ignitible Liquids

Liquids have physical properties that cause them to behave differently from most gases or solids. The most common ignitible liquids used as fire accelerants have unique characteristics that manifest themselves in both fire development and evidence from fire scenes. Key properties of common ignitible liquids include the following:

- Liquids flow downgrade and tend to form pools or puddles in low areas.
- Almost all hydrocarbon liquids are lighter than water, are immiscible, and may display "rainbow" coloration (sheen floating on water). Certain other common ignitible liquids (e.g., alcohol and acetone) are water soluble.
- Almost all commonly used ignitible liquid accelerants tend to form flammable or explosive vapors at room temperature.
- The vapors of most commonly used ignitible liquid accelerants are heavier than air and tend to flow downward into stairwells, cellars, drains, pipe chases, elevator shafts, and so on.
- Many ignitible liquids used as fire accelerants are readily absorbed by structural materials and natural or synthetic substances.
- Many ignitible liquids are powerful solvents, which tend to dissolve or stain many floor surfaces, finishes, and adhesives.
- Common ignitible liquids used as fire accelerants do not ignite spontaneously.
- Ignition of a given ignitible liquid vapor requires that the vapor be within its flammable or explosive range at the point where it encounters an ignition source at or above its ignition temperature.
- When an ignitible liquid is poured onto a floor and ignited, two major things occur:
  - Many types of synthetic surfaces (e.g., vinyl) or surface treatments mollify (soften) beneath the liquid.
  - At the edges of the pool, burning vapors adjacent to the liquid edge cause many floor surfaces, such as wood, to char while certain others, such as vinyl, melt and then char. As the liquid pool boils off, its edge recedes. Floor surface charring (or melting and charring) follows the receding liquid edge. The floor area under the ignitible liquid is protected from the effects of burning until the liquid boils off that section.
- Experiments indicate that the greatest temperatures in an ignitible liquid-accelerant fire occur above the center of the burning liquid pool. Scientific experiments have shown that maximum concentrations of ignitible liquid residues are more often found at the edges of the burn pattern and minimum concentrations toward the center. Some arson investigators believe that this is controversial and so take samples from both the edges and the center.
- Ignitible liquids with high vapor pressure, such as alcohol or acetone, tend to "flash and scorch" a surface, while ignitible liquids with higher boiling components, such as kerosene or turpentine, tend to "wick, melt, and burn," leaving stronger patterns. The amount of ventilation available to the fire is a factor in burn pattern appearance.

As a general rule, all fire scenes should be thoroughly photographed inside and outside before any debris excavation is done. As pieces of potential evidence are recovered, the investigator should photograph each in place and fix its location in a crime scene sketch.

Collection of Liquid Samples for Ignitible Liquid Testing

In cases in which a container is suspected of containing an ignitible liquid used as an arson accelerant or in which such a liquid was a factor in an accidental case, there is a series of recommendations for evidence recovery:

- Find out whether investigators recognize the odor of the liquid to allow later testimony about odor recognition.
- Collect a sample of the liquid into a sterile glass container with a hard plastic cap by pouring or drawing the liquid into a sterile pipette or eyedropper. It is good practice to place a piece of aluminum foil on top of the container before screwing down the cap.
- Remove the container from the fire scene, pour a small amount onto a safe surface, and attempt to ignite it to allow later testimony about ignition properties.

If an investigator has evidence to suspect that a specific ignitible liquid was used to set a fire (from the odor in the debris or an accelerant can left behind), he or she should inform the chemist who is working on the project.

Sterile, 3- or 4-oz (90- to 118-mL) pharmacy bottles with hard plastic caps (not glued cap liners) are recommended for collection of suspected pure ignitible liquid samples. Using a sterile eyedropper or pipette, the investigator should collect about 2 or 3 oz (60 to 90 mL) of the liquid, place a piece of aluminum foil over the bottle top, screw down the cap, and place a label on the side of the bottle.

Sterile gauze bandages are recommended for skimming suspected ignitible liquid residue sheen (rainbow colors) off the surface of suppression water in the pour pattern area. If sanitary napkins must be used as a substitute absorbent, the nondeodorized, individually wrapped type should be used.

A pure ignitible liquid sample should never be placed in a metal can. If the can should become heated, internal vapor pressure could pop off the cover.

The type of glass bottles, pipettes, and eyedroppers described can usually be obtained from a forensic supplier, a forensic laboratory, or a pharmacy. A plastic fishing tackle box makes a convenient container in which to keep a supply of pipettes, glass bottles, gauze pads, and evidence labels.

Collection of Liquid Evidence Absorbed by Solid Materials

Accelerant residue sampling at a fire scene can be done in a way that maximizes laboratory identification of accelerant residues. Most of the laboratory procedures for this type of evidence involve testing "headspace" vapor in various ways. *Headspace* is the zone inside a sealed evidence can between the top of fire debris and the bottom of the lid. Fire and arson chemists generally recommend that evidence containers be filled to two-thirds volume with the debris sample, leaving the top one-third volume as empty air headspace.

New, unlined, uncoated steel paint cans with V-groove lids are usually recommended for the collection, preservation, and analysis of fire debris that is suspected of containing ignitible liquid residue.

After taking the sample, the investigator should seal it in the can by firmly tapping the V-groove lid onto the can top, trying not to distort the sides of the can, which could lead to seepage of volatile gases. Some agencies require investigators to further secure the lid with tamper-resistant tape. The identification mark or evidence label should be placed on the side of the evidence can, not on the lid.

The latest generation of heat-sealed, plastic evidence pouches has eliminated earlier problems, according to a 1991 laboratory test conducted by the Bureau of Alcohol, Tobacco, Firearms and Explosives (ATF) laboratory. This type of evidence container has important advantages but remains puncture prone. The investigator should ask his or her laboratory for its recommendation. Whichever evidence container is chosen a sample container should be submitted for testing and comparison.

Training for obtaining samples of materials suspected of containing ignitible liquid residue is provided by state and federal arson investigation schools. The interFIRE VR training program contains specific training on expert methods of sampling from the seven types of common surfaces: carpet, wood flooring, vinyl flooring, soil, concrete, vinyl floor tile, and ceramic tile.

Training for sampling ignitible liquid residue from these seven common floor types can be obtained online at www.interfire.org under the Training Center. Specific collection techniques are taken from *A Pocket Guide to Accelerant Evidence Collection*, 2nd edition, which is available from www.arsonpocketguide.com.

The investigator should always keep in mind that ignitible liquids exposed to fire quickly burn away. Sampling for ignitible liquid residues involves identifying and sampling from the base of absorbent materials placed on a floor surface and cutting narrow splinter samples from seams and joints on wood flooring, thresholds, door casings, furniture, the edges of suspected pour patterns on carpet, and so forth. Whenever possible, the investigator should try to pulverize, shred, or splinter sample residue evidence material. Breaking a large solid sample into many smaller pieces dramatically increases the surface area from which to extract residue. Excess water should be drained.

It is advisable to fill the evidence container to two-thirds of its volume loosely and not to pack it down. The sample should be loaded vertically (i.e., "chimney roll") whenever the type of sample permits (e.g., carpet or pad samples, linoleum seams, paper, cardboard, or cloth). Sample specimens from seams and joints in wood flooring should be splintered or pulverized and loaded vertically in the container when possible. When collecting sample material that is best stacked vertically, it is advisable to lay the evidence can on its side while loading it.

A strong sample should be placed in the container that is closest to its size without packing down. The investigator should not dilute the sample by adding material that is not suspected of containing ignitible liquid residue just for the sake of filling the can, but use a smaller container instead.

If a strong odor of an ignitible liquid is present in the sample, a normal-sized sample should be taken whenever possible (at least one quart). When a weak odor or no discernible ignitible liquid odor is present, large-size samples should be taken as required (not more than one gallon each). If a small object such as a matchbook is found to contain an ignitible liquid odor, the object should be sealed in aluminum foil, then further sealed in a suitable evidence container. The chemist should be advised of the odor.

Fire scene procedure requires that, before collection of physical evidence, the condition and location of the evidence be documented with a photograph and then fixed in a line drawing. It is considered good practice to include a permanent feature (e.g., a radiator, wall, valve, or door casing) in each evidence photograph.

The basic rule with line drawings of fire scenes is to measure the location of any movable item of evidence from two or more fixed, permanent objects whenever possible. The investigator should always use a magnetic compass to orient the drawing to north.

Comparison Samples and Exemplar Products

Comparison samples should be taken whenever sampling liquid or solid materials are believed to contain ignitible liquid residue. This is especially true when trying to establish a second point of fire origin or when sampling from a surface believed to have a petrochemical base, as in the case of many common floor surfaces such as carpet or vinyl.

Comparison samples can often be successfully obtained away from the suspected pour pattern on the same carpet or surface by removing sample material from underneath large objects that sit on the floor, such as couches or cabinets. Taking a sample of uncontaminated carpet for laboratory testing helps to establish a baseline of chemical components to which the contaminated sample can be compared.

In cases in which an appliance or control is believed to have caused a fire, the investigator can often learn the model type and source by simply interviewing the occupants of the area of fire origin. In serious fire cases, obtaining an exemplar unit is often desirable. A good resource for product safety information is the U.S. Consumer Product Safety Commission (CPSC) website (www.cpsc.gov), which has information on product recalls by manufacturer and model. This information includes products that have been recalled for fire and electrocution hazards. Mechanical or electrical engineers are usually involved in examination to determine the cause of a failure.

Fire investigators are cautioned not to perform destructive examinations that exceed their training and experience. In these cases, the investigator should always consult with the insurance carrier and the person sustaining the loss about the need to perform any testing or plans to remove any evidence that could be destroyed or further damaged.

Collection of Electrical Equipment and System Components

Before examination of electrical equipment, the first priority is to make certain that there is no electrical current to the equipment. If there is any question, it is advisable not to touch the equipment. A qualified electrician should be called to the scene.

Fire-damaged electrical equipment, such as electrical panels or switchgear, appliances, controls, and the like, needs to be examined and fully documented on scene since movement will often degrade its evidentiary value.

IDENTIFICATION OF PHYSICAL EVIDENCE

Procedures for Identification

The first stage of this process is photographing the evidence where it was found and placing the evidence into a scene drawing by measuring its exact position from fixed objects such as walls or door casings.

It is an accepted practice to take an overall photograph of any important article of evidence to show its relationship to other major objects in the room or vicinity, a second photo to show the object in its immediate surroundings, and then a close-up photo of the item. A fourth photo of the evidence should be taken with a measurement scale in the photo to establish size.

The collection of an item of evidence must be accompanied by marking the evidence for positive identification. Crime scene procedure calls for marking, tagging, or bagging of evidence. It is best to directly mark on the evidence itself if this will not be destructive. Otherwise, the item can be tagged or placed in a bag. The investigator should be careful not to place a mark in a location where other evidence such as a latent fingerprint or tool mark may exist.

Marking the Evidence

When marking an item of physical evidence for identification the following data should be included:

- Date and time collected
- Case number
- Location
- Brief description of the evidence
- Where and at what time the item was discovered
- Name of the investigator(s) collecting the evidence

The materials used for marking the evidence should not be susceptible to removal, damage, or alteration.

Chain of Custody

Forensic and legal requirements mandate that evidence be positively identified and maintained in a chain of custody from the point where the evidence is collected right to the courtroom.

The integrity of any piece of evidence may be challenged, so the investigator needs to keep an accurate historical accounting of the evidence, which is referred to as chain of custody. This record details all the parties who have come into possession of the evidence. The inability to show the chain of custody for an evidentiary item can lead to challenges of admission as evidence in court. Exhibit 14.4 shows an example of an inventory report form used to record the chain of custody.

The strongest chain of custody has one link: the fire investigator who collected the evidence. If possible, the investigator who recovers the evidence should place it into storage, transport it to and from the laboratory, place it back into storage, and then bring it to court. If this isn't possible, then be certain to maintain a strict accountability for each person who has custody of the evidence in accordance with local requirements.

TRANSPORTING AND STORING EVIDENCE

At some point, physical evidence needs to be transported for examination or laboratory testing. The major concern at this point is maintaining the physical integrity of the evidence.

The investigator may use several methods to transport evidence. Personal delivery is the preferred method. When this is not possible, the evidence may be shipped via the post office or a common carrier. Specific requirements vary

BURNSVILLE POLICE DEPARTMENT **PROPERTY AND INVENTORY REPORT**

UOC TITLE ☐ EVIDENCE ☐ RECOVERED ☐ FOUND ☐ OTHER C.F. No.____

Receiving Officer | Date/Time | Citation No. | EVIDENCE TECH: USE

Arresting Officer

Property taken into custody at: (Address)

SUSPECT Full Name D.O.B. / /

Address | Phone No. (Home) | Phone No. (Work)

City/County/State/Zip

Full Name ☐ VICTIM ☐ OWNER ☐ FINDER D.O.B. / /

Address | Phone No. (Home) | Phone No. (Work)

City/County/State/Zip

CASE DISPOSITION:
☐ EXC. CLRD. ☐ PENDING ☐ ARREST ADLT. ☐ JUV. REFERRAL ☐ INACTIVE ☐ ASST. & ADVSD.

PROPERTY PLACED IN:
☐ Property Room ☐ Locker No. ☐ Prisoner Inventory ☐ Other (explain)

BIKE / INFO	Brand	Model	☐ Boys ☐ Girls	Tire Size	Serial No. and/or P.I.N.	Speed
	Colors – Frame	Fenders	Seat	Grips		Est. Value

Item #	**PROPERTY** Itemize; Describe; List Serial Nos.	Est. Value

OFFICER DISPOSITION: ☐ Other ☐ Return to Owner ☐ Hold ☐ Destroy ☐ Prisoner Transfer

CONTROLLED SUBSTANCES: Type____ Weight ____ Quantity/Count

☐ Deliver to BCA by: Date____ New BCA Case ☐ Yes ☐ No

CHAIN POSSESSION:

Item	Time	Date	Place	Delivered By	Received By

CLAIMANT'S RECEIPT

I certify that I have received: items #____ and that I am the lawful owner or claimant.

I certify that I have received: items #____ and that I am the lawful owner or claimant.

Released By:____ Name ____ Name ____

Address ____ Address ____

Date ____ Signature ____ Signature ____

FORM 44100-23 Rev. 10/99 **CASE FILE COPY**

EXHIBIT 14.4 Property and Inventory Report Form
(Source: Burnsville Police Department)

depending on the type of evidence. The laboratory should be consulted on the safest way to ship a given category of evidence. Some materials, such as explosive devices and flammable liquids, cannot be shipped under normal circumstances.

Some form of return receipt is necessary for all shipments of evidence, and a letter of transmittal is generally appropriate as well. The transmittal letter should generally include a detailed list of the evidence submitted, the nature and scope of the testing requested, and possibly the circumstances surrounding the event in question.

Evidence should be stored under conditions that maintain it in the best possible condition, guarding it against excessive heat, humidity, or other sources of contamination.

TESTING PHYSICAL EVIDENCE

After an item of evidence has been collected, it may be submitted for examination or testing. Testing is often performed to establish chemical composition or for failure analysis. Some of the tests that are conducted on evidence include the following:

- *Gas chromatography (GC):* Used to separate mixtures into their individual components
- *Mass spectrometry (MS):* Used in conjunction with gas chromatography to further analyze individual components
- *Infrared spectrophotometer (IR):* Used to identify certain chemical species
- *Atomic absorption (AA):* Used to identify individual elements in nonvolatile substances
- *X-ray fluorescence:* Used to analyze for metallic elements

Other tests may be conducted to determine flash point, flammability, burning characteristics, heat and smoke release rates, and other factors. The fire investigator should be familiar with the purposes and standards established for each test. The investigator should confirm that the testing facility is conducting the appropriate test in accordance with the established standards before the testing is conducted.

ACTIVITY 

Contact a nearby testing laboratory and find out what methods of testing they use for fire debris evidence.

ACTIVITY 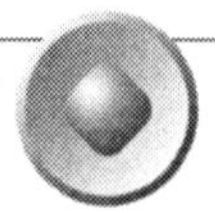

Review standard operating procedures on how your agency or firm packages, labels, and stores evidence that it takes from a fire scene. Are these procedures adequate? What recommendations for improvement would you suggest?

Finally, after the evidence has been gathered, tested, and stored, the question arises as to how long the item needs to be maintained. Many factors influence the length of storage. Legally, some cases—such as a murder-related investigation—may have no statute of limitations, and so evidence may be brought from storage many years after the fact. Even after a conviction in a criminal case, the defendant is entitled to appeal, and the evidence should be maintained during the time allotted for an appeal.

Civil cases also have lengthy time frames during which evidence may be called into court. In general, the fire investigator should maintain all evidence until proper written authorization to dispose of the evidence has been received from all who are concerned in the investigation. After that, items of evidence should be returned to the owner.

interFIRE TRAINING

A copy of interFIRE VR is available at www.interfire.org. Table 14.1 provides an interface between NFPA 921 and the interFIRE VR training program. It brings the user from the NFPA 921 section of interest to the corresponding area in interFIRE. The user can apply the interFIRE information or find more information on what NFPA 921 offers.

TABLE 14.1 NFPA 921/interFIRE VR Training

921 Section	Knowledge/Skill	*interFIRE Tutorial Student Activity*	*interFIRE Scenario Student Activity*	*interFIRE Resource Section Student Activity*	*www.interFIRE.org Student Activity*
Chapter 16—"Physical Evidence"	*Demonstrate the ability to recognize, document, collect, preserve, and transport physical evidence*	Information on the process of evidence collection is a focus of this program. Review these Tutorials containing supplemental information on recognition, documentation, collection, preservation, and transportation of physical evidence: **Module: The Process of Investigating a Fire** (organization and sequence protocol) **Module: Before-the-Fire Practices** • Define Official Responsibilities • Equip Your Investigative Team **Module: Roll-Up Practices** • Preserve Scene and Physical Evidence • Secure Scene and Witnesses **Module: Preliminary Scene Assessment Practices** • Size Up the Structure **Module: Fire Scene Examination** • Guiding Principles • Document the Scene • Handling Injuries and Fatalities • Determining the Need for Additional Resources • Collect and Preserve Evidence • Use Canine Units • Submit Evidence for Testing	interFIRE's Scenario enables you to conduct a fire scene examination of an actual fire-damaged building and practice real fire investigation functions such as location and field examination of evidence; collecting, identifying, and specifying what types of tests should be done by the program's forensic lab. Investigative work is retained in a file where it can be reviewed as the investigation progresses. **Exercises** Focus on collecting evidence. Your preliminary physical examination and ruling out of potential accidental causes should give you an idea of where to begin evidence collection to establish your case. Try using the canine unit to search different areas of the residence. Sample at each alert and specify lab tests. Remember to request lab reports for trace evidence, such as fingerprints left on items, not just accelerant tests. Now turn your attention to other physical evidence connected to canine accelerant alerts. For example, if you are leaning toward an incendiary cause determination, examine ingress and egress points to the residence and consider collecting doorknobs for latent fingerprint examination. Finally, document or, in some cases, collect any potential accidental sources of ignition that are beyond your training or capability to rule out for additional laboratory examination. Elimination of potential causes of ignition is required for isolation of a single cause. In the follow-up investigation, read your lab results and discuss how they inform your investigation. Get together with other students and compare your evidence logs, the choices you made, and the conclusions they helped you draw. Did you miss any key evidence? How or why? What could you do differently next time?	The Resource Section contains additional information on proper methods for documenting evidence commonly found at fire scenes. Beginning with the *browse* view (file drawer view) select the file "Fire Scene Examination." **We recommend a review of the following documents:** • Evidence Collection & Handling • General Evidence Collection Guidelines • The Evidence Collection Kit • Laboratory Services • ATF National Laboratory • Accelerants • Excerpts from *The Pocket Guide to Accelerant Evidence Collection*	All of the reference and photographic information available in the CD-ROM Resource Section is available online at www.interfire.org. Additionally the following "Featured Articles" containing information and examples of **Physical Evidence** are available on the website: • *Full-Scale House Experiment for interFIRE VR* • NIST *Fire Dynamics at Cherry Road* (May 2000) Check the website frequently for additional information on this topic.

▸▸▸ QUESTIONS FOR CHAPTER 14

1. True or false: In legal and law enforcement organizations physical evidence is also known as *real evidence.*
   A. True
   B. False
2. True or false: The recommended container for the collection of liquid and/or solid evidence is a common plastic bag.
   A. True
   B. False
3. True or false: An investigator should label all evidence for storage immediately on returning to the office.
   A. True
   B. False
4. True or false: The diagramming and photographing of physical evidence at a fire scene should be done before the evidence is moved or disturbed.
   A. True
   B. False
5. True or false: At the beginning of a fire scene investigation, the entire scene should be considered physical evidence and therefore should be protected and preserved.
   A. True
   B. False
6. True or false: To avoid cross-contamination, all collection tools should be properly cleaned, and new latex gloves should be used during the collection of each subsequent item sample evidence at a scene with multiple areas of fire origin.
   A. True
   B. False
7. True or false: The fire service plays a vital role in preserving physical evidence to promote fire prevention and arson prosecution.
   A. True
   B. False
8. True or false: The most important physical evidence is most often located at or near of the point of fire origin.
   A. True
   B. False
9. True or false: Samples of a suspected ignitible liquid can be properly placed in an evidence can for laboratory analysis.
   A. True
   B. False
10. True or false: Any sample that is obtained by canine alert must be confirmed by laboratory analysis to be considered valid.
    A. True
    B. False

11. True or false: The responsibility for the preservation of the fire scene and physical evidence lies solely with the fire investigator.

    A. True

    B. False

12. Name three of the six ways in which evidence should be identified.

13. Which of the following terms refers to the documentation needed to establish the proper control of an item of evidence between the fire scene and the courtroom?

    A. Preservation

    B. Chain of custody

    C. Evidence log

    D. Chain of integrity

14. True or false: If there is a possibility that a test or examination may alter a piece of evidence, it should be performed only after all potential parties in a litigation should be notified and provided an opportunity to observe the test.

    A. True

    B. False

15. True or false: Evidence containers should be filled to two-thirds capacity with sample material for laboratory testing.

    A. True

    B. False

Origin Determination

OBJECTIVES

Upon completion of Chapter 15, the user will be able to

- Follow recommended techniques for determining the origin of a fire
- Analyze heat movement and intensity patterns
- Explain the need for fire scene reconstruction

The origin of the fire is the location at which the fire started. If the origin cannot be identified, the cause cannot be determined. This chapter examines fire damage assessment, scene assessment, scenario development, interior and exterior examination, fire scene reconstruction, fire spread scenarios, and total burns as they relate to origin determination.

Review Chapter 17, "Origin Determination," in NFPA 921.

POINT OF ORIGIN AND AREA OF ORIGIN

A fire's *point of origin* is the exact physical location where a heat source and a fuel come into contact with each other and a fire begins, such as where a lighted match comes into contact with a container of gasoline. The fire's *area of origin* refers to the room, building, or general area in which the point of origin is located. Determining the point and area of origin of a fire is essential to determining its cause. The task may be easy or difficult, depending on the amount of evidence, facts, or data that has been gathered; the reliability of the evidence; and the amount of destruction.

point of origin

The exact physical location where a heat source and a fuel come into contact with each other and a fire begins. (Source: NFPA 921, 2004 edition, Paragraph 3.3.115)

area of origin

The room or area where the fire began. (Source: NFPA 921, 2004 edition, Paragraph 3.3.9)

Although they are treated as two separate topics, origin determination and cause determination are closely related. If the investigator cannot reliably identify the area of origin, it is very difficult to reliably determine the fire's cause, or ignition scenario. Put another way:

No area of origin = no cause

To determine the origin of a fire, the investigator must be systematic in examining the fire scene. By being systematic and using the scientific approach, the investigator can reduce any potential errors or oversights that may creep into the process. By employing similar techniques on every fire, the investigator develops a standardized approach that should address the requirements of a complete fire scene investigation.

A recommended approach is for the investigator to work from the outside of a fire in toward the area or point of origin (or from the area of least damage to the area of greatest damage). It is important to use this method rather than beginning at what appears to be the point of origin and then working outward. However, there are times when the investigator can vary this procedure based on information that is available at the time of the investigation.

By working inward from the area of least damage to the area of greatest damage, the investigator can define the fire area. Then, by analyzing fire patterns, the investigator can identify potential areas of origin that are supported by facts. If the opposite procedure is used (working outward from areas of most damage to least damage), then the investigator may be tempted to develop a hypothesis as to what the area of origin is and collect data to support that theory. That would not be an objective analysis.

FIRE DAMAGE ASSESSMENT

Assessing fire damage is an ongoing process. It is a process of comparison and evaluation. As was stated earlier, the building and its damage should be evaluated systematically, working inward from the outside, from the area of least damage to the area of most damage. By using a systematic approach at every fire scene, the investigator develops habits that ensure a complete and thorough fire investigation.

Fire damage assessment is often conducted through the reconstruction of the fire scene. Fire pattern analysis involves a comparison of areas that are damaged to those that are not damaged. For example, the investigator can evaluate the degree of damage to walls, floors, ceilings, and contents.

Some of the methods used in documenting the fire scene during fire damage assessment include notes, diagrams, sketches, photography, vector diagrams, and depth-of-char survey grid diagrams. (Refer also to Chapter 13, "Documentation of the Investigation," in this *User's Manual.*)

RECOMMENDED PROCEDURE FOR ORIGIN DETERMINATION

The recommended procedure, as outlined by NFPA 921, includes the following six steps:

1. Preliminary scene assessment
2. Preliminary fire spread scenario development
3. In-depth exterior examination
4. In-depth interior examination
5. Fire scene reconstruction
6. Fire spread scenario report

Preliminary Scene Assessment

The preliminary scene assessment provides an initial overall look at the structure. It enables the investigator to determine what resources may be needed—for example, equipment, people, and security. It also helps the investigator to identify safety concerns—utilities, unstable structures, weather—and identify areas that require further study. The preliminary scene assessment phase is important because it provides the investigator with a big-picture perspective of the building and damage. It is also very important in identifying safety concerns that need to be addressed. Following the preliminary scene assessment, the investigator should be able to determine the safety of the fire scene, the staffing requirements, and the equipment requirements for the investigation.

Building Identifiers. Beginning with the preliminary scene assessment, it may be beneficial to establish building identifiers and to use them throughout the investigation. These identifiers provide consistency in various aspects of an investigation—such as in the description of photographs, diagram preparation, report writing, and testimony. Building identifiers or a building marking system can include any of the following:

- Points on a compass (north side of the building, etc.)
- Front, back, left side, right side
- Similar method such as those used in the incident command (Side A, Side B, etc.) or for urban search and rescue

Whatever building-marking system is used, it should be consistent and clearly understandable to everyone involved. NFPA 1670, *Standard on Operations and Training for Technical Rescue Incidents,* has information on standardized marking systems.

Surrounding Areas. The preliminary scene assessment should include examination of surrounding areas. This involves looking for contents from the structure and fire patterns. Important witnesses can be identified, among them the reporting party, fire suppression crews, and neighbors.

Weather. The weather at the time of the fire is an important component of the preliminary scene assessment. Wind can play a role in the movement of smoke and fire. Weather information can be obtained from a number of sources, including the local airport, local media outlets, the National Weather Service, and Internet sources. Another source would be the weather observations of witnesses and fire fighters obtained through interviews.

It is important to evaluate the distance from the point at which the weather readings were taken and the fire scene. Weather conditions can vary dramatically from one location to another. The investigator should verify official reports with observations from local witnesses.

Structural Exterior. As part of the preliminary scene assessment, the investigator should walk around the entire exterior to evaluate the possibility of extension from an outside fire, to examine the construction of the building, to determine the occupancy or use of the building, and to identify areas that may require further study. During this and all subsequent phases, the investigator should constantly be evaluating the safety of the building and identifying any conditions that need to be corrected. These conditions include securing utilities, shoring unstable areas, removing water, cordoning off unsafe areas, and noting inhalation dangers.

Structural Interior. The investigator should then move to the interior of the building and conduct a preliminary assessment of conditions, moving from areas of least damage to areas of most damage. Interior damage should be compared with the location and type of damage on the exterior. Among the things to examine are the following:

- The contents of the structure
- How the contents were stored
- How the contents were positioned
- The living conditions
- The type of construction
- The composition of the surface coverings of walls and floors
- Ventilation aspects of windows, doors, and other openings, including ventilation openings
- Indicators of smoke and heat movement, such as patterns on surfaces
- Areas of fire damage
- Relative extent of damage (severe, moderate, minor, none) in each area

The investigator should note any apparent postfire alterations, which might include debris removal or movement, content removal or movement, alterations to electrical service, gas meter removal, and indications of prior investigations. The investigator should identify those responsible for these alterations and question them to determine, as closely as possible, the conditions before the alterations were made.

Preliminary Fire Spread Scenario Development

Following the preliminary scene assessment, the investigator should begin to develop a preliminary fire spread scenario based on the gross evidence available (fire spread, fire damage, burn patterns, and witness statements). The purpose of this critical phase of the investigation is to provide a preliminary hypothesis.

Note that this is a *preliminary* scenario development. The investigator must not become focused solely on this hypothesis and exclude any others that may be raised during the ongoing investigation. This preliminary scenario should be changed or discarded if the evidence uncovered during the investigation does not support it.

In-Depth Exterior Examination

Once the preliminary scene assessment has been completed and the preliminary scenario has been developed, it is time for the investigation to move into a more comprehensive stage. This stage is when the efforts of the first two phases pay off, in that the investigator has identified the areas that need to be examined further. This in-depth examination can be made by using observations, photographs, sketches, or other means.

A detailed exterior surface examination is critical because there are issues that may be brought up later in the investigation that may be important and related to the exterior. For example, the location and condition of a window (open or closed) may have affected the ventilation and the subsequent fire growth and spread. If this observation is not documented, then its value in determining the fire spread is lost.

Another critical reason to conduct an in-depth exterior examination is to look for additional fires that may have occurred at the scene. Investigators may mistakenly zero in on the first fire they find or the area with the greatest damage, not realizing that there could be other areas where fires occurred.

As with the preliminary scene assessment, the investigator should start from the exterior of the building and work inward, to the interior. Even if the fire clearly originated in the interior, this strategy should be used to document various building features, including, but not limited to, the following:

- *Prefire conditions:* The state of repair of the building and its components, the conditions of structural elements, damage to the structure, condition of the structure's fire suppression and detection systems
- *Utilities:* Type, location, meter readings
- *Doors, windows, and other openings:* Their location; condition (open, closed, or broken, and for how long prior to the fire); means of securing
- *Explosions:* The presence or absence of evidence that explosions may have affected the exterior components
- *Fire damage:* Overall damage; damage to natural openings (windows, doors); damage resulting in unnatural openings (holes made by the fire or explosion, holes made during suppression efforts)

The exterior evaluation of fire damage should include an analysis of the fire suppression efforts. These may have affected the fire patterns and therefore could have had a significant impact on the direction of the fire and the subsequent damage that occurred. It is important to interview fire fighters to determine what actions were taken on the fireground, such as forcible entry, ventilation, overhaul, and other related activities.

In-Depth Interior Examination

Many fires start in the interior of a structure. Even if the fire did start in the exterior, the conditions inside the building should be documented, especially any damage that occurred as a result of the fire spreading to the interior. Furthermore, there may be issues that come up later in the investigation relating to conditions inside the building. This in-depth interior examination is probably the investigator's only chance to document the following conditions:

- *Prefire conditions:* Housekeeping; fuel loads (location, quantity); electrical appliances; fire suppression and detection equipment and its condition
- *Utilities:* Location, condition, configuration, damage
- *Explosion damage*

The in-depth interior examination is an important phase because a great deal of information on the fire damage can be gathered, which can be instrumental in determining the area of origin.

Fire Scene Reconstruction

Adequate debris removal and reconstruction are essential to accurately observe, analyze, and document fire patterns. In this phase of the investigation, the investigator attempts to recreate the conditions that existed prior to the fire, as shown in Exhibit 15.1. By removing the debris carefully and systematically and then placing the remaining contents back in their proper place, a great deal can be learned.

The fire scene reconstruction phase is important because it can allow for a strong visualization of how the fire developed and spread, based on pattern analysis as related to the location of the contents. During the previous phases of the investigation (preliminary scene assessment, preliminary scenario development, in-depth exterior examination, and in-depth interior examination), the investigator can generally narrow down an area of origin. Therefore, the reconstruction efforts can be limited to the probable area of origin as opposed to the entire area of damage.

EXHIBIT 15.1 Fire Scene Reconstruction

To reach the evidence, it is necessary to remove the debris in a systematic and controlled manner. This ensures that there is no misinterpretation of patterns that became obscured by debris or by evidence contained in the debris. The investigator must remember that safety is of paramount concern during this phase. Debris removal can make a structure unstable. When removing debris, the investigation team should remove only as much of the debris as is necessary. Fire suppression crews should attempt to minimize contents and debris before starting the investigation. If this is not possible, the investigator should document the conditions prior to the debris removal.

The debris should be moved only once. It should be removed in a systematic fashion, and the process should be fully documented, including location, condition, and orientation of any contents that are uncovered. The investigation team should work together, deciding what is important and what is not before beginning the process.

Following the debris removal, the contents should be placed back in their original location for scene reconstruction. If the location of an item cannot be definitively determined, the item should not be included. It is better not to have it there than to place it erroneously and perhaps draw an incorrect conclusion.

Mathematical calculations and computer fire models are sometimes used in reconstructing the fire dynamics or development. However, these methods require a certain level of expertise. They are also highly dependent on the correct information being entered to produce a valid solution (a situation known in programming as "garbage in, garbage out"). If the investigator is not comfortable using these tools, or does not have someone available who has the proper level of expertise, it is better not to use them.

The 2004 edition of NFPA 921 contains a form to assist in collecting information that can be used in calculations and computer fire modeling. See Figure A.15.3.2(i) in NFPA 921.

Fire Spread Scenario

Once the preceding five steps have been completed, it is time for the investigator to determine how the fire spread once ignition occurred.

The fire spread scenario (hypothesis) should be supportable by the evidence (data) gathered through the fire investigation process. Any evidence that contradicts the fire spread scenario proposed by the fire investigator must be resolved. If the hypothesis is not supported by the facts, the investigator may have to return to one of the earlier steps in the scientific method and repeat the process from that point to ensure that a sound conclusion is reached. (See Chapter 2 in this *User's Manual,* "Basic Methodology.")

The final phase in the process is to define the area of origin, using all of the information obtained. Once the area of origin has been pinpointed, the potential ignition sources in the area should be evaluated to determine the cause of the fire. If no area of origin can be reliably established, then it is difficult to reliably determine the cause of the fire.

TOTAL BURNS

A total burn—a fire that has completely consumed all of the fuel (contents and/or building)—is a difficult but not impossible fire scene to investigate. Although it is usually not possible to determine the origin and cause of a total burn, there are other areas of interest that may need to be considered. Even though the analysis is difficult, valuable data can still be recovered from a total burn site. The investigator should therefore attempt to follow the procedures outlined in this chapter to gather data, as follows:

- Interview witnesses.
- Obtain information on the building from records.
- Walk the perimeter.
- Identify any objects found that are relevant.
- Develop a site plan and indicate the location of the found objects.
- Inspect the interior and attempt to develop a floor plan using clues obtained from debris.
- Remove the debris in a systematic method, layer by layer.
- Evaluate the contents to determine if they are appropriate for the occupancy.
- Evaluate heat damage to the contents.

An attempt should be made to obtain photographs of the fire in its early stages from first responding units such as police officers. Sometimes, onlookers at the scene have captured the fire in its early stages on video or still cameras, and every attempt should be made to obtain this evidence.

ACTIVITY 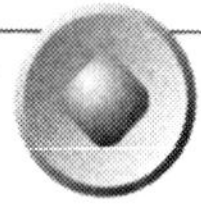

Review an experienced investigator's report on origin determination and discuss the findings with the investigator. ◀◀◀

interFIRE TRAINING

Table 15.1 provides an interface between NFPA 921 and the interFIRE VR training program. It brings the user from the NFPA 921 section of interest to the corresponding area in interFIRE. The user can apply the interFIRE information or find more information on what NFPA 921 offers.

TABLE 14.1 NFPA 921/interFIRE VR Training

921 Section	Knowledge/Skill	interFIRE Tutorial Student Activity	interFIRE Scenario Student Activity	interFIRE Resource Section Student Activity	www.interFIRE.org Student Activity
Chapter 17—"Origin Determination"	*Demonstrate the ability to coordinate information from evaluation of fire patterns, witnesses, fire science, and electrical systems to determine the fire's origin (17.1.1)* *Understand the effect overhaul has on potential evidence remaining after a fire and the need to protect fire scene evidence (17.7.3)*	Information and guidance on the process of origin determination is a focus of this program. Public fire investigators can favorably affect the outcome of fire scene evaluation by protecting and preserving the fire scene and its evidence. Four major Tutorials contain important information guiding origin determination. **Module: Roll-Up Practices** • Narrative and Flow Chart (17.7.3) • Guiding Principles • Preserve Scene and Physical Evidence • Contact the Incident Commander • Secure Scene and Witnesses **Module: Preliminary Scene Assessment** • Narrative and Flow Chart • Get Basic Incident Information • Size Up the Structure • Walkthrough with First Respondent **Module: Interviewing Witnesses** • Guiding Principles • Prioritizing Witnesses • Approaching the Witness **Module: Fire Scene Examination** • Guiding Principles • Document the Scene • Determine the Need for Additional Resources • Examine the Scene and Analyze Fire Flow • Reconstruct the Scene	The student is able to conduct an origin determination of the program's fire scene by coordinating information derived from the following (17.1.1): • Evaluation of the fire patterns • Interviews with public safety officials and witnesses from the scene • Analysis of the physics and chemistry of fire initiation, growth, and development • Evaluation of electrical appliances and circuits in and around the origin As the student progresses through origin determination, he or she will need to utilize all means of evidence documentation described in (17.1.1), such as: • Recording the Scene • Note Taking (statements and photograph notations) • Evidence Identification and Collection During this process, the student should perform a preliminary scene assessment (17.3); search the areas around the structure (17.3.2.1 and 17.3.2.2); note the weather (17.3.3); examine the exterior of the structure (17.3.4); examine the interior of the structure (17.3.5); note prefire conditions (17.5.1), condition of utilities (17.5.2), and status of all exterior doors and windows (17.5.3). The student who works the case further will be able to excavate the area of fire origin and wash down the floor (17.7.3). **Exercise** Write a paper or participate in a class discussion of the factors leading to identification of the area of fire origin in the 5 Canal Street fire.	Within the Roll-Up and Preliminary Scene Assessment files, review the following documents on topics covered in Chapter 17. **Roll-Up Practices** • Fire Fighters and Arson Detection • Hazard Identification • Overhaul Techniques • Evidence Preservation • Thirty-Three Questions for Fire Fighters • Fire Suppression Questionnaire **Preliminary Scene Assessment** • Working the Fire Scene: Search and Seizure Issues • Spoliation of Evidence: A Fire Scene Dilemma • Watch Out for Civil Tort Liability for Spoliation of Evidence	All of the reference and photographic information available in the CD-ROM Resource Section is available online at www.interfire.org. Additionally, the following "Featured Articles" and other references are available on the website containing information and examples of origin determination: • *Full-Scale House Experiment for interFIRE VR* • *Fire Modeling: An Introduction for Attorneys* • *Glass Breakage in Fires* Check the website frequently for additional information on this topic.

QUESTIONS FOR CHAPTER 15

1. In determining the origin of the fire, what four main areas are covered in the coordination of information?
2. When might an investigator be able to determine an area of origin from a single piece of evidence?
3. Explain why there is normally no single item or observation that is sufficient to determine the origin of a fire.
4. Describe how observing and documenting fire patterns that indicate fire movement and intensity provide the necessary data to determine the point of origin.
5. When assessing fire damage, what key information should be gathered from each category listed below?
   A. Notes
   B. Photography
   C. Vector diagrams
   D. Depth-of-char survey grid diagrams
6. What is the purpose of a preliminary scene assessment?
7. True or false: Once the preliminary scene assessment has been completed, the exterior of the structure should be analyzed in detail.
   A. True
   B. False
8. Describe the purpose of fire scene reconstruction and the methods used to accomplish it.
9. What must an investigator do when there are contradictions to the fire spread scenario?
10. True or false: A total burn—a fire that burns unimpeded until it self-extinguishes due to a lack of fuel—is not worthy of investigation.
    A. True
    B. False

Fire Cause Determination

▶▶▶ OBJECTIVES

Upon completion of Chapter 16, the user will be able to

- Identify the four elements of a fire cause
- Understand evaluation of hypotheses
- Explain the process of elimination

Generally, the primary role of an investigator is to determine the cause for a fire or explosion. However, determining the fire cause is not limited to identifying the ignition source, such as smoking materials, an electrical arc, or a match. There are four elements that need to be identified for the determination of a fire cause: the ignition (heat) source, the fuel (material) first ignited, the oxidizer present, and the ignition factor (ignition scenario). When discussing hypotheses or making conclusions regarding the fire cause, it is incumbent on the investigator to identify and discuss these four elements.

fire cause

The circumstances, conditions, or agencies that bring together a fuel, ignition source, and oxidizer (such as air or oxygen) resulting in a fire or a combustion explosion. (Source: NFPA 921, 2004 edition, Paragraph 3.3.55)

FOUR ELEMENTS OF FIRE CAUSE

Ignition Source

One of the first decisions that the investigator must make in testing any hypothesis regarding potential heat sources is determining whether those ignition sources are *competent.* A competent ignition source is one that has sufficient energy, temperature, and time to raise the first fuel to its ignition temperature. If the heat source cannot meet these criteria, it can effectively be eliminated as a potential ignition source, and the hypothesis should be discarded.

Why is this important? To support a hypothesis, it is necessary to determine whether the heat source that was found in the area was competent to ignite the first fuel under the existing conditions. If this heat source cannot raise the first fuel to its ignition temperature, then the heat source was not the ignition source for the fire. For example, a piece of paper that moves quickly through the flame from a candle does not ignite despite the energy and temperature from the flame being higher than the ignition properties of the paper. The key element here is the duration. This time of contact between the flame and the paper (fuel) is not sufficient to allow ignition to occur. This can happen in real-world situations as well. The mere presence of a heat source at the origin does not necessarily make that heat source "the ignition source" for the fire.

First Fuel Ignited

In some cases, it can be difficult to separate the fuel first ignited from the major fuel package. For example, if flammable vapors present in a basement were ignited by the pilot flame on a water heater, the combustible fuel load in the basement (i.e., contents and structure) could be ignited by the subsequent explosion and fire. To determine the root cause of the incident, it is important to differentiate the two from each other.

Oxidizer Present

In most fires, the identification of the oxidizer present is seldom articulated, because it is atmospheric oxygen. This does not mean that the oxidizer is not an essential element of the fire cause. There can be fires that have an oxidizer other than atmospheric oxygen as the initial or primary oxidizer, an example being a fire initiated by the interaction of a fuel with chlorine. Despite the presence of atmospheric oxygen, it is the chlorine that is the fire cause oxidizer. The investigator is reminded not to skip this step in fire cause determination, despite its seeming intuitive that atmospheric oxygen is the primary oxidizer, because it is the exceptions that can lead to a faulty conclusion.

Review Chapter 5 in NFPA 921 for further discussions of oxidizers other than atmospheric oxygen.

Ignition Sequence

Simply put, the ignition factor is the sequence of events that brought together a heat source and a fuel under the right conditions for a sustained combustion to occur:

$$\text{Heat source} + \text{first fuel} + \text{event} = \text{fire}$$

In formulating the sequence, it is important to review several factors affecting the sequence. The fuel must be sufficiently heated to generate combustible gases. Remember, it is the gases that burn, not the fuel itself. (See Table 16.1.)

TABLE 16.1 Sources and Forms of Ignition

Source	*Form*
Short circuit	Spark or arcing
Water heater pilot	Open flame
Cigarette	Smoldering material

Review Section 5.3, "Ignition," in NFPA 921.

Heating and Proximity. A number of factors can influence whether or not a fuel is heated sufficiently. Some of these include proximity and duration of exposure. If the fuel and the heat source are not within a critical distance (proximity) of each other, insufficient heat is transferred from the source to the fuel. The critical distance varies according to the heat transfer mechanism and heat flux. For example, if a candle is 10 feet from a newspaper, it is unlikely that ignition will occur, because the fuel is beyond the critical radiant and convective heat flux. However, if the newspaper is held 2 inches over the flame, it will be heated sufficiently to ignite.

Experiment with a lighted candle in a safe environment. Hold a paper over the flame at varying distances and see whether it ignites and the time it takes to ignite. Now hold the paper at the same distances to the side of the flame and record your results. What causes the different results?

Duration. If the fuel and heat source are not close enough for a sufficient amount of time (duration) to allow the heat transfer to occur, then ignition does not follow. Using the same example given above, if the newspaper is passed quickly within inches of the flame, ignition might not occur because there was not sufficient time for the newspaper to be heated by the flame to start giving off combustible gases.

Repeat the previous candle experiment with a piece of cardboard and then a piece of wood. Note that the three pieces of fuel were all made of cellulose. What accounts for the different results?

Thermal Inertia. The ignition process is directly related to thermal inertia. Thermal inertia can be thought of as the amount of energy it takes to raise a material's temperature to a point at which it gives off combustible gases that can then be ignited.

thermal inertia

The properties of a material that characterize its rate of surface temperature rise when exposed to heat; related to the product of the material's thermal conductivity (k), its density (ρ), and its heat capacity (c). (Source: NFPA 921, 2004 edition, Paragraph 3.3.152)

Consider, for example, a fire that occurs in a deep-fat fryer in a kitchen. Determination of the cause requires more than just the identification of the ignition source and fuel. A "grease fire in a deep-fat fryer" does not constitute the cause. The "cause" requires the identification of the mechanism of failure, such as contacts that were defective or a high-limit thermostat that failed.

Remember that the ignition factor is the explanation of how the heat source and the first fuel came together. In a fire caused by someone having dropped a lit match into a wastebasket filled with newspaper, the four elements of the fire cause would be:

- The lit match as the heat source
- The newspaper as the first fuel
- Atmospheric oxygen as the oxidizer
- The dropping of the match into the wastebasket as the ignition factor

EVALUATION OF HYPOTHESES

Fire cause determination involves a process of hypothesis development and testing to arrive at a supportable theory about the source of ignition and the material that was first ignited. This process requires that all potential sources of ignition in the area of origin be evaluated. That is, they are either cognitively or experimentally tested.

Testing the Hypotheses

To cognitively test a hypothesis is to evaluate the theory mentally by applying rigorous reasoning. It is a process of questioning. Are there sufficient facts to support the theory? Does the theory make sense given the supporting facts? Are the fire dynamic theories sound? Is there cause and effect? The process continues until the hypothesis is found to be valid or fails the testing.

Experimentally testing the hypotheses can occur on many levels, from testing of supporting facts to testing of fire development. The tests can range from simple to complex, such as a field flammability test, a bench scale test of a material's ignition temperature, a heat release test of a piece of furniture, or a full-scale compartment fire.

Only when a hypothesis is not supportable to a sufficient level of certainty (refuted) is it effectively eliminated. The process of elimination, therefore, is simply a process of developing and testing hypotheses.

Determinations Using the Process of Elimination

It is occasionally possible to make a credible determination about an ignition source even in the absence of physical evidence of the heat source. This can be accomplished by using the same hypothesis development and testing process noted earlier.

The determination of the ignition source (which is one element of the fire cause) in the absence of physical evidence is referred to as the *process of elimination.* The process of making determinations in the absence of physical evidence of the cause is long-standing in the fire investigation community. However, this process still requires an analysis that reaches a conclusion based on the scientific method. There is no exception to the scientific method.

Although it may occasionally be possible to determine a fire cause in the absence of physical evidence of an ignition source, there are some limitations on the use of this procedure. Sometimes it is appropriate, and other times it is inappropriate. The proper application for the process of elimination referred to in NFPA 921 is contingent on the origin being "clearly defined." The terminology *clearly defined* indicates the importance of the identification of the origin. Ideally, the origin should be obvious.

Generally, the identification of the origin becomes more difficult as the damage increases, such as in fully developed (post-flashover) compartment fires. Where the origin is not determined, the use of the process of elimination is inappropriate.

Large fires or fires with heavy damage can be problematic if the determination of the area of origin is based solely on fire pattern analysis without investigating the supporting facts related to the production of those patterns. The investigator has to determine the preexisting conditions, the fire growth, and the fire suppression to understand the complexity of the processes that led to the development of the fire patterns. Fire pattern analysis that is unsupported by documentation of factors that affect fire pattern development can lead to a faulty conclusion. It is advisable to retest the hypotheses about the fire pattern analysis.

Because the process of elimination may allow for a determination of cause in the absence of physical evidence of the ignition source, it is justifiably important that the origin be absolute. The investigator should be aware that there are risks in using the process of elimination. A critical error can occur if the origin is misidentified. In that event, the real ignition source will not be located because the real origin is not being investigated. Furthermore, in the case in which the suspected origin is incorrectly identified and the ignition source is wrongly determined to be an open flame, the cause may be inappropriately classified as incendiary. This is why the accurate determination of the origin is critical.

Another consideration before identifying an ignition source in the absence of the origin is that the concluding cause must be consistent with all other known facts. For example, if a countertop is identified as the area of origin, then all potential ignition sources within the area must be considered to determine what specifically caused the fire. The process to eliminate potential ignition sources may entail inspecting or testing each electrical outlet and each electrical appliance. The procedure may further require collecting additional empirical data from the fire scene. This may include determining whether any smoking materials were present, or if there were any cooking appliances that may have served as a potential heat source. Above all, the suspected ignition source must be competent in relation to the first fuel that was ignited. The identification of the first fuel may assist in developing hypotheses about the ignition scenario or the elimination of incompetent ignition sources.

The introduction of an open flame seldom leaves physical evidence for the determination of that heat source. When all other potential sources of ignition in the discrete area of origin have been credibly eliminated, the hypothesis that the fire resulted from the introduction of a competent ignition source may be supported. In many of those incidents, there is other evidence that helps to support that hypothesis.

Process of Elimination in Accidental Fire Causes

Although the process of elimination methodology is most often associated with application to incendiary fires, this methodology is also prevalent in the identification of ignition sources in accidental fires. In these circumstances, after determining an origin, an investigator may conclude that the heat source that was present at the origin is the ignition source, merely because of its presence. The identification of an appliance as "the" ignition source is usually accomplished by determining the mechanism of failure. Concluding that the appliance was the ignition source because of its proximity to the origin is rarely justified. Whether it was the ignition source is not easily determined by visual examination and if that is done the investigator should be expected to differentiate its condition as the ignition source versus a victim of the fire.

interFIRE TRAINING

Table 16.2 provides an interface between NFPA 921 and the interFIRE VR Training Program. It brings the user from the NFPA 921 section of interest to the corresponding area in interFIRE. The user can apply the interFIRE information or find more information on what NFPA 921 offers.

TABLE 16.2 NFPA 921/interFIRE VR Training

921 Section	Knowledge/Skill	*interFIRE Tutorial* Student Activity	*interFIRE Scenario* Student Activity	*interFIRE Resource Section* Student Activity	*www.interFIRE.org* Student Activity
Chapter 18—"Cause Determination"	*Ability to accurately identify the circumstances and factors necessary for the fire to have occurred (18.1)* *Determination of the cause of a fire loss encompasses four major investigative areas:* *1. Cause of the fire (circumstances that bring together a fuel, ignition source, and an oxidizer, causing a fire or explosion)* *2. Factors responsible for fire spread (exposures, construction, fire protection, etc.)* *3. Cause of deaths/ injuries (adequacy of alarm systems, condition of egresses, toxicity of environment, etc.)* *4. Degree of human fault (incendiary or negligence)*	Information and guidance on the process of cause determination is a focus of this program. The fire service and public fire investigators play a major role in protection of evidence allowing cause determination by restraining fire fighting tactics and minimizing overhaul activities. Five major Tutorials contain information on cause determination: **Module: Roll-Up Practices** • Preserve Scene and Physical Evidence **Module: Preliminary Scene Assessment** • Size Up the Structure • Walkthrough with first respondent **Module: Fire Scene Examination** • Guiding Principles • Document the Scene • Handling Injuries and Fatalities • Examine the Scene and Analyze Fire Flow • Reconstruct the Scene • Collect and Preserve Evidence • Eliminate Accidental Causes • Make On-Scene Cause Determination **Module: Interviewing Witnesses** • Guiding Principles • Corroborate Information **Module: Follow-Up Investigation**	This fire scene contains physical and circumstantial evidence that would enable most experienced fire investigators to render an opinion as to the cause of this fire. Forensic and nonforensic expert examination of specific evidence are available to corroborate the investigators findings. After the student isolates the fire origin at the 5 Canal Street scene, the ignition factor (cause) is determined by evaluation of the source and form of the heat of ignition (18.3) and identification of the first material ignited (18.4). To isolate one ignition factor as the cause, the student must first systematically examine and eliminate all other potential ignition sources in the area of fire origin (18.2). **Exercise** On closing the scene investigation at 5 Canal Street, an origin and cause determination must be made. To make the cause determination, select the factors that support your determination. In your group, put up the list of possible accidental causes and discuss each one according to what you determined in your investigation. Reach a group consensus on what, if any, of these items is the cause of the fire. Do the same for the incendiary causes. For each cause, discuss whether the statement is true or false for the investigation at 5 Canal Street (i.e., were there canine hits in the dining room?). If the answer is true, then discuss whether that fact is proper support for cause determination. Remember, not every point that is true can be used to support cause determination. Discuss this and reach a group consensus on each point.	The program contains a great deal of information on fire cause determination. Refer to the file "Fire Scene Examination" to review the following documents on origins and cause: **Determination** • Cause Determination • Thermal Effects on Materials (photo catalogue) • Origin and Cause: Legal Standard of Truth • Reflections on the Origin & Cause **Accidental Fire Causes** • Fire Scene Electrical Checklist for the Non-Electrical Engineer **Incendiary Fire Causes** • Incendiary Fire Basics • Incendiary Fire Indicators • The Arson Set: Excerpts from *The Pocket Guide to Accelerant Evidence Collection*	All of the reference and photographic information available in the CD-ROM Resource Section is available online at www.interfire.org. Additionally, the following "Featured Articles" and other references are available on the website containing information and examples of fatal injury fire investigations: • *Smoke Detector Technology and the Investigation of Fatal Fires* • *Heat Release Rate: A Brief Primer* • *Temperatures in Flames and Fires* Check the website frequently for additional information on this topic.

QUESTIONS FOR CHAPTER 16

1. True or false: When a fire is set with a match, the fire cause cannot be determined unless the match is found.
   A. True
   B. False
2. True or false: When the origin of a fire is clearly defined, it is possible to make a credible determination regarding the cause of the fire even when there is no physical evidence of that cause available.
   A. True
   B. False
3. List the four elements of the fire cause.

CHAPTER 17

Analyzing the Incident for Cause and Responsibility

OBJECTIVES

Upon completion of Chapter 17, the user will be able to

- Understand the four classifications of fire cause
- Understand the differences between cause and responsibility

Determining cause is not limited to identifying the ignition sources, such as smoking materials, or the fuel, such as an upholstered chair. It also involves the steps that brought the two together and the conditions under which the fire was able to start and propagate. This chapter focuses on classification of cause, source and form of ignition, first material ignited, ignition factors, and determination of responsibility.

cause

The circumstances, conditions, or agencies that brought about or resulted in the fire or explosion incident, damage to property resulting from the fire or explosion incident, or bodily injury or loss of life resulting from the fire or explosion incident. (Source: NFPA 921, 2004 edition, Paragraph 3.3.22)

FEATURES OF THE INVESTIGATION

Although the focus of an investigation is typically on the fire cause, there are other significant features that are often addressed during an investigation. Some or all of these features may be addressed in varying degrees, depending on the investigator's responsibility. In fact, in some cases, the investigator's responsibility may emphasize or be limited to issues other than isolating the "cause of the fire." These features fall into four classifications:

1. *Cause of the fire or explosion:* This area identifies the elements of a cause: the heat source, the first material ignited, and the conditions or circumstances that allowed the heat source and fuel to create a fire or explosion.
2. *Cause of damage to property resulting from the incident:* This area considers the factors that are responsible for fire spread from the origin. These factors may include the combustibility of contents or construction materials, the adequacy or inadequacy of passive (e.g., fire walls and fire doors) and active fire protection systems (e.g., water sprinklers), and structural compliance to model fire or building codes.
3. *Cause of bodily injury or loss of life:* This area identifies factors related to human injuries or deaths. The factors may include analysis of fire alarm or smoke detection systems and the analysis of human behavioral response to the fire. (See Chapter 10, "Fire-Related Human Behavior," of this *User's Manual.*)
4. *Degree to which human fault contributed to any of the above:* This area concerns factors related to human contribution to the any of the features listed above. For example, combustible materials might have been stored too close to a heat source.

CLASSIFICATION OF CAUSE

The classification of cause relies on the development and testing of hypotheses using the scientific method. (See Chapter 2, "Basic Methodology," in this *User's Manual.*) Cause classification is the culmination of identifying the four elements of the fire (ignition source, fuel first ignited, oxidizer present, and ignition factor), and then classifying them according to the generally accepted definitions. At this time, the determination of the cause and the classification of the cause are separate activities.

There are four general categories to which an explosion or fire cause can be classified:

1. Accidental
2. Natural
3. Incendiary
4. Undetermined

Accidental

Accidental fires are those that are not the result of a deliberate (intentional) act. This also encompasses fires that are ignited deliberately but become hostile, such as intentionally ignited brush or trash fires that spread beyond the intended confinement. Traditionally, the accidental classification is used to classify fires ignited by juveniles who are below the legal age for being responsible.

Natural

The natural fire cause classification is for fires that occur without human intervention. This category includes fires that result from natural phenomena such as lightning, wind, and earthquakes.

Incendiary

These are fire causes that result from deliberate acts—that is, fires that are ignited or result from deliberate actions in circumstances in which the person knows that there should not be a fire. Fires resulting from reckless or negligent acts are not specifically referenced but may be included on the basis of the circumstances in which "the person knows the fire should not be ignited." Intent is therefore a key element of this classification.

Undetermined

The undetermined classification is an appropriate category for fires that have not yet been investigated, for fires that are under investigation, and for fires that have been investigated and the cause is not proven to an acceptable level of certainty.

DISCUSSION OF CLASSIFICATION

The first three classifications (accidental, natural, and incendiary) are classifications in which the hypothesis is supported by the data. If the hypothesis is not supported by the facts (data), the investigator returns to one of the earlier steps in the scientific method and repeats the process from that point to ensure that a supportable conclusion is reached. (See "Basic Methodology," Chapter 2 in this *User's Manual.*) If the cause cannot be determined because the data are insufficient to support the hypothesis, then the cause must be classified as "undetermined." With the development of new data, the investigator can reevaluate the original classification of undetermined and change the classification.

Even after a classification has been made, if new evidence (data) should become available that does not support the original hypothesis, then the investigator should retest the original hypothesis with the new data. If appropriate, the investigator should reach a new cause determination.

It is not necessary that all the elements of the cause be identified for the cause to be classified. As was discussed earlier in this chapter, by using the process of elimination, it may be possible to make a credible determination in the absence of the physical evidence of the ignition source. There also may be circumstances in which the available evidence at the fire scene allows for the identification of the ignition source and the first fuel, but the cause cannot be classified. In these circumstances, the appropriate classification is undetermined.

In some circumstances, such as those in the examples below, there is insufficient information to classify the cause. The investigator cannot use evidence or analysis of intent or motive to classify the cause. (See Chapter 20, "Incendiary Fires," in this *User's Manual.*) In those cases, the cause is undetermined.

Suspicious Determinations

At no time is a fire cause to be classified as "suspicious." This is not an acceptable classification for a fire cause. *Suspicious* is not an appropriate term to describe the unexplained. Although the untrained layperson, including news media, might refer to a fire as *suspicious,* this term is inappropriate and should not be used by the fire investigator.

Arson

Arson is a term that denotes a crime and is therefore determined by judicial process. A fire classified as incendiary can have sufficient probable cause to be called arson by the law enforcement community. Depending on local laws, not all incendiary fires are arsons.

DETERMINING RESPONSIBILITY

Determining responsibility is a process that occurs after the fire cause has been determined. It is through determining responsibility that codes and standards can be changed, fire safety can be addressed, and civil or criminal litigation actions are initiated. This generally requires that the investigator conduct a failure analysis.

responsibility

The accountability of a person or other entity for the event or sequence of events that cause the fire or explosion, spread of the fire, bodily injuries, loss of life, or property damage. (Source: NFPA 921, 2004 edition, Paragraph 3.3.126)

Failure analysis considers what factors contributing to any injuries, loss of life, and property damage may have resulted from the root incident. The information gathered through a failure analysis of an incident can prove invaluable information in helping to ensure that these losses are not repeated in the future. When looking at an incident from a failure analysis point of view, it is important to understand that it is often a chain of events that contribute to the cause.

failure analysis

A logical, systematic examination of an item, component, assembly, or structure and its place and function within a system, conducted in order to identify and analyze the probability, causes, and consequences of potential and real failures. (Source: NFPA 921, 2004 edition, Paragraph 3.3.50)

Some examples of failure analysis that can be conducted during an investigation are as follows:

- Did the sprinkler system operate? If so, why did it not control the fire? Was sufficient water provided for the hazard being protected?
- Were there sufficient exits for people to escape from the fire? How did people react when they tried to escape?
- Did the fire alarm sound? If not, why not? Were the smoke detectors properly placed to react to the fire development?

Some of the most tragic fires have resulted in significant building code changes:

- One of the significant contributing factors to the loss of life in the Cocoanut Grove fire in Boston, Massachusetts, in 1942 was the lack of proper exits.

- The two hotel fires that struck Las Vegas in the 1980s resulted in a widespread movement in the lodging industry to call for the installation of sprinklers.
- The fire at Seton Hall University that killed three students in 2000 was the impetus for the installation of sprinklers in all student housing in the state of New Jersey.

All of these incidents, and others, resulted in dramatic changes because failure analysis identified significant contributing factors. Going beyond the initial determination of the cause of the fire brought out other factors relating to responsibility.

▶▶▶ QUESTIONS FOR CHAPTER 17

1. Which term is NOT an accurate description of a fire cause?
   - **A.** Incendiary
   - **B.** Natural
   - **C.** Suspicious
   - **D.** Undetermined
2. True or false: The nature of responsibility in a fire or explosion incident may be in the form of an act or omission.
   - **A.** True
   - **B.** False

CHAPTER 18

Failure Analysis and Analytical Tools

OBJECTIVES

Upon completion of Chapter 18, the user will be able to

- Describe methods available to assist in the analysis of a fire incident
- Document information collected during the incident
- Organize information into a rational and logical format

As part of determining fire cause and origin, the fire investigator may conduct a failure analysis when issues related to performance of equipment or systems are involved. Such activities normally include evaluating time-based information and aspects of equipment or systems performance. The methods used are timelines and systems analysis.

TIMELINES

Understanding the timeline of any incident is key to creating the failure analysis because it assists the investigator in determining the sequence of events that occurred. A timeline is a comprehensive listing of specific events, no matter how inconsequential, that can be verified and attributed to a stated degree.

The actual timeline can be either a graphic or narrative representation of all or some activities related to a fire incident that have been arranged and formatted in chronological order. Timelines may include events that occurred before, during, and after the fire. Estimates of fire size or conditions are often valuable in developing timelines. With the use of fire dynamics, fire conditions can be related to specific events. Information that is gathered from detection and suppressions systems may be useful in determining the spread of the fire as well as establishing a time for the events. Remember that the value of the timeline is directly related to the accuracy of the information it includes, and the reliability of a timeline is a function of the level of confidence that can be placed in specific elements that it includes. Not all information can be related with a specific time and may have to be listed as a *time interval* during which the event occurred. A variety of components are used to develop timelines. These include incidents described as either *hard times* or *soft times.*

Hard Times

When developing a timeline, incidents that can be related to a known exact time are generally referred to as *hard times.* This means that the time of occurrence is specifically known. For example, a fire department's incident history records can provide the specific times when units were dispatched, arrived on scene, and so forth. Such data serve as benchmarks in developing the timeline. Sources of hard times that will identify a specific point in time or serve as benchmarks for other events are listed in NFPA 921.

Soft Times

Other times can be considered *soft times.* For example, a witness might not be able to precisely state the time a certain event such as a flashover occurred or the time when an occupant was sent to leave a building but might be able to relate such observations to another event, such as the arrival time of the fire department or the moment when a bus or train went by. Through interviews and further gathering of data, an investigator may be able to narrow down the range of occurrence for such soft times to relatively specific periods of time.

Benchmark Events

Some events are particularly valuable as a foundation for the timeline or may have significant relation to the cause, spread, detection, or extinguishment of a fire. These areas are referred to as *benchmark events.*

It is important when developing a timeline that includes different hard time sources to note any discrepancies between the clocks and synchronize the times. Such data should be double-checked to ensure that errors were not made in synchronizing all such sources of data.

Macro and Micro Time

A variety of timelines may be required to effectively evaluate and document sequence events precipitating the fire, events during the actual fire incident, and post-fire activities. Depending on complexity, these may be evaluated in a macro or micro timeline.

A *macro* evaluation of events can cover months or even years and may incorporate activities that occurred a long time before the fire, such as those related to building construction, modification of codes, and/or code enforcement activities. Conversely, a *micro* evaluation is used to look at small or narrow segments of the macro timeline in detail. Examples of micro incidents may be travel times for a witness to walk between rooms or the dispatch and on-scene times.

Other Timelines

Parallel timelines can be used to look at multiple events that occur simultaneously. Graphic timelines can be very helpful in putting together the sequence of events that transpired, and approaches involving matrices may be helpful when numbers of events occur simultaneously, as illustrated in Table 18.1.

TABLE 18.1 Sample of an Abbreviated Matrix That Can Be Used in Developing a Timeline

Time	*Engine 1*	*Engine 2*	*Ladder 1*	*Battalion 1*
0604	Dispatched	Dispatched	Dispatched	Dispatched
0605	Responding	Responding	Responding	Responding
0608	On scene, assuming command			
0610	Making entry with hoseline			
0611		On scene	On scene	
0613	Attacking fire in NW bedroom	Advancing to second floor with hoseline for primary search and rescue	Extending ground ladder to roof for ventilation	
0614				On scene, assuming command
0615	Fire knocked down in NW bedroom	Victim located in hallway, second floor	Accessing roof	Requesting ambulance
0616	Conducting primary search, first floor	Victim removed	Making ventilation opening	

Note: The investigator might want to add columns for Police Department or perhaps non-fire service witness accounts.

SYSTEM ANALYSIS

Analytical approaches involving systems analysis take into account characteristics, behavior, and performance of a variety of elements—including human activities and mechanical features of equipment—and integrate these to provide as complete a picture of events surrounding an incident as possible.

Simple examples of systems analysis include evaluations of incidents that investigators make every day. An example of such an informal systems analysis can take the form of a properly functioning stove left on by a homeowner, which leads to a kitchen fire. One aspect of such an incident involves human factors, and the second aspect involves the properties of the stove and its surroundings. Together, these make up the system whose properties led to the incident. A variety of tools are available for the investigator to use in analyzing such incidents. These are discussed in the next section.

Fault Trees

Fault trees (also known as decision trees) illustrate series of events and decisions that must take place for a specific outcome to occur. When this type of graphic logic or reasoning is applied to a situation, a solution may become more readily apparent, and incorrect solutions may be eliminated from consideration. When developing a fault tree, the investigator should use deductive reasoning. A fault tree diagram places, in logical sequence and position, the conditions and chains of events that are necessary for a given fire or explosion to occur.

Readers might be familiar with similar decision trees used in electronics or computer programming with "and" and "or" gates at the point where decisions must be made. (Refer to NFPA 921, Figure 20.3.1.1.) For an "and" gate, all events or conditions must be present. For an "or" gate, any one of several events or conditions must be present. Fault trees identify the conditions and chain of events of a fire or explosion. If the conditions or events did not happen in an order that would lead to the event, then the proposed scenario is not possible. The investigator may apply probabilities to the events. The end result may produce multiple feasible scenarios for the fire scene. There is software for developing fault trees.

Failure Mode and Effects Analysis (FMEA)

Failure mode and effects analysis (FMEA) is another graphical method or technique used to determine the causes and effects involved with an event or subevent leading to a fire. By identifying specific components associated with potential ignition sources or fire spread, it may be possible to identify specific predecessor events occurring or activities preceding an incident.

The development of an FMEA table can be involved or quite simple, depending upon the complexity of the incident or the depth of analysis required. When compiling the information on an FMEA, the investigator should consider the environmental conditions and the process status for each item or action. Probabilities or degrees of likelihood can be assigned to each occurrence. The investigator should remember that the accuracy of the table depends on identifying the system components and human actions that are relevant to the incident.

MODELING OF FIRES

Models or approaches commonly used to analyze fire incidents are generally of two types: mathematical and engineering models and graphic models.

Mathematical/Engineering Models

A widely used group of modeling techniques involves the application of engineering models. Data are developed that incorporate both known and approximated properties of materials and systems, specific features and components of a fire incident, and physical property estimates defined to a stated degree of certainty. To these assumed "fact sets" or "modeling input data," accepted engineering and analytical techniques are applied to develop descriptions of what may or may not have occurred under a clearly stated set of conditions.

Mathematical models can be as simple as those used when flammable gas concentrations are calculated on a hand calculator to determine the upper and lower explosive limits for a given gas in a given space. Other mathematical models may be quite complex, such as those used for fire growth modeling. These involve the applications of zone or finite element techniques that must be run by a qualified modeler on a well-equipped personal computer. See Exhibit 18.1, which is a graphic depiction of heat release for NFPA standard fire growth rates for slow, medium, fast, and ultrafast fire growth rates.

It is important to understand that these techniques are only tools and that repeated calculations should be carried out to bracket the conditions that most

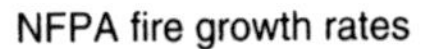

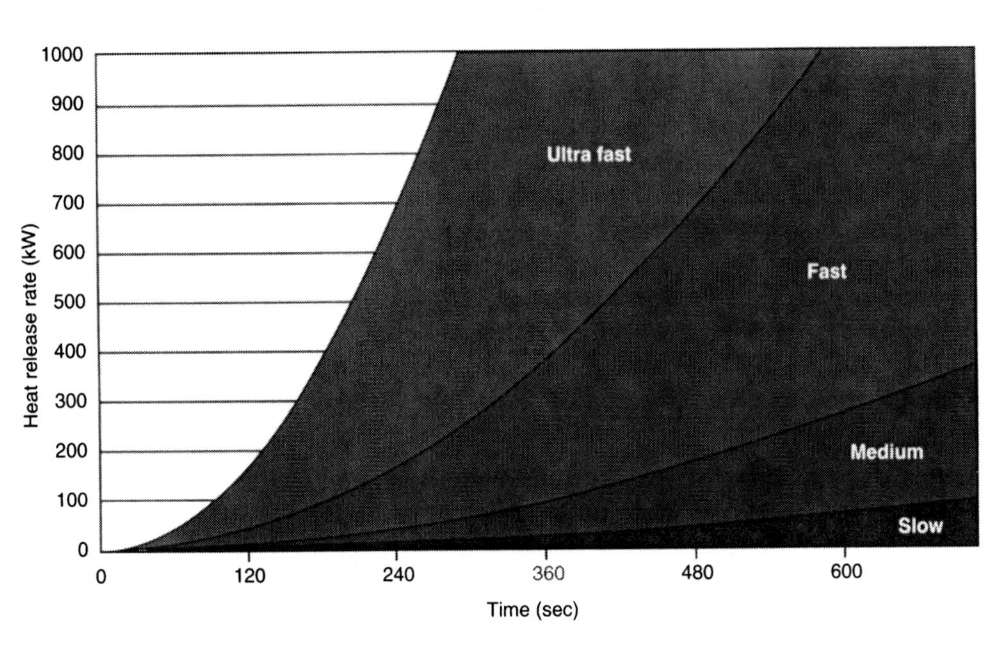

EXHIBIT 18.1 Fire Growth Curves

likely existed with a given incident. This type of modeling can be used to determine a probable scenario given a specific set of boundary conditions. Models do not necessarily provide a definitive solution. As with all evidence, modeling results should be considered in light of other data that has been gathered during the fire investigation. Models can frequently be used to help support or disprove a particular hypothesis.

Mathematical modeling must be conducted by an individual who has sufficient training and experience to carry out the type of analysis required. Examples of typical sorts of engineering modeling techniques that can be used effectively in fire investigation work are discussed below.

Heat Transfer Models. Heat transfer models allow the investigator to determine how heat was transferred from a source to a target by one or more of the common heat transfer modes: conduction, convection, or radiation.

Heat transfer models can be used to test hypotheses such as the competency of a given heat source to act as an ignition source in a given fire causation scenario. Other useful applications of heat transfer modeling are to explain damage or ignition to adjacent buildings via fire spread, ignition of secondary fuel items, and transmission of heat through building elements.

Flammable Gas Concentrations. By being able to determine what the concentration of gas was within a given space, the investigator can support or disprove the theory of whether the presence of a flammable gas during an incident was a contributing factor to the sequence of events that occurred. Flame concentration results can support or refute theories about the location and size of a leak or the competency of an ignition source.

Hydraulic Analysis. When a fire is not controlled by an operating sprinkler system, it is important to determine whether the sprinkler system functioned as designed and intended and whether the design was adequate for the particular occupancy and furnishings present.

The analysis of a sprinkler system and its water supply should determine whether the system and water supply were matched to the hazard being protected. Questions should be asked to determine whether there were any faults in the system, such as closed valves or faulty sprinkler heads. It might be necessary to evaluate why an apparently functional system did not control a given fire. This involves an analysis of the water supply characteristics, water flow through the piping network, and sprinkler discharge characteristics. It may also involve developing a fire growth or heat release model to assess the potential effectiveness of the sprinkler system under various circumstances.

Hydraulic analysis can also be used on systems such as carbon dioxide, gaseous suppression agents, dry chemicals, and fuel distribution systems.

Thermodynamic Chemical Equilibrium Analysis. Fires and explosions that are believed to be caused by reactions of known or suspected chemical mixtures can be investigated by a thermodynamics analysis of the probable chemical mixtures and potential contaminants. Thermodynamics analysis can be used in incidents involving chemical reactions and to determine the feasibility of a given scenario.

Structural Analysis. Structural analysis can provide the investigator with clues to why a building failed by collapse at a given point during a fire. This type of analysis can be critical in helping to point toward where the fire was able to have the most significant impact on the strength of the building.

Egress Analysis. The reasons why a fire victim did not escape from a given fire scene is a critical question for a fire investigator to answer. The cause for such a failure may be related to egress design features (or lack) of a building or to maintenance issues. The investigator should obtain data such as the location of the exits, egress routes, travel distance, and egress route widths to help in this analysis. Computer-based fire models such as CFAST have modules that can provide egress data for the specific fire situations that they can be used to model.

Fire Dynamics Analysis—Fire Growth Modeling. A number of computer models can assist the investigator in attempting to determine the growth of the fire. Fire endurance models can calculate the probable fire endurance of specific elements such as doors, flooring, ceiling, and walls after a fire had reached flashover conditions.

The investigator is reminded and cautioned that these models are only tools. Although they can provide additional data on a fire, modeling results must be weighed in relation to other information gathered during the investigation and the reliability of the input data and assumptions made.

A number of variables that can influence fire modeling results include the following:

- Fire load characteristics
- Ventilation openings (size and open or closed)
- HVAC flow rates
- Heat release rates

The two primary types of models currently used to assess fire growth are zone models and field, computational fluid dynamics (CFD) models. Zone models have the following characteristics:

- Can be run on a personal computer
- Divide room into two zones
- Are well accepted, validated by peer review

Field, computational fluid dynamics (CFD) models have the following characteristics:

- Require larger computing capacity
- Divide the room into thousands of small cells
- Are expensive
- Are time intensive
- Require a high level of expertise
- Provide the highest degree of computational accuracy currently available

Graphic Models

The second general class of models sometimes used to represent what has occurred during a given fire incident are graphic models. These include drawings of all sorts as well as physical models and computer animations, which are being used increasingly today. This class of models is frequently used by both the investigation team and forensic evaluation personnel.

Among the reasons that an investigation team uses graphics are to better understand an incident location, to assist in interviewing witnesses, and to better define and identify materials and systems and their involvements in an incident.

In forensic applications, graphics can be used to help a judge and/or a jury to better understand important features of a fire scene and underlying scientific and engineering principles that caused a particular outcome in a given fire. Graphic models should not be confused with mathematical models, described earlier. Mathematical models are based on calculations, whereas graphic models are based on geometric representations of a scene or set of facts.

FIRE TESTING

Fire testing involves taking known facts from a fire scenario and exposing them to a well-defined set of conditions. The results can be applied to better understand what did or did not happen under a given set of circumstances.

Examples of these may include determining the burn-through time of a wall or door using recognized fire endurance testing techniques such as those found in the ASTM E-119 standard or determining the heat release characteristics of a piece of cushioned furniture to assess whether it could have been responsible for fire growth and damage patterns noted. Application of the types of data that are developed in testing can provide invaluable information to support or refute various hypotheses being evaluated.

EXHIBIT 18.2
Fire Progression
(Source: Fire Cause Analysis)

(a)

Exhibit 18.2 shows a series of images of 4-minute fire growth progression prior to extinguishment. Fire was initiated with a single match igniting a single sheet of newspaper.

As with computer modeling, it is important to make certain that test conditions are consistent with the fire scenario being studied and that testing is conducted as accurately as possible to obtain valid results. It is rarely possible to absolutely duplicate conditions from a given fire. It is generally possible to replicate enough of the important facts and data to conduct tests that will provide important, needed insights into what actually happened.

Exhibit 18.3 shows temperature development during ASTM E-1537-99 fire testing of a couch. Measurements were made at locations 4 ft (1.2 m) above the couch and directly above the couch back.

DATA REQUIRED FOR MODELING AND TESTING

To conduct valid modeling and testing, it is important that the investigator gather data that are as accurate and complete as possible. As in all other endeavors, the "garbage in, garbage out" concept applies here. It should be anticipated that during court proceedings, test results will be subjected to a Daubert review, and the results that are presented will be only as valid as the accuracy of the information from which they were derived and the care that was taken to develop them.

Important information to be obtained includes structural information, materials and contents, and ventilation information.

(b) (c) (d) (e)

EXHIBIT 18.2 ***Continued***

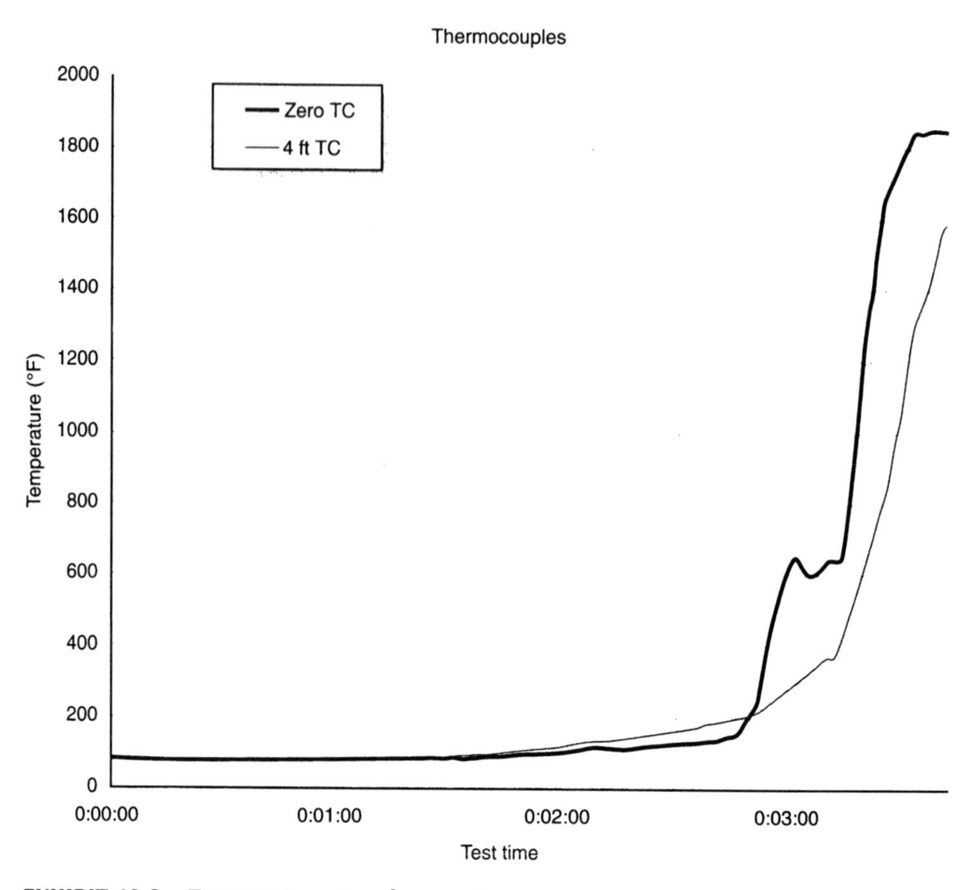

EXHIBIT 18.3 Temperature Development

See Figure A.15.3.2(a) in NFPA 921 for a suggested fire scene data collection form.

Structural Information

Structural information, such as the room dimensions and size of structural components as they apply to the fire information, should include the following:

- Length, width, and height of rooms and buildings
- Wall thickness
- Slopes of floors and/or ceilings
- Construction materials
- Construction features present

Materials and Contents

A meaningful analysis of a fire requires understanding of the heat release rate, the fire growth rate, and total heat released. The information should include the following:

- Type of contents, including materials
- Location of contents
- Configuration and condition of contents

Ventilation

Understanding ventilation conditions is important to the validity of a fire test or model. The position and conditions of openings should be included, as well as the following:

- Location of openings
- Size of openings
- Status (open or closed)
- Area of usable opening (fully open, partially open, closed)
- Ventilation effects, including wind and HVAC
- Fire department operations

Obtain and read the NIST report on the Pulaski Building fire by Harold E. Nelson available at http://fire.nist.gov/bfrlpubs/fire94/PDF/f94057.pdf. What conclusions can be drawn from the timeline activities documented in the report? ◖◖◖

▶▶▶ QUESTIONS FOR CHAPTER 18

1. True or false: One means of identifying the quality of data on a timeline is by showing events related to the fire incident as hard times or soft times.
   A. True
   B. False
2. Significant entries that are particularly valuable as a foundation for a timeline or that have significant relation to the cause, spread, detection, or extinguishment are referred to as what type of event?
   A. Benchmark
   B. Historical
   C. Micro
   D. Parallel
3. On what does a macro evaluation of a timeline usually focus?
   A. One discrete segment of the total timeline
   B. Two or more series of events in a timeline
   C. Activities that occurred before the fire
   D. The entire timeline of events
4. What kind of reasoning is used when developing a fault tree logic diagram for a fire or explosion event?
   A. Inductive
   B. Deductive
   C. Analytical
   D. Subjective
5. True or false: Failure mode and effects analysis (FMEA) is a technique used to identify basic sources of failure within a system and is best utilized for small fires.
   A. True
   B. False
6. Which type of analysis is used to predict fire phenomena and characteristics of the environment?
   A. Egress analysis
   B. Structural analysis
   C. Fire dynamics analysis
   D. Thermodynamic chemical equilibrium analysis
7. When using fire dynamics analyses, which category divides each room into two or more spaces?
   A. Specialized fire dynamics routines
   B. Field models
   C. Computational fluid dynamics
   D. Zone models
8. True or false: A meaningful analysis of a fire requires understanding of the heat release rate, fire growth rate, and total heat released.
   A. True
   B. False
9. True or false: When conducting a fire test or using a fire model, it is important to understand the ventilation conditions to validate the test.
   A. True
   B. False

Explosions

OBJECTIVES

Upon completion of Chapter 19, the user will be able to

- Identify the types of explosions
- Identify characteristics of explosion damage
- Identify factors controlling the explosion effects
- Analyze and establish the origin of the explosion
- Establish the fuel source of the explosion
- Determine the ignition source
- Analyze the damage effects

As is discussed in Chapter 21 of NFPA 921, an explosion is the rapid equilibration of a high-pressured gas within its environment. The operative term is *rapid.* It must be so rapid that the energy contained in the high-pressured gas is dissipated in a shock wave. The factors necessary to define an explosion include the evidence that indicates the nature of the movement or shattering of materials as a result of the generation and sudden escape of gas that has been confined under pressure. Although noise usually accompanies an explosion, noise is not an essential element in defining the term *explosion.* The scene itself can be evidence.

The factors influencing the effects of the explosion as well as the nature of the explosion are easily identified. They include the fuel, the manner of containment, and the manner of venting. First, the fuel factors include the type, quantity, and configuration and location of the fuel. Second, the containment factors include the size and shape of the containment vessel and the type and strength of the materials of the containment vessel. Third, the venting factors include the type and amount of venting that is present.

TYPES OF EXPLOSIONS

As indicated in Exhibit 19.1, the two major types of explosions are mechanical and chemical. Types are distinguished by the source and the mechanism that produces the explosion.

Mechanical Explosions

Mechanical explosions do not involve changes in the basic chemical nature of the substance(s) in the vessel. Rather, mechanical explosions are explosions of a high-pressure gas, which produces a physical reaction, as in the rupture of a gas storage cylinder.

A boiling liquid expanding vapor explosion (BLEVE) is the most frequent type of mechanical explosion that the user of this manual will encounter. Exhibit 19.2 shows the damage from a BLEVE. In short, a BLEVE is a subtype of mechanical explosion involving containers of liquids that become flammable when under pressure and at temperatures above their atmospheric boiling point. The user should note that the vessel(s) may range from a small butane lighter to a railroad tanker. A BLEVE may occur as a result of the internal temperature of the liquid or vapor being increased by exposure to a fire. Exposure to the fire increases pressure within the vessel, leading to rupture. The release of the liquid results in instant vaporization. If the contents of the vessel are ignitible, then the vapor almost always results in a fire. The ignition of the vapor usually occurs from the original heat that caused the BLEVE or from another electrical or mechanical source. A common example of a BLEVE that does not involve an ignitible liquid is the rupture of a steam boiler. Finally, a BLEVE may result from overfilling, runaway reaction, vapor explosion, or mechanical failure.

ACTIVITY

Contact your local member of the National Propane Gas Association (NPGA) (www.npga.org). Identify yourself as a student of NFPA 921 seeking an opportunity to review brochures and videos on the characteristics of propane.

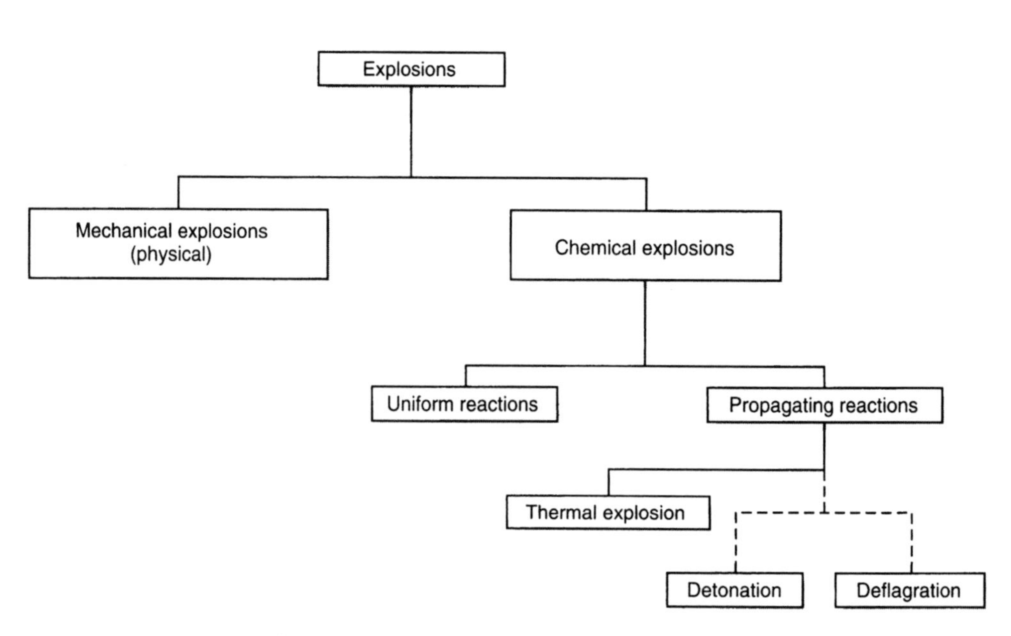

EXHIBIT 19.1 Types of Explosions

EXHIBIT 19.2 BLEVE Damage

Chemical Explosions

Chemical explosions are those in which a chemical reaction is the source of the high-pressure gas. In this reaction, the fundamental chemical nature of the fuel is changed. Although chemical explosions can involve solid combustibles or explosive mixtures of fuel and an oxidizer, most typical are the propagating reactions that involve gases, vapors, or dust mixed with air.

The most common chemical explosions are those caused by the burning of combustible hydrocarbon fuels. A combustion explosion is characterized by the presence of a fuel (such as dust) with air as an oxidizer. The elevated pressures are created by the rapid burning of the fuel and the rapid production of combustion by-products and gases. The velocity of the flame front propagation through the fuel determines whether the combustion reaction is classified as a deflagration or detonation.

Uniform reactions are reactions that occur more or less equally throughout the material. This classification would include ordinary chemical reactions that form gaseous products at a rate faster than they can be vented. Propagating reactions are reactions that initiate at a specific point in the material and propagate as a reaction front through the unreacted material.

Thermal explosions are the result of exothermic reactions occurring within confinement without provisions for dissipating the heat of reactions. This can accelerate to the point at which high-pressure gases are generated and an explosion occurs.

A *deflagration* is a reaction that propagates at subsonic velocities, several feet per second. A deflagration can be vented successfully. The difference between a detonation and a deflagration is the magnitude as to pressure versus time for the system involved in the combustion reaction.

A *detonation* is a reaction that propagates at supersonic velocities, more than 1100 ft per second (335 m per second). Detonations cannot be vented due to the speed of the reaction. A significant difference between deflagration and detonation is time.

Finally, the fuels involved in combustion explosions may be flammable gases, vapors, ignitible (flammable and combustible) liquids, combustible dust, and smoke and flammable by-products of incomplete combustion (such as in a backdraft explosion).

In a nuclear explosion, the high pressures within the primary system as well as the secondary system (such as steam generators and boilers) are created by the enormous heat produced by the fission or fusion of atoms. NFPA 921 does not cover the investigation of nuclear explosions.

CHARACTERISTICS OF EXPLOSION DAMAGE

The terms *low-order damage* and *high-order damage* are the preferred terms to characterize explosion damage. The user should note that the differences in damage are more a factor of the rate of pressure rise and the strength of the confining vessel or structure than of the maximum pressures within the system.

Low-Order Damage

Low-order damage is produced by a slow rate of the pressure rise. The characteristics of low-order damage include walls bulged out or not laid down, roofs lifted slightly and returned to approximate position, windows dislodged with the glass intact, and thrown-out debris that is generally not only large but also a short distance from the structure.

High-Order Damage

High-order damage is a result of a rapid rate of pressure rise. The characteristics of high-order damage include the shattering of walls, roofs, and structural members, resulting in small, pulverized debris, and debris that is thrown great distances from the structure (hundreds of feet).

EFFECTS OF EXPLOSIONS

Explosions have four major effects: blast pressure wave effect, shrapnel effect, thermal effect, and seismic effect.

Blast Pressure Front Effect

Blast pressure front effect is a result of large quantities of gas being produced by the explosion of the material. The gases move outward at high speed from the point of origin. The blast pressure front creates a positive pressure phase and a negative pressure phase based on the direction of the forces in relation to the point of origin. The outward movement of the gases and displaced air is known as the positive pressure phase. The movement of air rushing toward the area of origin as result of low pressure is known as the negative pressure phase.

The positive pressure phase with the expanding gases moving outward from the point of origin is more powerful than the negative pressure phase and is responsible for most of the pressure damage. The negative pressure phase results from the rapid outward movement of the positive pressure phase that leaves a

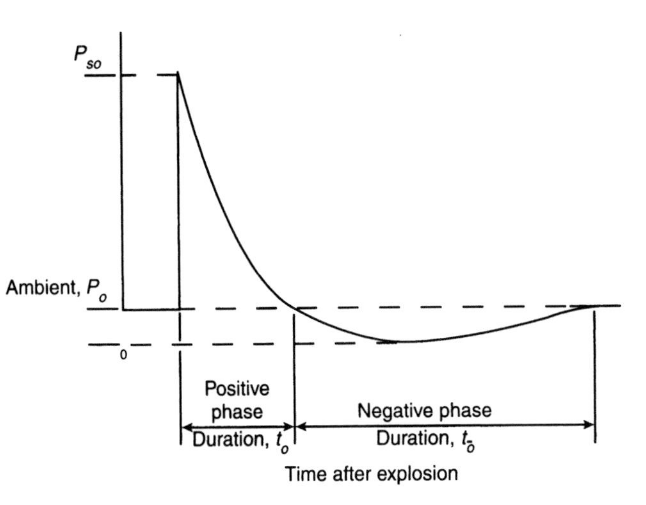

EXHIBIT 19.3
Blast Pressure Front Effect

low air pressure behind it. Air moves into the low pressure area toward the point of origin. (See Exhibit 19.3.)

The air movement may cause secondary damage by propelling artifacts toward the point of origin, which may become obscured by the debris. The user should note that although the negative pressure phase is weaker than the positive pressure phase, additional damage may be caused as a result of structural components previously being weakened by the positive pressure phase. The shape of the blast front would normally be spherical expanding evenly in all directions from the epicenter (Exhibit 19.4). However, under actual conditions, confinement changes the shape and force of the front. Furthermore, the change in direction of the blast front may be the result of either the venting path or redirection when the front is reflected off a solid object. Regardless, the force of the blast pressure front decreases with distance from the epicenter, assuming no propagating reactions.

The correlation between the rate of pressure rise and the damage effects of the explosion is shown in Table 19.1.

Where the pressure is less rapid, the venting effect has an important impact on the maximum pressure that develops. For example, in fugitive gas explosions in residential or commercial buildings, the maximum pressure is limited to a pressure that is slightly higher than the pressure that the components of the building enclosure can sustain without rupture. In a well-built residence, this pressure seldom exceeds 3 psi (21 kPa).

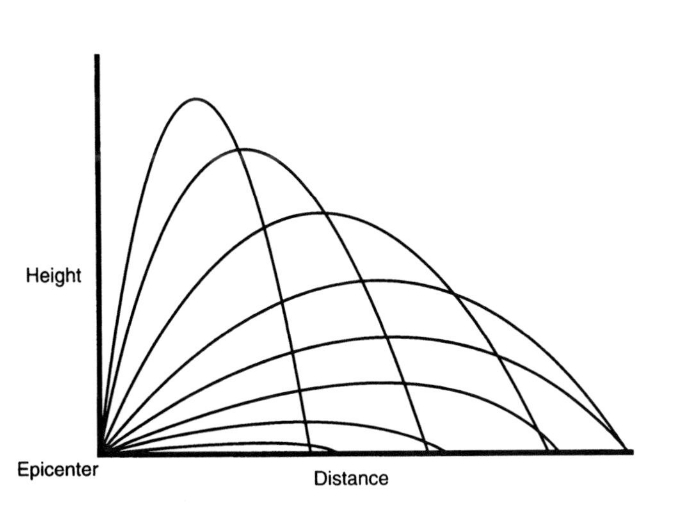

EXHIBIT 19.4
Epicenter

TABLE 19.1 Rate of Pressure Rise Versus Damage

Rate of Pressure Rise	*Damage*
Slow	Pushing or bulging type of damage Weaker parts of the structure will rupture first Characteristic of low-order damage
Rapid	Shattering of confining vessel or container Debris will be thrown great distances Characteristic of high-order damage

Shrapnel Effect

The user should recall that the effects of explosions may be observed in four categories or groups. The first category was identified as the blast pressure front effect. The second grouping is identified as the shrapnel effect. Shrapnel is the name for the pieces of debris that result from the rupture of vessels or containers as a result of the blast pressure fronts. The shrapnel may cause personal injury or damage a great distance from the source of the explosion. The shrapnel may also sever electric utility lines, fuel gas lines, or storage containers. The distance that the shrapnel will be propelled is greatly dependent on the initial direction as well as on the weight and aerodynamic characteristics of the projectile.

Thermal Effect

Thermal effect is the term for the third grouping of effects of explosions. All explosions release quantities of energy that can ignite nearby combustibles. The secondary fires may increase the damage as well as injuries from the explosion. Unfortunately, it is often difficult to determine which occurred first, the fire or the explosion. The thermal damage from a chemical explosion depends not only on the nature of the explosive fuel but also on the duration of the high temperature. Table 19.2 provides an analysis of explosion type in resulting thermal effect. The user should note the part that time plays in thermal effects.

Seismic Effect

The transmission of tremors through the ground is known as *seismic effect.* Such an effect is a result of the blast wave expansion causing structures to be knocked to the ground. Accordingly, small explosions result in negligible seismic effects.

TABLE 19.2 Thermal Effects from Various Explosions

Explosion Type	*Effect*
Combustion	Releases heat energy Can cause secondary fires
Chemical	Releases great quantities of heat
Detonating	Produces extremely high temperatures of short duration
Deflagration	Produces lower temperatures of longer duration

However, larger explosions may produce damage to structures, underground utility services, pipelines, tanks, and cables.

FACTORS CONTROLLING EXPLOSION EFFECTS

The factors that control the effects of an explosion are as follows:

- Type and configuration of fuel
- Nature, size, volume, and shape of containment vessel
- Location and magnitude of ignition source
- Venting of the containment vessel
- Relative maximum pressure
- Rate of pressure rise

The nature of the noted factors and their variables can produce a wide variety of physical effects. The variables that may affect the characteristics of a blast pressure front as it travels from the source include blast pressure front modification by reflection and blast pressure front modification by refraction and blast focusing. The blast pressure front may amplify due to its reflection from objects in its path. The reflection can cause the overpressure to increase as much as eight times at the surface of reflection. This phenomenon is negligible with deflagrations.

A blast pressure front that encounters a layer of air at a significantly different temperature may cause the front to bend or to refract. This occurs because the speed of sound is proportional to the square root of temperature and air. A low-level temperature inversion may cause refraction, resulting in a focus on the ground adjacent to the center of the explosion. This effect is also negligible with deflagrations.

SEATED EXPLOSIONS

The crater or area of greatest damage may be characterized as the seat of the explosion. The seat may be of any size, depending on the amount and strength of the explosive material. The seated explosion is generally characterized by high pressure and rapid rates of pressure rise. The types of fuels that may generate seated explosions include explosives, steam boilers, highly confined fuel gases and liquid fuel vapors, and BLEVEs in small containers.

explosion seat
The crater or area of greatest damage located at the point of initiation (epicenter) of an explosion. (Source: NFPA 921, 2004 edition, Section 21.6)

Definition

Explosives generally have a highly centralized epicenter or seat. Accordingly, because explosives have a high-velocity positive pressure phase, they usually produce craters or localized areas of great damage. Boiler and pressure vessel explosions exhibit effects similar to explosives but with lesser localized overpressure adjacent to the source. Fuel gases are ignitible vapors that, when combined with small vessels, may also produce seated explosions. Finally, a BLEVE produces a seated explosion if the confining vessel is of a small size (such as a can or barrel) and the rate of pressure release is sufficiently rapid.

NONSEATED EXPLOSIONS

A nonseated explosion occurs when the fuels are dispersed or diffused at the time of explosion with moderate rates of pressure rise and subsonic explosive velocities. Supersonic detonations may also produce nonseated explosions, depending on the conditions. Fuel gases, such as natural gas and liquefied petroleum, usually produce nonseated explosions. This is due to the confinement of the gases in large containers such as rooms. Explosions from the vapors of flammable fuel or combustible liquids are nonseated explosions. A subsonic explosive speed as well as the magnitude of the area that they cover result in small, high-damage seats. Dust explosions most often occur in confined areas with wide dispersal, such as grain elevators, processing plants, and coal mines. The large areas of origin also preclude the development of explosion seats. Finally, smoke explosions or backdrafts usually involve a widely diffused volume of combustible gases and particular matter. Accordingly, the explosive velocities are subsonic, thus limiting the production of pronounced seats.

GAS AND VAPOR EXPLOSIONS

See Table 21.8 in NFPA 921 for properties of common flammable gases. Review NFPA 68, *Guide for Venting of Deflagrations.*

Fuel gases or the vapors of ignitible liquids are the most commonly encountered explosion. The more violent explosions involving lighter-than-air gases are less frequently encountered than are explosions involving gases or vapors with vapor density greater than 1.0 (heavier than air). Ignition temperatures in the range of 700°F to 1100°F (370°C to 590°C) are common. Thus, fuel gas–air mixtures are the most easily ignitible fuels that may result in an explosion. Note that minimum ignition energies begin at 0.25 mJ (an extremely low level of energy). Exhibit 19.5 shows the damage resulting from a vapor explosion.

Interpreting Explosion Damage

The explosion damage to structures is related to several factors:

- Fuel–air ratio
- Vapor density of the fuel
- Turbulence effects
- Volume of the confining space
- Location and magnitude of the ignition source
- Venting
- Strength of the structure

Fuel–Air Ratio. The entire volume of air need not be occupied by a flammable mixture of gas and air for there to be an explosion. Relatively small volumes of

EXHIBIT 19.5 Vapor Explosion Damage

explosive mixtures capable of causing damage may result from gases or vapors collecting in a given area. A mixture at or near the lower explosive limit (LEL) or the upper explosive limit (UEL) of a gas or vapor usually produce less violent explosions than do those near the optimum concentration. The optimum concentration is usually just slightly rich of stoichiometric. This is a result of lower flame speeds and lower maximum pressures in a less-than-optimum ratio of fuel to air. Thus, explosions of mixtures near the LEL do not tend to produce large quantities of postexplosion fire in that most of the available fuel is consumed during the explosive propagation. It follows that explosions of mixtures near the UEL tend to produce postexplosion fires because of the fuel-rich mixtures. The delayed combustion of the remaining fuel may produce a postexplosion fire. In short, the stoichiometric mixture (slightly fuel rich) produces the most efficient combustion and therefore the highest flame speeds, rates of pressure rise, and maximum pressures and consequently the most damage.

Flame speed may vary widely, depending on temperature, pressure, confining volume, confining configuration, combustible concentration, and turbulence. *Burning velocity* is the rate of flame propagation relative to the velocity of the unburned gas ahead of it. Fundamental burning velocity is an inherent characteristic of a combustible as a fixed value. It is the velocity at which a flame reaction front moves into the unburned mixture as it chemically transforms the fuel and oxidant into combustion products. It is only a fraction of the flame speed. The *transitional velocity* is the sum of the velocity of the flame front (caused by the volume expansion of the combustion products due to the increase in temperature), and any increase in the number of moles and any flow velocity due to the motion of the gas mixture prior to ignition. The burning velocity of the flame front can be calculated from the fundamental burning velocity as reported in NFPA 68 at standardized conditions of temperature, pressure, and composition of unburned gas.

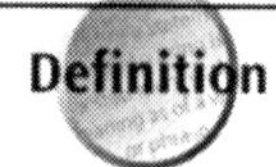

flame speed

The local velocity of a freely propagating flame relative to a fixed point.

burning velocity

The velocity at which a flame reaction front moves into the unburned mixture as it chemically transforms the fuel and oxidant into combustion products. It is an inherent characteristic of a combustible and is a fixed value. It is only a fraction of the flame speed.

transitional velocity

The sum of the velocity of the flame front caused by the volume expansion of the combustion products due to the increase in temperature and an increase in the number of moles and any flow velocity due to motion of the gas mixture prior to ignition.

Vapor Density. Air movement from both natural and forced convection is a dominant mechanism for moving gases in a structure. The vapor density of the gas or vapor may affect the movement of the fugitive gas as it escapes from its container or system. Heavier-than-air gases and vapors (vapor density greater than 1.0 such as ignitible liquids and LP-Gases) have a tendency to flow to lower areas. In contrast, lighter-than-air gases (such as natural gas) tend to rise and flow to upper areas. Where lighter- or heavier-than-air gases are involved, the operation of heating and air-conditioning systems and temperature radiance may cause mixing and movement that can reduce the effects of vapor density. Field tests have confirmed that vapor density effects are minimized by most air exchange rates that exist in older homes. Thus, vapor density effects are greatest in still air conditions.

Full-scale testing of flammable gas concentrations has shown that near-stoichiometric concentrations of gas develop between the location of the leak and either the ceiling for lighter-than-air gases or the floor for heavier-than-air gases. Users should note that a heavier-than-air gas that leaks at floor level may create a greater concentration at floor level, but with time, the gas will slowly diffuse upward. The inverse relationship is true for a lighter-than-air gas leak at ceiling height.

Turbulence. Turbulence within a fuel–air mixture increases the flame speed and thus increases the rate of combustion and the rate of pressure rise. The turbulence may vary according to the size and shape of the combining vessel and thus affect the severity of the explosion. The investigator should be aware of any mixing or turbulent sources such as fans and forced-air ventilation and watch for chevrons or other structural components that could create turbulence.

Nature of Containment. The size, shape, construction, volume, materials, and design of the confining space greatly affect the nature of the explosion damage. A long, narrow corridor or a long production oven filled with a combustible vapor–air mixture when ignited at one end will be very different in its pressure distribution, rate of pressure rise, and effects on the structure than if the same volume of fuel–air were ignited in a cubicle compartment. In general, the smaller the volume of the vessel, the higher the rate of pressure rise for a given fuel–air mixture, and, as noted previously, the more violent the explosion.

Location and Magnitude of Ignition Source. The highest rate of pressure rise will occur if the ignition source is in the center of the confining structure. Although the energy of the ignition source has a minimal effect on the course of the explosion, unusually large ignition sources such as explosive devices can significantly increase the speed of pressure development and thereby cause transition from deflagration to detonation.

Venting. The nature of the venting of the containment vessel for gas, vapor–gas, or dust-fueled explosions has varying effects on explosion damage. A length of pipe may rupture in the center if it is sufficiently long. A room may experience destruction or merely movement of the walls and ceiling depending on the number, size, and location of doors and windows. (See Exhibits 19.6 and 19.7.) Furthermore, the venting of a vessel structure can cause damage to materials in the path of the venting. The user should note that if the explosion is typed as a detonation, venting effects are minimal, due to the high speeds of the blast pressure fronts. In short, the blast pressure fronts are too fast for any venting to relieve the pressure.

Underground Migration of Fuel Gases

Lighter-than-air and heavier-than-air fuel gases that have escaped from underground piping systems migrate underground to enter structures, resulting in fires or explosions. The fugitive gases may permeate the soil, migrate upward, and dissipate harmlessly in the air, but if the surface of the ground is obstructed by rain, snow, freezing, or paving, the gases may migrate laterally and enter into structures through means of disturbed soils.

Fugitive gases may enter buildings through any opening into the structure, such as sewer lines, electrical and telephone conduits, or drain tiles, or through basement and foundation walls. The user should note that natural gas and propane have no natural odors of their own. Foul-smelling compounds (such as ethyl mercaptan) are added to gases. Odorant verification should be part of any explosion investigation.

EXHIBIT 19.6 Gas Explosion Damage to Wall

EXHIBIT 19.7 Gas Explosion Damage to Window

Multiple Explosions

Secondary explosions or cascade explosions may result when gas and vapors have migrated to adjacent stories or rooms, resulting in pockets. Thus, when ignition or explosion takes place in one story or room, subsequent explosions may occur in adjoining areas or stories. The migration or pocketing of gases may produce areas with different air–fuel mixtures. Accordingly, the dynamics of air–fuel mixtures may result in a series of vapor–gas explosions depending upon the ratio of the air–fuel mixture.

DUST EXPLOSIONS

Violent explosions can be fueled by dust that is dispersed within the air. This is true for combustible materials as well as for materials not normally considered to be combustible. For example, dust explosions may occur from the following materials: agricultural products, carbonaceous materials, chemicals, dyes and pigments, metals, plastics, and resin.

Particle Size

The rate of pressure rise generated by combustion is largely dependent on the surface area of the dispersed dust particles. The finer the dust, the more violent the explosion. The total surface area and consequently the violence of the explosion increase as the particle size decreases. In general, the explosion hazard can exist when the particles of the dust are 420 microns or less in diameter.

Concentration

The user must note that minimum concentrations may vary with the specific dust, but unlike most gases and vapors, there is generally no reliable maximum limited concentration. The reaction rate is controlled more by the surface-to-air mass ratio than by a maximum concentration. The minimum concentration may be from as low as 0.015 oz/ft^3 to 2.0 oz/ft^3 (20 g/m^3 to 2000 g/m^3), the most common concentrations being less than 1.0 oz/ft^3 (1000 g/m^3).

The combustion rate and maximum pressure decrease if the mixture is fuel-rich or fuel-lean. The rate of pressure rise and the explosion pressure are low at the lower explosive limit and at the high fuel-rich concentration. Thus, as in gases and vapors, the pressure rise and the maximum pressure that occur in dust explosions are high if the dust concentration is at or close to the optimum mixture. The rate of combustion and thereby the rate of pressure rise are greatly increased when there is turbulence within the suspended dust–air mixture. Furthermore, the size and shape of the confining vessel affect the nature of turbulence and thereby can have a great effect on the severity of the dust explosion.

Moisture

Increasing the moisture content of the dust particles increases the minimum energy required for ignition and the ignition temperature of the dust suspension. The initial increase in ignition energy and temperature is low, but as the limiting value of moisture concentration is approached, the rate of increase in ignition energy and temperature becomes high. The user should note that the moisture content of the surrounding air has little effect on the propagation reaction once ignition has occurred.

Sources of Ignition

The variables that affect the ignition sensitivity of a dust include the ignition temperature, the minimum energy, and the minimum concentration. The Bureau of Mines has classified thousands of samples to identify ignition sensitivity as well as the explosion severity by which the index of explosivity can be established. Sources of ignition include: open flames, smoking materials, light bulb filaments, welding operations, electric arcs, static electricity discharge, friction sparks, heated surfaces, and spontaneous heating.

The actual ignition temperature for most material dust ranges from 600°F to 1100°F (320°C to 590°C). Minimum ignition energies are higher for dust than for gas or vapor fields and are generally within the range of 10 to 40 mJ, significantly more than those for most flammable gas vapors.

Multiple Explosions

Dust explosions usually occur in a series within industrial and agribusiness operations. (See Table 19.3.) The initial explosion is usually less severe than the secondary explosion. The user should note that the first explosion puts additional dust into suspension, which then results in additional explosions. The structural vibrations in the blast front from the first explosion will propagate faster than the flame front, thus placing the dust ahead of it into suspension. The secondary explosion may progress from one area to another or from one building to another building.

TABLE 19.3 U.S. Grain Dust Explosion Statistics

Subject	1995	1996	1997	1998	1999	2000
Number of Explosions	14	13	16	18	7	8
Deaths	1	1	1	7	0	1
Injured	12	19	14	24	19	12
Losses in millions	6.7	29.6	11.4	29.8	4.4	8.2
Types of Explosions						
Corn	3	5	8	9	7	2
Soybeans	1	2	2	1	0	0
Wheat	1	0	3	2	0	1
Barley	0	1	0	0	0	0
Beet pulp	0	1	0	0	0	0
Corn starch	2	0	0	1	0	0
Distillers feed	1	0	0	0	0	0
Mixed feed	2	1	0	0	0	0
Rice	3	1	0	0	0	0
Wheat flour	0	1	1	1	0	0
Other	1	1	2	3	0	5

Source: Department of Grain Science and Industry, Kansas State University

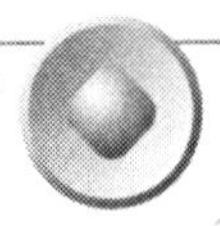

Contact the following entities for videos, statistics, and additional information:

- Grain Elevator and Processing Society (GEAPS), Minneapolis, MN, (612) 339-4825. Standards/training/statistics
- U.S. Department of Labor (OSHA). Standards/training
- Department of Grain Science and Industry, Kansas State University, (785) 532-4080. *Deadly Dust* (1980), *Deadly Dust II* (1990), and *Deadly Dust III* (1995)

BACKDRAFT OR SMOKE EXPLOSIONS

It is common for fires in airtight rooms to become oxygen depleted. This results in the generation of flammable gases due to incomplete combustion. The heated fuels accumulate in areas within the structure where there are insufficient oxygen and insufficient ventilation. The opening of a window or door results in the mixture of the fuels with air. The fuels can then ignite and burn sufficiently fast to produce low-order damage (less than 2 psi/13.8 kPa).

OUTDOOR VAPOR CLOUD EXPLOSIONS

The release of gas, vapor, or mist into the atmosphere may result in a cloud forming within the fuel's flammable limits and, subsequently, ignition culminating in an outdoor vapor cloud explosion. The phenomenon is referred to as *unconfined vapor–air explosion* or *unconfined vapor cloud explosion*. It is most frequently related to catastrophic failure of vessels and tankers in low-lying areas.

EXPLOSIVES

Explosives are categorized into two main types: low explosives and high explosives. The user should not confuse these categories with low-order and high-order damage.

explosive

Any chemical compound, mixture, or device that functions by explosion. (Source: NFPA 921, 2004 edition, Paragraph 3.3.45.)

Low explosives are characterized by deflagration (subsonic blast pressure wave), a slow rate of reaction with the development of low pressure. Examples of low explosives are smokeless gunpowder, flash powders, solid rocket fuels, and black powder. Low explosives are designed to work by pushing or heaving effects.

High explosives are designed to produce shattering effects due to their high rate of pressure rise and extremely high detonation pressure. Thus a characteristic of high explosives would be the detonation propagation mechanism. The high, localized pressures are responsible for creating localized damage near the center of the explosion.

A deflagration (fuel–air) explosion usually results in structural damage that is uniform and omnidirectional with relatively widespread evidence of burning, scorching, and blistering. The rate of combustion of a solid explosive is extremely fast in comparison to the speed of sound. Thus pressure does not equalize throughout the explosion volume, and high pressures are generated near the explosive. The location of the explosion should be evidenced by the crushing, splintering, and shattering that are produced by the higher pressure. However, major distance from the source of the explosion may leave little evidence of intense burning or scorching except where shrapnel may have landed on combustible materials. Common examples of high explosives are dynamite, water gel, TNT, ANFO, RDX, and PETN. The extremely high detonation pressure may reach 1,000,000 psi (6,900,000 kPa).

Table 19.4 shows some explosion investigation tips. Only investigators with appropriate training should conduct an investigation relating to possible explosive instances. Investigators who do not have this training should coordinate their investigation with appropriate experts.

TABLE 19.4 Investigation Tips

Damage	*Fuel–Air*	*Solid Explosive*
Structural	Uniform and omnidirectional	Nonuniform. Crushing and shattering near the location of the explosion
Fire	Widespread burning, scorching, and blistering	Localized around the source of the explosion

INVESTIGATING THE EXPLOSION SCENE

The objectives of an explosion scene investigation are no different from those of a fire investigation: Determine the origin, identify the fuel and ignition source, determine the cause, and establish the responsibility for the incident.

A systematic approach is important in an explosion investigation. The following investigative procedures are quite comprehensive for the large incident, but the same principles should be applied to small incidents. Again, extensive damage would require the coordination of additional experts, including, but not limited to, an explosion expert and a structural expert. The investigator should note the necessity for absolute control of the scene to eliminate as much as possible the potential for contamination or loss of evidence and artifacts.

Securing the Scene

First responders to the explosion should establish and maintain physical control of the structure and surrounding areas. No unauthorized person should enter the scene or have contact with any blast debris, regardless of how remote it is from the scene. Evidence may be small and easily disturbed or removed by traffic in and around the scene. Caution must be used to prevent cross-contamination of the scene by investigators or authorized personnel wearing clothing or footwear that may contain explosive residue from other scenes or sources such as an explosives firing range. Furthermore, securing of the scene may minimize injuries to unauthorized personnel.

Establishing the Scene. The outer perimeter of the incident scene should be established at one and one-half times the distance of the farthest piece of debris found. Blast debris might be propelled great distances, in which case the scene perimeter should be widened accordingly.

Obtaining Background Information. All investigations should have sufficient background data to establish a timeline for the purpose of analysis.

First, all information should be obtained relating to the incident itself. The information would include a description of the incident site, systems and operations involved, conditions, and sequence of events that led to the incident, including material safety data sheets. Most important, the investigator must identify not only the locations of any combustibles and oxidants that were present and what conditions existed at the time of the incident but also information relating to what combustibles and oxidants and what hazardous conditions currently exist at the site.

Investigators should examine witness accounts, maintenance records, operational logs, manuals, weather reports, previous incident reports, and other relevant records. As in any failure analysis, the most recent changes to equipment, procedures, and operating conditions may be especially significant. Blueprints of the building and drawings and prints of the process can assist in the proper documentation of the scene.

Establishing a Search Pattern. The scene should be searched from the outer perimeter inward toward the area of greatest damage. The final determination as to the explosion's epicenter should be made only after all of the scene has been examined. The search pattern may be spiral, circular, or grid-shaped. The scene itself, with its particular circumstances, often dictates the nature of the pattern. The search pattern should overlap so that no evidence is lost at the edge of any search area, regardless of the methodology. The number of actual searchers needed depends on the size and complexity of the scene. Consistent procedures as to the identifying, logging, photographing, marking, and mapping of evidence must be maintained. The location of evidence may be marked with chalk, spray paint, flags, stakes, or any other marking means. The evidence should be photographed, and if no involvement of potentially responsible parties is identified, the evidence may be moved and secured accordingly.

All fire investigation safety recommendations listed in Chapter 10 also apply to an investigation of explosions.

Safety at the Scene. Structures that have been involved in an explosion are often structurally damaged. Accordingly, the possibility of collapse of floor, walls, ceiling, roof, or the entire building is great. Involvement of a structural engineer or construction engineers should be considered. Construction equipment may provide temporary support mechanism, minimizing risk to the investigators. All toxic materials in the air need to be neutralized. Accordingly, material safety data sheets should be consulted to identify the appropriate personal safety equipment (PSE) to be used.

A thorough search of the scene should be conducted for any secondary devices before the investigation is initiated. If undetonated devices and explosives are found, the area should be evacuated, isolated, and explosives disposal should be notified.

Initial Scene Assessment

The investigator should make an initial assessment of the type of incident. If the investigator determines that the explosion was fueled by explosives or an explosive device, the investigator should discontinue the scene investigation, secure the area, and contact the appropriate entities. (Review Table 21.13.3.1, "Typical Explosions Characteristics," in NFPA 921.) The following tasks are sequenced to assist the investigator in the initial scene assessment:

1. Identify whether the incident was an explosion or fire.
2. Determine high- or low-order damage.
3. Identify seated or nonseated explosion.
4. Identify the type of explosion.
5. Identify the potential general fuel type.
6. Establish the origin.
7. Establish the fuel source and explosion type.
8. Establish the ignition source.

Identifying Explosion or Fire. The first task is to determine whether the incident was a fire, explosion, or both and, if both, which came first. The investigator should look for signs of overpressure such as displacement or bulging of walls, floors, ceilings, doors and windows, roofs, structural members, nails, screws, utility service lines, panels, and boxes. The investigator should also analyze the extent of heat damage to the structure and its components to determine whether the damage can be attributed to fire alone.

Determining High- or Low-Order Damage. The investigator should determine whether the damage indicates high-order or low-order damage. (Review Section 21.3 of NFPA 921.) This will assist in the classification of the type, quantity, and mixture of the fuel involved.

Identifying Seated or Nonseated Explosion. The investigator should determine whether the explosion was seated or nonseated. (Review Section 21.6 of NFPA 921.) This will assist the investigator in classifying of the possible fuel involved.

Identifying the Type of Explosion. The investigator should identify the type of explosion involved: mechanical, chemical, or BLEVE.

Identifying the General Fuel Type. The investigator should identify the types of fuel that were potentially available at the explosion scene. This is accomplished by identifying the condition and location of utility services (fuel gases) and sources of ignitible dusts or liquids. The investigator should analyze the nature of the damage in comparison to the damage patterns consistent with the following: lighter-than-air gases, heavier-than-air gases, liquid vapors, dust, explosives, backdrafts, and BLEVEs.

Establishing the Origin. The investigator should attempt to establish the origin of the explosion as soon as possible. The origin is usually identified as the area of most damage—a crater or localized area of severe damage in the case of a seated explosion. The investigator should note that if it is a diffused fuel–air explosion, the origin is consistent with the confining volume or room of origin.

Establishing the Ignition Source. The investigator should attempt to identify the ignition source. Potential sources include hot surfaces, electrical arcing, static electricity, open flames, sparks, and chemicals in which fuel–air mixtures are involved. If explosives are involved, the ignition source may be a blasting cap or pyrotechnic device. The investigator should note artifacts from the ignition sources that may have survived the explosion.

Detailed Scene Assessment

Once the investigator has the general information from the initial scene assessment, he or she should commence a more detailed study of the blast damage. The following tasks comprise the necessary activities for a detailed scene assessment:

- Identify the damage effects of the explosion
- Identify preblast and postblast fire damage
- Locate and identify articles of evidence
- Identify force vectors
- Analyze the origins (epicenter)
- Analyze the fuel source
- Analyze the ignition source
- Analyze as to cause

Effects of Explosion. A detailed analysis of the explosion overpressure damage should be made. The articles that are damaged should be identified as having been affected by one or more of the following forces: blast pressure front—positive phase, blast pressure front—negative phase, shrapnel impact, thermal energy, and seismic energy.

The investigator should examine the type of damage as to whether the debris was shattered, bent, broken, or flattened as well as a change in the pattern. At a distance from a detonation explosion center, the pressure rise is moderate, and the artifacts resemble those of a deflagration explosion. Items in the immediate vicinity of the detonation center exhibit splintering and shattering.

The scene should be examined carefully to identify any fragments of foreign material.

A review of Tables 21.13.4.1.5(a) and 21.13.4.1.5(b) in NFPA 921 should be made to assist in estimating the peak blast overpressure from the structural damage. Table 19.5 may assist in the analysis.

Damage to personnel from explosion blast pressures is usually as a result of acceleration in the high velocity air steam with subsequent impact against a rigid surface, rather than compression in the airwave itself. Some threshold values for physiological effects are shown in Table 19.6.

Preblast and Postblast Damage. Debris that has been burned and propelled away from the point of origin may indicate that a fire preceded the explosion. Glass fragments with smoke residue and soot found some distance from the structure may indicate a fire of some duration followed by an explosion. Glass fragments that are clean and debris that is not burned but found some distance from the structure may indicate an explosion prior to the fire.

TABLE 19.5 Structural Damage and Required Pressure

Structural Element	*Failure Modality*	*Peak Overpressure psi/kPa*
Glass windows, large and small	Occasional shattering	0.5–1/3.4–6.9
Frame failure, corrugated steel or aluminum paneling	Connection failure followed by buckling	1– 2/6.9–13.8
Wood panels, standard house construction	Usually failure occurs in main connection allowing whole panel to be blown in	1–2/6.9–13.8
Concrete or cinder block wall panels, 8 in. or 12 in. thick (not reinforced)	Shattering of the wall	2–3/13.8–20.7
Self-framing steel panel building	Collapse	3–4/20.7–27.6
Oil storage tanks	Rupture	3–4/20.7–27.6
Wooden utility poles	Snapping	5/34.5
Rail cars	Overturning	7/48.3
Brick wall panel, 8 in. or 12 in. thick (not reinforced)	Shearing and flexure failures	7–8/48.3–55.2

TABLE 19.6 Physical Damage and Required Pressure

Physiological Effect	*Peak Overpressure (psi/kPa)*
Knock personnel down	1.0–3.0/6.9–20.7
Ear drum rupture (50%)	6.3/43.4
Lung damage	10/68.9
Fatality threshold	14.5/100
50% fatality	20.5/141.3
99% fatality	29.0/200

Articles of Evidence. The methodology used by an investigator to document scene artifacts may include locating, identifying, noting, logging, photographing, and mapping of physical evidence.

The probability of physical evidence being propelled both inside and outside of the structure may result in the evidence being found imbedded in walls, resting in adjacent vegetation, inside adjacent structures, and within the body and clothing of victims. Photographs must be taken of the injuries to the victims as well as any materials removed from them during medical treatment. Hardhats, gloves, boots, and respirators as well as clothing and materials removed from the victims should be preserved for further examination.

The condition and position of damaged structural components—walls, ceilings, floors, roofs, foundations, support columns, doors, windows, sidewalks, driveways, and patios—should be noted. The condition and position of buildings contents such as furnishings, appliances, heating or cooking equipment, manufacturing equipment, clothing, and personal effects should also be noted.

The condition and position of the utility equipment, such as fuel gas meters, regulators, fuel gas piping and tanks, electrical boxes and meters, electrical conduits and conductors, heating oil tanks, parts of explosive devices and fuel vessels, should be examined.

Force Vectors. The investigator should document the debris that has been propelled away from the area of origin as well as the following parameters: the direction of travel, the distance of travel, the material propelled, and the material's size, weight, and configuration. This process assists the investigator in identifying the trajectories of the artifacts involved.

Analyze Origin (Epicenter). In *explosion dynamics analysis,* the general path of the explosion force vectors is followed from the least to the most damaged area. The process may require more than one explosion dynamics diagram to identify the debris movement: a large-scale analysis indicates the general area or room for further analysis as to the origin, and a small-scale diagram analyzes the explosion dynamics of the area of origin itself. (See Exhibit 19.8.)

It is necessary to plot the directions of debris movement and the relative force necessary for the movement of each major piece of debris. The investigator should note that the explosion dynamics analysis may be complicated by secondary explosions. The user should also recall that secondary explosions (dust explosions) are often greater than the primary and thus cause more dam-

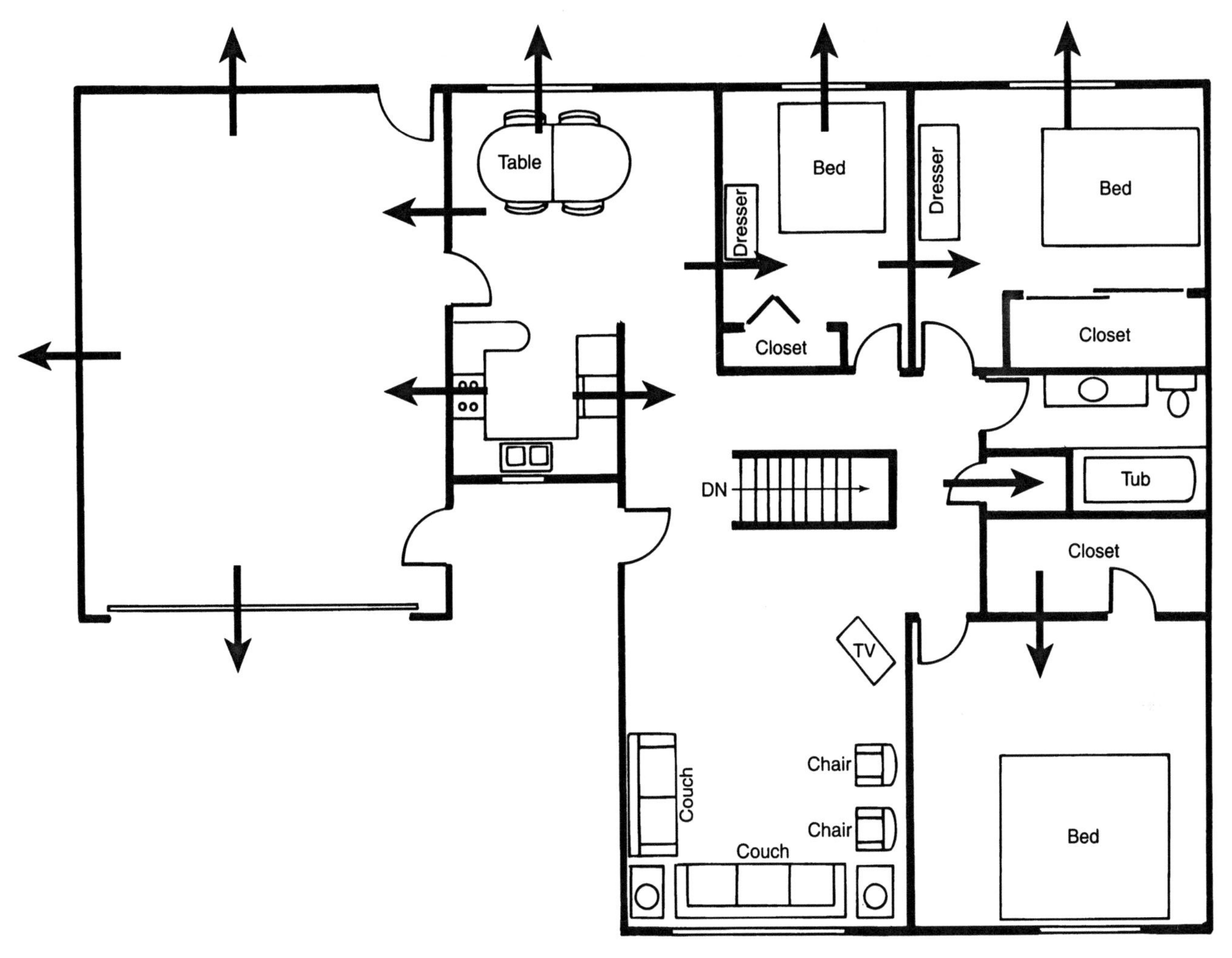

EXHIBIT 19.8 Explosion Dynamics

age. The analysis of the explosion dynamics is based on debris movement away from the epicenter of explosion in a roughly spherical pattern. The investigator should recall that the farther an object is from the epicenter, the less force the object is subjected to.

explosion dynamics analysis

The process of using force vectors to trace backward from the least to the most damaged areas following the general path of the explosion force vectors. (Source: NFPA 921, 2004 edition, Section 21.14)

Analyze Fuel Source. All available fuel sources should be considered then limited to one fuel source that meets all the physical damage criteria. Clinical analysis of debris, soot, and soil as well as air may assist in identifying the fuel. Gas chromatography, spectrography, and other chemical tests of samples may identify the fuel. Once the fuel has been identified, the investigator can determine the source. Further, as has been noted, odor verification should be part of any explosion investigation involving flammable gas. Although stain tubes may be used, the collection of a sample for submission to a lab is the most accurate means.

Analyze Ignition Source. A careful evaluation of every possible ignition source should be made. The factors to consider include minimum ignition energy of the fuel, ignition energy of the potential ignition source, ignition temperature of the fuel, temperature of the ignition source, location of the ignition source in reference to the fuel, presence of both fuel and ignition source at the time of ignition, and witness accounts of conditions prior to and at the time of the explosion.

Establish Cause. An analysis to determine the simultaneous presence of the fuel and the ignition should include the following techniques: timeline analysis, damage pattern analysis, debris analysis, relative structural damage analysis, correlation of blast yield with damage, analysis of damage items in structure(s), and the correlation of thermal effects.

QUESTIONS FOR CHAPTER 19

1. True or false: An explosion is the sudden conversion of potential energy into kinetic energy with the production and release of gases under pressure.
   A. True
   B. False
2. True or false: Noise is an essential element in defining an explosion.
   A. True
   B. False
3. True or false: A timeline is a graphic or narrative representation of events arranged in chronological order.
   A. True
   B. False
4. True or false: A pyrotechnic device that initiates an explosion, including wires and components parts, will not survive the explosion itself.
   A. True
   B. False
5. True or false: A BLEVE frequently occurs when the temperature of the liquid and vapor within a confining vessel is raised by an exposure fire to the point at which the increasing internal pressure can no longer be contained and the confining vessel explodes.
   A. True
   B. False
6. True or false: The blast pressure front occurs in two distinct phases, based on the direction of the forces in relation to the origin of the explosion.
   A. True
   B. False
7. True or false: Within the classification of hazardous reactions, explosions may be classified as either chemical or physical (mechanical) explosions.
   A. True
   B. False
8. True or false: Because of the magnitude of damage involved, dust explosions most often result in pronounced seats of explosion.
   A. True
   B. False
9. True or false: Explosions that occur in fuel–air mixtures at or near the lower explosive limit (LEL) always produce less violent explosions than do those that occur in fuel–air mixtures at or near the upper explosive limit (UEL).
   A. True
   B. False
10. True or false: The vapor density of the gas–vapor fuel has a direct correlation with the relative elevation of the structural explosion damage above floor level as a result of the explosion.
    A. True
    B. False

11. *Deflagrations* are propagating reactions, a subtype of chemical explosions, and as such are defined in which of the following definitions? (There may be more than one correct choice.)

A. The energy is transferred from the reaction zone to the unreacted zone on a reactive shockwave.

B. The velocity of the reaction is always greater than the speed of sound.

C. Structural damage is uniform and omnidirectional.

D. Relatively widespread evidence of burning, scorching, and blistering result.

12. Identify the two major types of explosions from the choices below.

A. Explosions that originate from strictly physical phenomena, such as a boiler explosion

B. Explosions that originate due to the rapid combination of ammonium nitrate and fuel oil (ANFO)

C. Explosions in which a chemical reaction is a source of the high-pressure gas

D. Explosion that is the result of rapid expansion of gas due to the Rickover Effect

13. An analysis of samples taken from commercial "natural gas" may include a mixture of which of the following gases? (There may be more than one correct answer.)

A. Butane

B. Nitrogen

C. Propane

D. Ethane

E. Methane

F. All of the above

14. Dust explosions occur in a wide variety of materials. Regardless of the material, which of the following factor(s) is/are fundamental for an analysis of a dust explosion? (There may be more than one correct answer.)

A. Particle size of the dust

B. Moisture of the dust particle

C. Turbulence within the suspended dust–air mixture

D. Concentration of the dust in air

E. Ignition temperature of the dust

15. It is helpful to characterize the explosion incident on the basis of the damage that has occurred to the structure. The following recommended terms have been used to classify the incidents.

A. High-grade damage

B. Medium-effect damage

C. High-order damage

D. Low-order damage

16. There are numerous factors that control the effects of an explosion. Identify the appropriate factors.

A. Type of fuel

B. Quantity of the fuel

C. Configuration of the fuel

D. Size and shape of the containment vessel

E. Type and strength of materials used in the construction of the vessel

F. Type and amount of venting present

G. BDE

H. All of the above

17. A boiling liquid expanding vapor explosion will produce a seated explosion if which two of the following conditions occur?

A. The relief valve lifts at 125 percent of the designated pressure.

B. The rate of pressure release from the vessel is sufficiently rapid.

C. The confining vessel is of a small (sufficiently small) size.

D. Fire fighters place a water stream on the vapor portion of the vessel rather than on the liquid portion of the vessel.

18. The distance that shrapnel may be propelled from an explosion depends on which of the following projectile perimeters? (There may be more than one correct answer.)

A. Temperature of the projectile

B. Initial direction of the projectile

C. Weight of the projectile

D. Aerodynamic characteristics of the projectile

E. All of the above

19. Chemical explosions produce great quantities of heat. Which of the following statements is true concerning thermal effects?

A. Detonating explosions produce extremely high temperatures of very limited duration.

B. Deflagration explosions produce lower temperatures but for much longer periods.

C. All of the above.

20. The following parameters that are true for dust explosions are also true for gas–vapor explosions. Identify the correct relationships.

A. As in most gases and vapors, there is generally a maximum limit of concentration. A maximum concentration controls the reaction rate.

B. The rate of pressure rise and the maximum pressure that occur in a dust explosion are higher if the pre-explosion dust concentration is at or close to the optimum mixture.

C. The combustion rate and maximum pressure decrease if the mixture is fuel-rich or fuel-lean.

D. The rate of pressure rise and total explosion pressure are very low at the lower explosive limit and at the high fuel-rich concentration.

Incendiary Fires

OBJECTIVES

Upon completion of Chapter 20, the user will be able to

- Identify incendiary fires
- Document evidence regarding origin and cause
- Use indicators to develop ignition hypotheses
- Record and examine other evidentiary factors

An incendiary fire is a fire that has been deliberately ignited under circumstances in which the person knows that the fire should not be ignited. This chapter covers the indicators that point to incendiary fires, including indicators not directly related to combustion, and other evidentiary factors.

INCENDIARY FIRE INDICATORS

The indicators that are discussed in this section can be data to be studied for their possible support of the hypothesis that the fire is incendiary.

Multiple Fires

Multiple fires are fires with no obvious connection between them that would have allowed one fire to ignite the fuel in another area. A fire in the basement that spread through the walls of a balloon-frame house to the attic would not be classified as an incident with multiple fires. The fire spread naturally through a vertical opening and ignited combustible fuels in a location that was remote from the initial fire. The investigator must determine that a separate fire was not the natural outgrowth of the initial fire.

Other "natural" means of fire spread that could cause multiple fires include the following:

- Conduction, convection, or radiation
- Flying brands
- Direct flame impingement
- Falling flaming materials (drop down)
- Fire spread through shafts
- Fire spread within wall or floor cavities
- Overloaded electrical wiring
- Utility system failures
- Fuel gas or dust explosions

The investigator should not confuse the site of a previous fire with that of a more recent fire. This could cause the investigator to develop the hypothesis that there were multiple simultaneous fires. The fire history of the structure should be obtained and examined. It is important to conduct a full scene examination to determine whether there were separate fires. If this full scene examination is not conducted, valuable evidence might not be located.

Trailers

Arsonists attempt to spread fire to other areas by linking them together with combustible fuels or ignitible liquids called *trailers.* One fire will, in turn, ignite other areas via these trailers.

Trailers can leave distinctive patterns on horizontal surfaces such as floors. When the floor is cleared of debris, the pattern may be easily discernible. However, it is important to determine that these patterns are not the result of other mechanisms or materials such as open areas bordered by protected areas or the effects from full-room involvement. At full-room involvement, radiant heat can be expected to create burn patterns on floors that can be misinterpreted as trailer burn patterns. (See Chapters 3 and 4 in this *User's Manual* for additional information.)

Materials that can be used as trailers include the following:

- Ignitible liquids
- Clothing
- Paper
- Straw

Many common household goods (cleaning fluids, gasoline, etc.) are ignitible liquids, and their presence on the fire scene is not necessarily indicative of use as a trailer. It is not the fuel that constitutes a trailer but the manner in which the fuel was used.

Lack of Expected Fuel Load

When the observable fire damage is not consistent with the observable fuel load, further investigation is warranted. The investigator should attempt to quantify the fire damage that would be caused by the observable fuel load. The absence of fuels is not enough to classify the fire cause as incendiary.

Examples of areas and spaces that routinely have low or limited fuel loads include (among others) corridors, stairways, hallways, and vacant homes. If the origin for a fire is in a low- or limited-fuel area, the investigator should look for physical evidence of fuels such as an ignitible liquid and take samples. However, burning in these areas might not be unusual if the burning represents fire extension or movement from another area, particularly if the adjacent space has developed past flashover.

Lack of Expected Ignition Sources

Another indicator that deserves further investigation is when a fire origin does not have a readily apparent ignition source. The investigator may have to look closely through the debris in the search for ignition sources that have burned, melted, or been consumed. Closets, crawl spaces, and attics are typical areas, rooms, and spaces in which a limited number of heat sources are present.

Exotic Accelerants

Mixtures of fuels with Class 3 or Class 4 oxidizers and thermite mixtures may be considered exotic accelerants. These types of accelerants can cause exceedingly hot fires and generally leave residues that may be visually or chemically identifiable. Indicators of high temperature accelerants (HTA) include the following:

- Rapid rate of growth
- Brilliant flares
- Melted steel or concrete

With the lack of scientific proof that HTAs are an indicator of incendiary fires, other reasons for the cause should be considered. These might include ventilation, fire suppression tactics, or type and configuration of fuels.

Unusual Fuel Load or Configuration

A firesetter might hope to create a fire that will burn more aggressively or effectively by moving contents or materials into a configuration to allow for more rapid fire growth or fire spread than would be expected if the contents were spaced farther apart. This could also be done in an attempt to provide more complete burning of the fuels. Witnesses might be able to provide information as to the position or location of contents prior to the fire.

The types of fuels can be evaluated to determine whether they are the types expected in a given occupancy, as fuels may be added to an area to assist in fire growth or spread.

An unusual fuel load or configuration should not be assumed to be related to the fire cause. The investigator should seek an explanation for the configuration if it is truly unusual.

Burn Injuries

For all classifications of fires, all known burn injuries to persons should be analyzed to determine their relationship to fire ignition and the investigative hypothesis. The investigator should also attempt to determine the type of burn injury (e.g., one resulting from a hot object or open flame). Because burn injuries might be sustained while setting an incendiary fire, local hospitals should be contacted for identification of recent burn victims. Some jurisdictions require the reporting of burn injuries. A detailed interview may be helpful to determine the origin, cause, or spread of the fire.

Incendiary Devices

If incendiary devices are used, remains of the device can often be found. Some incendiary devices are constructed as delay devices to allow the firesetter to safely leave the area.

If, during the investigation, the investigator finds a device that has not activated, he or she should not move it. Adequate precautions and safeguards should be taken, including the notification of trained ordinance personnel if an active or live device is found.

Almost any appliance or heat-producing device can be used as an incendiary device. If there are no other obvious ignition sources, then efforts should be made to determine whether a device was used. Some examples include (but are not limited to) the following:

- Combination of cigarette and matchbook
- Candles
- Wiring systems
- Electric heating appliances (Exhibit 20.1)
- Fire bombs/Molotov cocktails (Exhibit 20.2)
- Paraffin wax–sawdust incendiary device (fireplace starters)

Exhibit 20.3 shows a time delayed incendiary device using a cigarette and matches. Exhibit 20.4 shows another time-delayed device: a plastic jug full of gasoline with a birthday candle used as the heat source.

EXHIBIT 20.1 Heater Used as Incendiary Device

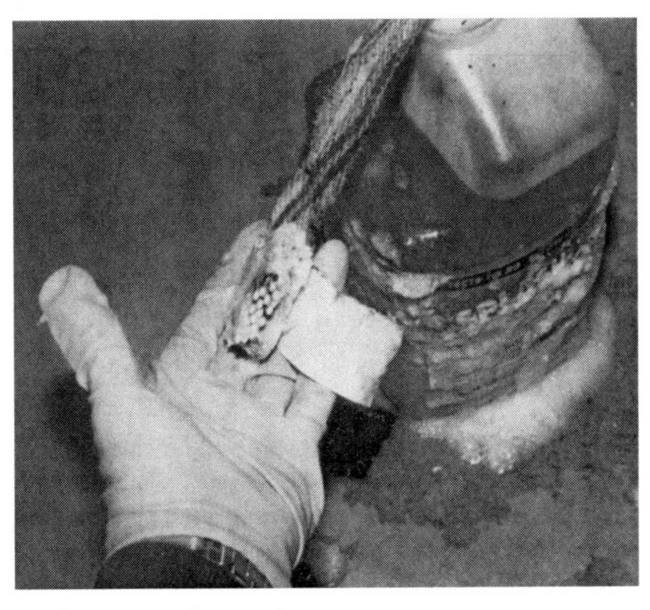

EXHIBIT 20.2 Fire Bomb

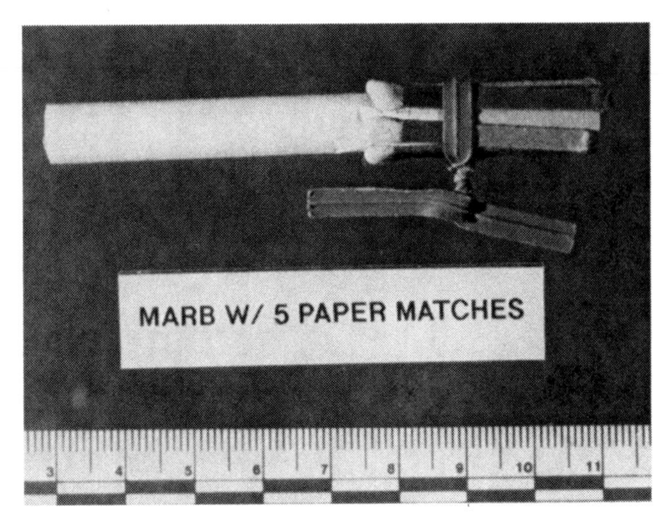

EXHIBIT 20.3 Time-Delayed Incendiary Device: Cigarette and Matches

There may be occasions when the firesetter may try to mask the true cause of the fire by attempting to use an appliance as the "obvious" cause of the fire. For example, a firesetter might pour an ignitible liquid into a coffeemaker, causing a fire to occur. The investigator should not assume that the appliance malfunctioned, causing the fire. Further testing or evaluation may be warranted to determine the true cause of the fire. The investigator should be aware of spoliation issues when conducting this type of investigation.

Because arsonists may set multiple fires, the investigator should inspect the building to determine whether other fires were set but did not extend. The arsonist might have used similar devices in these other fires that can provide valuable clues.

Fire Growth and Damage

If a fire spread more quickly than can be explained by the expected fuel load or beyond the area where it would normally be expected to be confined, then the investigator should look more closely at what contributing factors could have

EXHIBIT 20.4 Time-Delayed Incendiary Device: Jug of Gasoline

caused this to happen. Fire growth is related to a large number of variables, including the volume of the compartment, the height of the ceiling, the heat release rates of fuels, the location of the initial fuel package within the compartment, and ventilation. In the absence of physical evidence, the investigator is cautioned against using subjective terms such as *excessive, abnormal,* or *unusual* to support an incendiary fire cause determination.

POTENTIAL INDICATORS NOT DIRECTLY RELATED TO COMBUSTION

Certain indicators can assist the investigator in developing an ignition hypothesis, questions for witnesses, or avenues for further investigation. These indicators typically are not related to the determination of the fire cause but tend to show that somebody had prior knowledge of the fire.

Remote Locations with Blocked or Obstructed View

A firesetter might start a fire in a remote location or one obscured from public view. However, accidental fires can also start in remote locations. Therefore, no conclusions should be made regarding the cause determination based on the location of the origin. More important to the investigator is obscurations that occur just prior to the fire, such as paper-covered or painted windows. These can provide the investigator with information for establishing a timeline as well as for questioning witnesses.

Fires Near Service Equipment and Appliances

To make a fire appear to be accidental, a firesetter might set a fire near appliances, hoping that the appliance will appear to be the ignition source for the fire. Each appliance should be carefully evaluated to determine whether it was the ignition source. The investigator should remember issues related to spoliation before any destructive evaluation.

Removal or Replacement of Contents Prior to the Fire

Contents are sometimes removed or replaced with items of lesser value prior to a fire. Careful documentation of the remains and the debris may be useful in establishing a fraudulent insurance claim, even in situations in which the fire cause cannot be determined.

The determination that contents have been removed or replaced requires verification that the items were present some time prior to the fire. Assessment concerning the absence, removal, or replacement of any item requires verification from the corroborated witness statements, inventory, or sales receipts.

Personal items or irreplaceable items may be removed prior to a fire. The items that were removed may provide the investigator with avenues for questioning and the identification of suspects. Examples include jewelry, photographs, pets, tax records, business records, and firearms.

Items and contents that may be replaced depend on the occupancy of the building. Some examples are as follows:

- *Residential:* furniture, clothing
- *Industrial/commercial:* machinery, equipment, stock, merchandise
- *Vehicles:* tires, batteries

If the contents of the building are abnormal to the occupancy, this could be an indication that further investigation is necessary.

Blocked or Obstructed Entry

To allow the fire more time to grow, the firesetter might place obstructions that will hinder or slow the fire-fighting operations. Any unusual obstructions that deny fire vehicle access should be noted and evaluated.

Sabotage to the Structure or Fire Protection Systems

The term *sabotage* refers to intentional damage or destruction. Firesetters often develop conditions that lead to rapid and complete destruction of a building and its contents. To accomplish this goal, the firesetter might sabotage the fire protection systems in order to delay notification to occupants and the fire department.

Fire Protection Systems. Fire protection systems include the following:

1. Detection systems
    - Heat
    - Smoke
    - Flame
    - Security
    - Video
2. Suppression systems
    - Sprinkler systems
    - Standpipe systems
    - Fire department connections
3. Special extinguishing systems
    - Carbon dioxide
    - Foam
    - Halon
4. Water mains and hydrants

If the fire suppression or detection systems are operating normally, they should detect or suppress the fire before it has a chance to cause any significant damage. In addition, they could serve to alert the fire department. If the fire suppression system or fire detection system failed, the investigator should inspect the systems for any signs of tampering, such as the following:

- Improper construction
- Lack of maintenance

- System shutdown
- Equipment or structural assembly failure

It is also important to determine whether these conditions existed prior to the fire. Inspection and maintenance records could provide valuable clues regarding the pre-fire status of the systems.

Some methods of disabling these systems include the following:

- Removing or covering smoke detectors
- Obstructing sprinkler heads
- Shutting off control valves
- Damaging threads on standpipes, hose connections, and fire hydrants
- Placing debris in Siamese connections and fire hydrants

Building Damage. A firesetter may intentionally damage the building for several reasons. Two of these are: (1) to hinder the ability of the fire fighters to effectively fight the fire and (2) to provide avenues for the fire to spread beyond its area(s) of origin.

Examples of the first could include the following:

- Cutting openings in the floors that fire fighters could fall through
- Jamming or barricading doors and windows to make it difficult for fire fighters to enter
- Sabotaging fire-rated doors, fire dampers, and the like so that they are in the open position during a fire and will not close automatically

Opening Windows and Exterior Doors

A fire needs oxygen to burn. To provide ventilation, the firesetter might open exterior windows and doors that would not normally be open. The objective of the firesetter could be to have the fire spread outside of the compartment of the area of origin. To accomplish this effectively, artificial avenues of fire spread might have to be created. It is important for the investigator to determine whether the occupants normally held the doors open or whether this might have been done to spread the fire. The occupants might normally prop open doors to allow for easier traffic flow. Wind direction might also be a factor that the firesetter attempted to take advantage of in spreading the fire.

OTHER EVIDENTIARY FACTORS

Other evidentiary factors are those indicators analyzed after the fire has been classified as incendiary to develop suspect profiles. It is through this analysis that trends or patterns can be detected. The key to this analysis is whether the fire setting is repetitive or not. *Serial firesetters* or *serial arsonists* are the terms used to identify individuals or groups who are involved in three or more firesets. There are three principal trends that may be identified through this analysis:

1. *Geographical or clusters:* Fire-setting tendencies are within the same geographical location or neighborhood.
2. *Temporal frequency:* A serial arsonist can choose the same time period or day of the week.

3. *Materials and methods:* Repetitive fire-setting behavior not only remains in the same geographical location but also uses similar fire-setting materials and methods.

Firesetters may start an incendiary fire to conceal another crime, such as a homicide or burglary. The issue of which crime occurred first has more to do with motive than with the cause of the fire. Determining motive, however, can aid investigators in their approach to the investigation. Some indicators revealed by an investigator include the following:

- Financial stress
- History of code violations
- Fires at additional properties owned by a single individual or group
- Overinsured property

Timed Opportunity

Firesetters sometimes take advantage of conditions or circumstances that add to the chances of successful destruction of the property. A timed opportunity can also increase their chances of not being apprehended. Examples of a few timed opportunities are as follows:

1. Natural conditions
   - Floods
   - Snowstorms
   - Hurricanes
   - Earthquakes
   - Electrical storms
   - High winds
   - Low humidity
   - Extreme temperatures
2. Civil unrest
3. Fire department unavailable
   - Calling in a false alarm
   - Parades

Motives for Fire-Setting Behavior

Motive is defined as an inner drive or impulse that is the cause, reason, or incentive that induces or prompts a specific behavior. The use of motive indicators in the fire investigation process should be used only to help identify potential suspects. These indicators should not be used to determine or classify the fire cause.

Motive Versus Intent. *Intent* is generally necessary to show proof of a crime and refers to the state of mind that exists at the time a person acts or fails to act. *Motive* is the reason that an individual or group may do something and is not generally a required element of a crime.

Classification of Motive. The classifications discussed in this chapter are based on Douglas et al., *Crime Classification Manual* (CCM). This manual helps the fire investigator to gain as much information as possible by identifying essential elements of analytical factors used to classify the motive of an offense.

The behaviors listed in the CCM may identify a possible motive leading investigators to possible suspects. These behaviors apply whether the fire is the result of a one-time occurrence or the action of a serial firesetter. There are three classifications of repetitive fire-setting behavior:

1. *Serial arson:* Involves an offender who sets three or more fires with a cooling-off period between fires.
2. *Spree arson:* Involves an arsonist who sets three or more fires at separate locations with no emotional cooling-off period between fires.
3. *Mass arson:* Involves an offender who sets three or more fires at the same site or location during a limited period of time.

The National Center for the Analysis of Violent Crime (NCAVC) has identified the following six motive classifications as the most effective in identifying offender characteristics for fire-setting behavior:

1. Vandalism
   - Willful and malicious mischief
   - Peer or group pressure
2. Excitement
   - Thrill seeking
   - Attention seeking
   - Recognition
   - Sexual gratification or perversion
3. Revenge
   - Personal retaliation
   - Societal retaliation
   - Instructional retaliation
   - Group retaliation
4. Crime concealment
   - Murder concealment
   - Burglary concealment
   - Destruction of records or documents
5. Profit
   - Insurance fraud
   - Eliminating or intimidating business competition
   - Extortion
   - Removing unwanted structures to increase property values
   - Escaping financial obligations
6. Extremism
   - Terrorism
   - Riot/civil disturbance

interFIRE TRAINING

Table 20.1 provides an interface between NFPA 921 and the interFIRE VR training program. It brings the user from the NFPA 921 section of interest to the corresponding area in interFIRE. The user can apply the interFIRE information or find more information on what NFPA 921 offers.

TABLE 20.1 NFPA 921/interFIRE VR Training

921 Section	Knowledge/Skill	interFIRE Tutorial Student Activity	interFIRE Scenario Student Activity	interFIRE Resource Section Student Activity	www.interFIRE.org Student Activity
Chapter 22—"Incendiary Fires"	*Identify and document evidence establishing the incendiary nature of a fire under investigation as a prelude to its classification* *NFPA 921 defines an incendiary fire as one intentionally ignited under circumstances in which the person knows the fire should not be ignited (22.1)*	Information and guidance on the process of evidence collection is a focus of this program. Review these Tutorials containing supplemental information on recognition, documentation, collection, preservation, and transportation of physical evidence: **Module: The Process of Investigating a Fire** (organization and sequence protocol) **Module: Before-the-Fire Practices** • Define Official Responsibilities • Equip Your Investigative Team **Module: Interviewing Witnesses** • Guiding Principles • Prioritizing Witnesses • Approaching the Witness **Module: Roll-Up Practices** • Preserve Scene and Physical Evidence	The case behind the fire at 5 Canal Street was designed to challenge the technical and investigative skills of every fire investigator. To "solve" this fire case the investigator must not only identify the fire's cause but must also sift through information obtained from witnesses, physical evidence, conditions at the fire building, databases, insurance documents, and more to uncover the true nature of this fire. The student should examine the fire scene and associated sources of evidence (witnesses, records, insurance, etc.) for "incendiary fire indicators" (i.e., "red flags" that often are associated with set fires): • Fire near an appliance? (22.3.2) • Multiple areas of fire origin? (22.2.1) • Unusual fuel load or configuration? (22.2.5) • Incendiary device? (22.2.7) • Presence of ignitible liquid in the area of origin? (22.7.3) • Signs of open windows and doors? (22.3.6) • Owner with fires at other properties? (22.4.6) **Exercise** Investigate each of these red flags at 5 Canal Street. Use both physical evidence and witness statements to determine if that red flag occurred at this fire. Pay special attention to inconsistencies between witnesses, especially the owner and the renters of the property. Close the scene investigation and select an incendiary cause determination. For each possible point of support, first discuss whether the statement is true or false for the investigation at 5 Canal Street (e.g., were there canine hits in the dining room?). If the answer is true, then discuss whether that fact is proper support for cause determination. Remember, not every point that is true can be used to support cause determination. Discuss this and reach a group consensus on each point. Check off those items used as support and then submit them. If you do not move on, read the feedback on your answers and discuss it. Then, return to the scene and investigate the points you got wrong in more detail.	The Resource Section contains vital information about the investigation of incendiary fires. Refer to the file "Fire Scene Examination" to review the following documents: • Origin and Cause Determination • Cause Determination • Thermal Effects on Materials (photo catalogue) • Burn Pattern (photo catalogue) • Legal Standard of Proof • Reflections on Origin and Cause **Origin and Cause: Accidental Fire Causes:** • Fire Scene Electrical Checklist for the Nonelectrical Engineer • Common Accidental Cause Burn Pattern (photo catalogue) **Origin and Cause: Incendiary Fire Causes:** • Incendiary Fire Basics • Incendiary Fire Indicators • Ignitible Liquid Classification • The Arson Set: Excerpts from *The Pocket Guide to Accelerant Evidence Collection*	All of the reference and photographic information available in the CD-ROM Resource Section is available online at www.interfire.org. Additionally, the following "Featured Articles" and other references containing information on incendiary fires are available on the website: • *Flammable and Combustible Liquid Spill Patterns* • *Full-Scale House Fire Experiment for interFIRE VR* • *Glass Breakage in Fires* • *Another Weapon for the Arson Investigator* • *Community Based Anti-Arson Programs* • *Arson Unit Technical Support* Check the website frequently for additional information on this topic.

❯❯❯ QUESTIONS FOR CHAPTER 20

1. Which means of fire growth is considered "natural" when examining a fire scene suspected of multiple fires?
   - **A.** Fires on different stories
   - **B.** Fires spread through shafts
   - **C.** Fires inside and outside a building
   - **D.** Fires in different rooms
2. True or false: Confirmation of multiple fires is a strong indication that the fire was incendiary.
   - **A.** True
   - **B.** False
3. Fire patterns from fuels intentionally distributed from one area to another are known as
   - **A.** Connectors
   - **B.** Accelerants
   - **C.** Linear
   - **D.** Trailers
4. What should an investigator do on finding an incendiary device that has not activated?
   - **A.** Notify trained personnel for handling
   - **B.** Examine device to determine fuel type
   - **C.** Collect device for use as evidence
   - **D.** Document the location and continue searching for other devices
5. Why is it important for an investigator to be able to specifically explain subjective terms that are used in reporting assessment of fire growth and damage?
6. How can igniting multiple fires be considered sabotage?
7. Identify three principal trends used by investigators to help identify serial firesetters or serial arsonists.
8. Which evidentiary factor used by serial arsonists involves setting fires to coincide with other events to increase the chances of not being apprehended?
   - **A.** Crime concealment
   - **B.** Temporal frequency
   - **C.** Timed opportunity
   - **D.** Sequential prospect
9. Which classification of fire arsonist behavior involves an offender who sets three or more fires at separate locations with no emotional cooling-off period between fires?
   - **A.** Mass
   - **B.** Spree
   - **C.** Serial
   - **D.** Extreme
10. What is the difference between motive and intent?

11. Which subcategory of revenge retaliation is generally not satisfied with a single fire?
    - A. Institutional
    - B. Personal
    - C. Group
    - D. Societal
12. True or false: Crime scenes such as murder or burglary that also involve fires usually destroy all physical evidence.
    - A. True
    - B. False

Fire and Explosion Deaths and Injuries

OBJECTIVES

Upon completion of Chapter 21, the user will be able to

- Identify the effects of fire and explosion on the human body
- Identify additional personnel required to investigate a fatal or serious injury
- Discuss several combustion products and their effects on healthy individuals

Fires and explosions often involve a complex investigation to put together the events of the incident. These investigations are even more complex when the fire or explosion involves serious injury or death. As with any investigation, it is important that the correct procedures be followed from the beginning.

FIRST RESPONDERS

The investigation of a fatal fire or explosion is a two-part investigation, involving the origin and cause of the fire or explosion and the cause and manner of the death. Therefore, the body of the victim is the most important piece of evidence in determining the cause and manner of the death. Without question, first responders to a scene are there to preserve life and property. They also need to be aware of the importance of preserving the scene as much as possible while still effecting fire suppression and essential rescue. If on discovery of a victim, there is no question that the victim is beyond help through medical intervention, then every effort should be made to leave the body in its original location. At a crime scene, it would be unthinkable to move a murder victim. The fire or explosion victim should not be considered any differently. An exception to this would be if it is determined that the body of the victim would be damaged further by allowing it to remain in its discovered position. It is preferable to have the body preserved than to have it lie under several floors of debris. Always remember that the body is a piece of evidence and should be treated as such.

NOTIFYING AUTHORITIES

When a death occurs, there are legal and procedural requirements for the notification of various parties including law enforcement, medical examiners, coroners, forensic laboratory staff, and possibly others. Often, these same parties are part of the investigative team.

DOCUMENTING THE FIRE SCENE

One of the most effective and quickest ways to document the scene is by photography, particularly if, because of the environment that it is in, the body must be moved early in the investigation. It is helpful to take a quick series of photos. The choice for this photography is color and a 35mm or larger format, or a good-quality digital medium should be used whenever possible. Instant photos and video are useful, but often lack the detail and flexibility that may be needed later.

To ensure complete documentation, the body and its surrounding area should be photographed before the scene is disturbed and throughout the process of debris and body removal. Attention should be given to photographing the body with respect to its relationship to such items as exits and their condition, alarm systems, fire or explosion patterns, and other physical characteristics of the scene. Possible fire patterns and blast effects on the body should be well documented by photography. The body is one of the pieces of evidence that is part of a fire or explosion scene. It should be photographed once it is placed in the body bag, when it is removed from same, and during the time that clothing is being removed. Close-up and scale photos of burns and other injuries should be taken. The location in which the body was resting at the fire scene should also be photographed.

Documentation of the scene should include sketches and diagrams as needed. Diagrams should detail and record the physical dimensions of the scene that cannot be documented with photos. The sketch should include the outline of the body for reference purposes. Often these sketches may be useful in court when photos of a victim are not admissible out of consideration for jury members.

ACTIVITY

Review Chapter 15 in NFPA 921 and Chapter 13 in this *User's Manual,* relating to documentation of a fire scene.

EXAMINING THE SCENE

All fire or explosion investigations require a team effort. In cases of death, the investigation expands the team to possibly include a homicide detective, the medical examiner, the coroner, forensic laboratory personnel, and a forensic pathologist. If the body is badly burned, the expertise of a forensic anthropologist and a forensic odontologist (dentist) are likely to be required.

The investigation often initially focuses on the area where the body is located. The investigators need to remember that important evidence can often be found some distance from the body. Fire suppression, the actions of the victim while alive, and the environment may position evidence away from the body.

To aid in a thorough examination, the investigators can use a grid system to divide the scene into sections. Each grid needs to be examined, and documented, and evidence needs to be identified. If the scene does not lend itself to a grid search, the investigators might wish to consider a spiral or grid search pattern. All methods that are employed should have some overlap for complete coverage.

The investigation of the debris and its location may reveal a sequence of events for the fire or explosion. Often, the area that must be searched is covered with multiple layers of debris, which must be removed in a systematic way. The examination of the fire or explosion scene can be compared to an archeological dig where each layer is removed and interpreted in relationship to the events. At times, the size of the evidence can require different examination techniques. Small pieces of evidence need to be located by removal of debris from a section and through careful screening of the debris.

During the removal of the body, items such as clothing and fire debris could be attached. These items should be left adhered to the body for later identification. Because the body is a piece of evidence, it is important that it be treated as such, with proper consideration for chain of custody and cross-contamination. The body should be placed in a new, unused, sealed body bag.

When the body has been badly burned or fragmented by the fire or explosion, great care should be taken to search for all remains, no matter how small. Burned bones and tissue can blend in with fire debris and may be overlooked. An experienced fatal-fire investigator can be helpful to the team in identification of these items.

The team might wish to bring a skilled forensic anthropologist directly to the scene to assist in skeletal identification as needed. (See Exhibit 21.1.) Sometimes it is necessary to determine whether bones recovered at a scene are animal or human.

EXHIBIT 21.1 Skeletal Identification: The skills of a forensic anthropologist are needed to examine the charred remains of the five people who died in this backseat area of a station wagon.

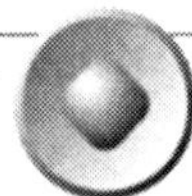

ACTIVITY

Review the members of the team that would respond to a fatal fire or explosion in your area. Does your agency have a written policy for fatal-fire investigation? If so, review it.

DEATH-RELATED TOXICOLOGICAL AND PATHOLOGICAL EXAMINATION

All victims of a fire or explosion should receive a full autopsy to determine the cause of the death. This examination must take into account several concerns.

One of the first steps in examination of the body should be full body x-rays. X-rays are useful in identification of any foreign matter in the body, such as a bullet or knife tip. (See Exhibit 21.2.) Also, x-rays may be helpful in the identification of the victim by determining any past injury such as a broken bone, surgical metal implant, or dental remains that can be compared to a known set of x-rays.

Testing for the carbon monoxide level of blood and tissue is one of the easiest and most common medical tests performed on the fire or explosion victim. This test often reveals much about the cause and sequence of events and should be performed whenever possible. Carbon monoxide is absorbed into the blood and tissue by breathing. It causes a cherry-pink coloration of the skin, which

EXHIBIT 21.2
Full-Body X-Ray
(Note the bullet in the pelvic area of this fire victim.)

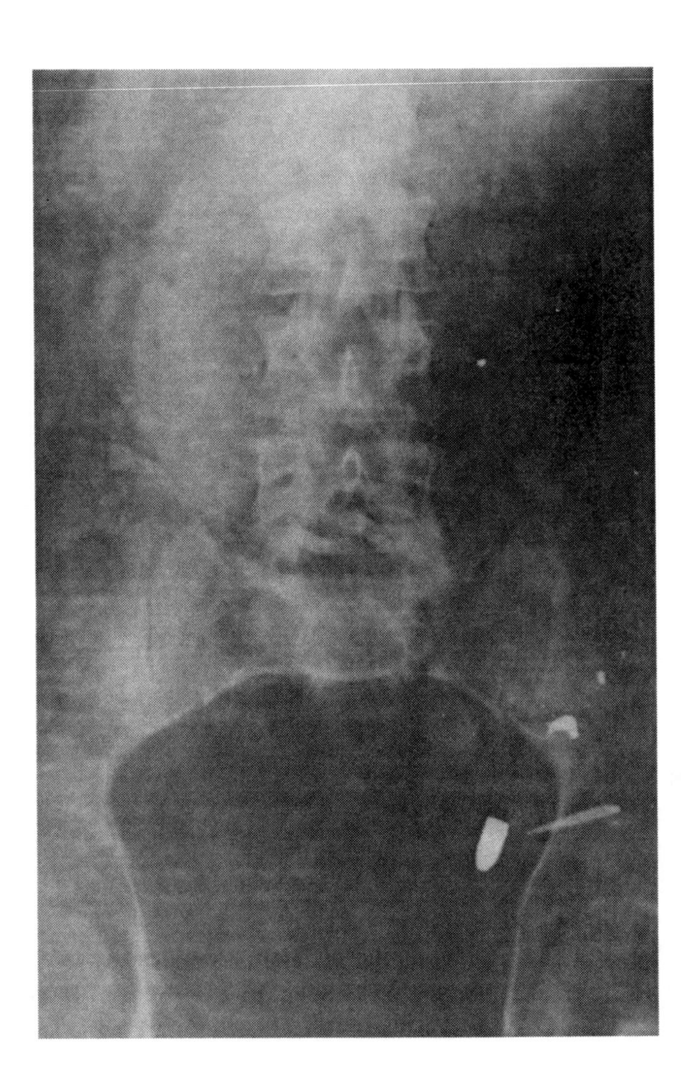

might not be visible in dark-skinned victims. The cherry-pink coloration could be visible in several areas on the body, including the lips and nipples, as well as in the blood that has pooled in the body (caused by postmortem lividity).

Lividity happens after death and is the pooling of the blood in the lower elevations of the body caused by the effects of gravity. Lividity becomes fixed in the body 6 to 9 hours after death.

In addition to the presence of carbon monoxide, the examination should look for soot or smoke in various areas of the body. The exterior of the body may show signs of soot around the nose and mouth. The investigator must be aware that these deposits are not always proof that the victim was breathing during the fire. A medical examination of the internal airways and the lungs often shows very clearly black soot deposits that are consistent with the fire victim's breathing in a smoke environment. Soot may also be discovered in the stomach of the fire victim.

The medical examination process should also include a search for other toxic products. With all fires, various combustion products are present. Such compounds as hydrogen cyanide and hydrogen chloride may be discovered in the blood and tissue. The same body elements may also reveal various drugs both medical and illicit, along with alcohol levels for the victim. Knowledge of these levels is useful to investigators as they try to interpret the ability of the fire or explosion victim to deal with the events as they take place.

PHYSICAL EFFECT OF FIRE ON THE BODY

The direct effects of the fire are often noted in burns to the body. This process takes place before and after death of the victim. Blistering of the skin (second-degree burn) can happen to a more limited degree after death. The effects of the heat after death begin to dehydrate the muscle tissue. As this tissue loses hydration, there is a noticeable shrinkage of the muscle, which in turn causes a constriction or tightening of the muscle. This is noticeable in the facial features of the victim and in the "pugilistic attitude."

The pugilistic attitude is a boxer-type stance observed in the hands and arms. This constriction of the muscles will also take place in the legs in the later stages of exposure. Often, investigators suspect that the victim was in a defensive stance as if warding off blows to the chest and head when instead it was a stance caused by the fire. This constriction of the muscles can even fracture bones in the arms or legs. The thermal effects directly on the bone also cause this fracturing.

During the examination of the body, special attention needs to be directed at the presence of any blood. Often, blood located on the body is an indication of non–fire-related trauma. The heat of the fire causes a dehydration of the tissue and, thus, seldom is an indication of blood present. One exception to this is small amounts of blood in the nose, ears, or mouth. Any blood that is found should be documented thoroughly.

Like all things in the fire scene, the body is part of the fuel load. Investigators at a fire scene need to remember to examine the burn patterns and extent of burns on a fire victim as they would any other part of the fuel load. The investigators should make sure that the patterns that are present are consistent with the other patterns in the area. When exposed to sufficient heat, the body is consumed to some degree. The skin and muscle tissues of the body are poor fuels but dehydrate and are consumed with time. Body fat is a combustible component of the body. The bones shrink and then shatter. There is a coloration change to the bone as it is heated. A darkened or black color evolves to a gray-blue and then to chalk white as it is heated.

The skull may shatter or crack from the effects of the fire. Most often, this is caused by a thermal expansion of the moisture in the brain cavity. The thermal expansion most often occurs along the suture lines of the skull. The investigator is warned not to assume that the fire caused any skull fracturing. Blunt trauma to the skull or the effects of a gunshot to the skull can cause fractures.

Much attention has been directed over the years to the phenomenon called spontaneous human combustion. Simply put, the human body does not spontaneously combust. Under the right conditions, the body fat of a fire victim may render out and, when absorbed by a wicking material such as a cotton shirt, serve as a fuel source for a small but concentrated flame similar to that of a candle. The energy from the flame may be insufficient to ignite adjacent combustible materials and will, with time, leave the body of the victim consumed in the torso area but allow the limbs and head (with their lower body fat concentration) to remain.

VICTIM IDENTIFICATION

One of the critical elements in the investigation related to the body is identification of the victim. As the body is discovered in the fire scene, it may be difficult for the investigator to determine whether the tissue mass is that of an animal or a human. The remains of large animals such as pigs, bears, and deer may resemble human remains. The skeletal remains of a bear paw compared to a human hand may be identifiable only to a trained anthropologist.

The bodies of children and infants pose special problems, as their skeletal structure is less developed than that of an adult. It may be impossible to determine the sex of the prepubertal victim based on examination of skeletal remains. In cremation, the body is exposed to temperatures over 1800°F (982°C) for 1 to 2 hours. Even after this exposure, there are remains of the skeleton. With a child, there is less development of the skeleton and less tissue mass. It is therefore possible in extreme cases to have remains that are not identifiable.

The effects of the fire on the body are varied, depending on many factors. In cases of moderate exposure, it may be possible to effect a visual identification of the victim. The investigator needs to bear in mind that the facial skin tightens and causes a more youthful appearance and that hair color may change.

The use of clothing and personal effects may provide some circumstantial evidence of identification. Fingerprints may be a possible source of identification. The pugilistic attitude causes the victim's fingers to curl into the palm, which may protect the fingertips sufficiently for printing and comparison.

X-rays of the dental remains and comparison to known sets are one of the surest ways of identification. The tissue mass of the head and the hardness of the teeth often help in their preservation in the fire. When the destruction to the head is extensive, the investigator needs to take care to recover dental remains for reconstruction and comparison.

With the recent advances in DNA, its use for identification is enhanced. Comparison of the DNA of the victim to a family member can be helpful.

After death, several changes begin to occur to the body. As was mentioned earlier, lividity sets in 6 to 9 hours after death. This pooling of the blood in its lower regions is caused by the effects of gravity. If the body is moved before lividity sets, the pooling of the blood changes. However, the investigator can recognize whether the body has been moved by examining the color of the tissue with respect to the body position. Rigor mortis begins to stiffen the joints of the body a few hours after the death. The effects of the rigor mortis begin to leave the body after about 12 to 24 hours.

There are several scenarios in which a body can be located in the fire scene:

1. An accidental fire that causes a fire-related death, such as a cooking fire that causes smoke inhalation
2. An accidental fire that causes a non–fire-related death, such as a cooking fire in which the victim dies leaping from the fire scene
3. An arson fire that causes an unintended death
4. An arson fire in which the death is intended
5. An arson fire that is used to cover a prior death, such as a suicide or murder

The ultimate questions related to the body are the cause and manner of the death. The *cause* of the death is regarded as the actual event or injury that brings about the cessation of life, such as smoke inhalation, burns, and gunshot. The *manner* of death is best regarded as the course of events that led up to the death, such as accidental suicide and others.

When trying to determine the cause and manner of death, it is helpful for the investigator to examine evidence and interpret it to determine the victim's activity prior to the fire. The physical location of the body (such as a bed), clothing, or items found on the body (such as a fire extinguisher) are telltale signs. Patterns of damage to clothing and/or the body should be considered in context with the total scene. Any inconsistencies should be examined. Burn patterns to body or clothing may be from attempts to extinguish the fire or to escape. Evidence found in the area may be a helpful indicator of past activity, such as the discovery of smoking material or food left out for cooking. Knowledge of the victim's prefire physical abilities is helpful in understanding the victim's movement or lack of movement.

MECHANICS OF DEATH

There are many products from the combustion process that can affect the victim of a fire or explosion. These products may include such items as carbon monoxide, carbon dioxide, nitrogen oxides, halogen acids, hydrogen cyanide, acrolein, benzene, soot, and ash. Breathing these products or direct skin contact can cause a variety of effects on the human body.

Carbon Monoxide

Carbon monoxide (CO) is a product of all fires and is caused by incomplete combustion of the fuel. Thus the level of CO produced by each fire varies on the basis of the completeness and type of the fuel package. CO is an anesthetic and an asphyxiate. When CO is inhaled, it binds with the hemoglobin in the blood to form carboxyhemoglobin. CO binds with the hemoglobin in the blood much more readily than oxygen does. Therefore, it is possible for the blood to form dangerous levels of CO with exposure to low concentrations of CO. When the fire survivor is removed from the CO environment, the levels of CO in the blood begin to decrease. However, this process can take many hours.

The stability of CO in the body is such that it can be tested for hours after the death. As a general rule, levels of CO in the body of 50 percent or greater are considered fatal. The actual levels can vary, with death possible in some cases at concentrations as low as 20 percent. The level of the CO in the body may be helpful to the investigator in making determinations about the death of the victim. When levels are below 20 percent, the death most likely was caused by other factors, such as thermal injuries or suffocation. When levels are 40 percent or greater, the toxic effects of the CO alone or in combination with other factors are likely a contributing cause of death. Studies have shown that about 80 percent of fire victims die from carbon monoxide poisoning.

There are other factors that can cause concentration of CO in the body. Smokers or people who have been exposed to automobile exhaust may have CO levels of 4 to 10 percent.

Hyperthermia

The thermal effects of the fire can result in death. Hyperthermia is caused when the temperature of the body is raised above fatal levels. Body temperatures above 109°F (43°C) are generally fatal in a few minutes.

The inhalation of hot gases and various toxic gases causes edema (swelling) and inflammation of the airway. The effects of hot gases are generally accompanied by facial burns or singed facial hair. The inhaling of soot may produce thermal injury and may carry toxic compounds or physical blockage to the airway. The fire may consume the oxygen to the point at which it poses a possible risk. Oxygen levels below 15 percent can cause disorientation and loss of judgment. At levels below 10 percent, unconsciousness occurs, followed by cessation of breathing.

POSTMORTEM TESTS AND DOCUMENTATIONS

Various tests and documentation during the postmortem examination provide the investigator with valuable information to aid in identifying the victim as well as establishing the cause and manner of the death. These tests may include the following:

- *Blood:* For level of CO, drugs, alcohol, or poisons
- *Internal tissue:* For level of CO, drugs, alcohol, or poisons
- *External tissue (near burns):* For determination of whether burns occurred before or after death
- *Stomach:* For contents, including the presence of soot
- *Airways:* For effects of the fire
- *Internal body temperature:* To assist in establishing a time of death
- *X-rays:* To assist in identification or location of foreign objects
- *Clothing or personal effects:* To assist in identification or check for the presence of ignitible liquids
- *Sexual assault evidence:* For possible motivation for setting a fire
- *Documentation by photography and sketching:* To record injuries and burns to the body and evidence recovered from the body

Contact your local medical examiner and see what medical tests they perform on fatal-fire victims.

INJURIES FROM FIRES AND EXPLOSIONS

Deaths from fires and explosions can occur well after the event. Serious injury fires and explosions should be investigated as if they were fatal investigations to ensure completeness. During the investigation of the injury scene, many items should be examined and documented. Often, physical evidence, both apparent and microscopic, is located in areas beyond the body.

Clothing should be collected as soon as possible to prevent its loss. Clothing may contain physical evidence (in pockets) or may have been a fuel source for the fire. Likewise, the furnishings of the area may be evaluated for burning properties.

As with any fire, potential ignition sources for the fire will be examined and documented.

Some areas require by law that notification be made to governmental agencies when serious fire or explosion injuries occur. These laws are similar to gunshot reporting laws.

Burn-Related Injuries

Burn injuries need to be documented and assessed to the following degree of burns:

- *First degree:* Reddening of the skin
- *Second degree:* Blistering of the skin
- *Third degree:* Full-thickness damage to the skin
- *Fourth degree:* Damage to the underlying tissue and charring of the tissue

Burns may not be distinguishable in appearance whether they are caused by scalding, chemical exposure, or hot gases or flames. There is a direct relationship to the forming of burns and the radiant heat flux.

The fire produces various by-products that affect the person exposed. Testing is needed to determine the levels of these by-products and to understand the effects. Narcotic gas such as carbon monoxide (CO), hydrogen cyanide (HCN), or the effects of hypoxia affect the response of the person who is exposed. Additionally, some of these same by-products and smoke can interfere with respiration, irritate eyes, and hinder the ability to see. The levels of these by-products begin to reduce when the person withdraws from the affected area.

Explosion-Related Injuries

The injuries related to an explosion scene are classified into four groups based on the blast effect.

Blast Pressure. Blast pressure injuries are caused from the concussion effects of the explosion. Damage to the internal organs is not uncommon depending on the level of the blast pressure. The pressure wave may also propel the victim into objects and cause blunt trauma injuries. With detonations, it may be possible for the body to experience severe injury or amputation.

Shrapnel. Shrapnel injuries from fragments from the blast center may cause serious effects on the body, such as amputation, lacerations, or blunt trauma.

Thermal. Thermal injuries may be caused by the explosive flame fronts, usually causing first- and second-degree burns. Thermal injuries may result from synthetic fabrics, which may melt on the victim. Cotton fabrics may scorch.

Seismic Effects. Seismic effects of the explosion may cause injury with the collapse of structures. These injuries are often seen as blunt trauma, lacerations, fractures, amputation, contusions, and abrasions.

interFIRE TRAINING

Table 21.1 provides an interface between NFPA 921 and the interFIRE VR training program. It brings the user from the NFPA 921 section of interest to the corresponding area in interFIRE. The user can apply the interFIRE information or find more information on what NFPA 921 offers.

TABLE 21.1 NFPA 921/interFIRE VR Training

921 Section	Knowledge/Skill	*interFIRE Tutorial Student Activity*	*interFIRE Scenario Student Activity*	*interFIRE Resource Section Student Activity*	*www.interFIRE.org Student Activity*
Chapter 23—"Fire and Explosion Deaths and Injuries"	*Fires or explosions resulting in serious injury or death require:* • *Special investigative effort (23.1)* • *Teamwork with medico/legal specialists (23.2)* • *Complete origin and cause examination* • *Death investigation*	interFIRE contains a Tutorial, "Handling Injuries and Fatalities" written by an internationally prominent pathologist. An addendum to this Tutorial entitled "Additional Information" presents specific recommendations for additional investigative steps. Tutorials containing additional information on injury/fatality fire scene investigation include: **Module: Roll-Up** • Preserve Practice Scene and Physical Evidence • Secure Scene and Witnesses **Module: Preliminary Scene Assessment Practices** • Get Basic Incident Information **Module: Fire Scene Examination** • Handling Injuries and Fatalities • Document the Scene • Determine the Need for Additional Resources • Examine the Scene and Analyze Fire Flow • Reconstruct the Scene • Collect and Preserve Evidence • Eliminate Accidental Causes • Use Canine Units **Module: Follow-Up Investigation** • Use Non-Forensic Labs and Technical Experts • Interpret Forensic Laboratory Results **Exercise** Meet with the medical examiner in your jurisdiction and discuss the content of the Tutorial section, "Handling Injuries and Fatalities." Relate this content to how your medical examiner operates at the scene and develop a joint plan for how you will handle fire scene fatalities at the location and at autopsy. Learn from your ME what signs he or she looks at when examining a fire death.	The investigator is required to perform a comprehensive investigation of this fire incident to establish the fire origin, its cause, and the person or persons responsible. The fire at 5 Canal Street resulted in non-life-threatening injuries to a passerby witness. This man stated that he was injured while attempting to rescue potential occupants of the burning home. In conducting this virtual reality investigation, you must determine his veracity on the basis of facts established during interviews with the victim, other witnesses, and evaluation of physical evidence.	The Resource Section contains valuable information on investigation of fire injuries and fatalities. Beginning with the *browse* view (file drawer view) select the file entitled, "General Fire/Arson Topics, Fatal Fire Investigation": • *Multiple Fatality Home Fires* • *NIST Simulator Provides New Picture of LODDs (Line of Duty Deaths)* • *U.S. Fire Fighter Deaths Reach 10-Year Peak* • *NIST Fire Dynamics at Cherry Road* • *Eight-Fatality Row-House Fire—Lessons Learned (USFA)* • *Five-Fatality Fire in Thornton, Colorado (USFA)*	All of the reference and photographic information available in the CD-ROM Resource Section is available online at www.interfire.org. Additionally, the following "Featured Articles" and other references containing information on incendiary fires are available on the website: • *Smoke Detector Technology and the Investigation of Fatal Fires* • *Establishing a Relationship between Alcohol and the Causalities of Fire* • *NIST Simulator Provides a New Picture of Line of Duty Deaths* Check the website frequently for additional information on this topic.

QUESTIONS FOR CHAPTER 21

1. True or false: X-rays should be taken of the entire body and of the clothing and associated debris found near the body.
   A. True
   B. False
2. True or false: Carbon monoxide causes a cherry-pink coloration to the skin of the fire victim and is visible on all skin types.
   A. True
   B. False
3. True or false: Evidence of smoke or soot in the lungs, bronchi, and trachea is one of the most significant factors in confirming that the victim was alive and breathing smoke during the fire.
   A. True
   B. False
4. True or false: The pugilistic attitude is a direct result of the fire.
   A. True
   B. False
5. True or false: Under certain circumstances, human bodies can spontaneously combust.
   A. True
   B. False
6. True or false: The remains of infants cannot be completely destroyed in a fire to the point where identification may not be possible.
   A. True
   B. False
7. True or false: Blood can accumulate fatal levels of carbon monoxide even when the victim is exposed to low-level concentrations in the air.
   A. True
   B. False
8. True or false: The presence, absence, and pattern of areas of lividity can help to establish the position of the body after death and can reveal whether it has been moved or repositioned after death.
   A. True
   B. False
9. True or false: External facial burns always accompany edema and inflammation caused by chemical irritants in smoke.
   A. True
   B. False
10. True or false: The human skull can fracture and explode from internal pressure caused by the expansion of moisture from the heat of the fire.
    A. True
    B. False
11. True or false: Heat can be transferred through clothing, causing burns to the skin without any identifiable damage to the clothing.
    A. True
    B. False

12. True or false: Carbon monoxide is a narcotic gas.
    A. True
    B. False
13. True or false: The carbon monoxide level of a fire survivor remains constant for several hours after removal from the fire environment.
    A. True
    B. False
14. True or false: Thermal injuries resulting from an explosion flame front are generally limited to first- and second-degree burns because of their very short duration.
    A. True
    B. False
15. List fundamental issues that an investigator may confront in investigating a death related to fire or explosion.

Appliances

OBJECTIVES

Upon completion of Chapter 22, the user will be able to

- Understand a variety of common appliances and their operation
- Know how to properly conduct an investigation of a fire involving an appliance

The material in this chapter expands on the information in Chapter 15, "Documentation of the Investigation," of NFPA 921. This chapter uses the methods outlined in that chapter, and several others for investigations of fires involving appliances.

RECORDING THE FIRE SCENE WHERE AN APPLIANCE IS INVOLVED

When conducting an investigation of a fire involving an appliance, the investigator should not disturb the appliance until it has been thoroughly documented. The investigator should review Paragraph 11.3.5.5, "Spoliation of Evidence," in NFPA 921, for proper procedures in handling the appliance and related evidence.

Once an origin area has been identified and one or more appliances have been found in the origin area, the appliance should be documented by photographs, diagrams, and measurements. Ascertaining the position of any controls, securing the appliance nameplate data, and gathering all component parts of the appliance must also be accomplished as part of the recording of the scene.

Read NFPA 921, Chapter 24, Section 24.1, "Appliances," through Section 24.3, "Origin Analysis Involving Appliances."

Photographs

The investigator should photograph the entire scene as outlined in Chapter 13 of this *User's Manual,* taking care to photograph the appliance from many different angles. He or she should record the entire area, including the appliance to establish its location in relation to combustibles, witness information, and landmarks in the room. Close-up photographs should also be taken to provide detailed information such as switch positions, and thermal protection devices. The appliance should not be moved until all other fire scene documentation and appliance documentation have been completed.

Diagrams and Measurements

The investigator should diagram the fire scene and locate the appliance on the diagram, being sure to include measurements locating the appliance in relation to fixed landmarks, combustible fuels, and other elements in the fire scene.

Documenting the Appliance(s)

On a sketch or photograph of the appliance, the following should be documented:

1. Controls
   - Dials
   - Switches
   - Power settings
   - Thermostat settings
   - Valve position
2. Branch circuit supply wiring to appliance
3. Bonding and grounding
4. Position of movable parts
   - Doors
   - Vents
5. Clocks (hand position)
6. Power supply
   - Battery
   - Electrical supply
7. Fuel supply
   - Gas: natural gas or propane
   - Electric
   - Fuel oil

Identifying Information

The investigator should obtain the following identifying information from the appliance:

- Manufacturer
- Model number
- Serial number

- Date of manufacture
- Name of product
- Warnings
- Recommendations
- Additional data located on the appliance

If it is necessary to move the appliance to gather this information, this should not be done until all other documentation has been completed as outlined in the lists.

Gathering Components

The investigator should gather together the various components that may have been moved during the fire or fire-fighting operations and reconstruct the appliance in its prefire location, if possible. The appliance should not be tested or operated at the fire scene. This could further damage the appliance and may be unsafe to do. At this point, all testing must be nondestructive such as testing for electrical continuity or resistance using a volt-ohm meter (See Exhibit 22.1.)

ACTIVITY

Read NFPA 921, Section 24.3, "Origin Analysis Involving Appliances," through Section 24.4, "Cause Analysis Involving Appliances." ◂◂◂

DETERMINING THE ORIGIN OF A FIRE INVOLVING AN APPLIANCE

This section applies and expands on the material in Chapter 6, "Fire Patterns," and Chapter 17, "Origin Determination," of NFPA 921.

Fire Patterns

Fire patterns should be used to establish that an appliance is located at the point of origin. The investigator should evaluate the fire patterns on the appliance in

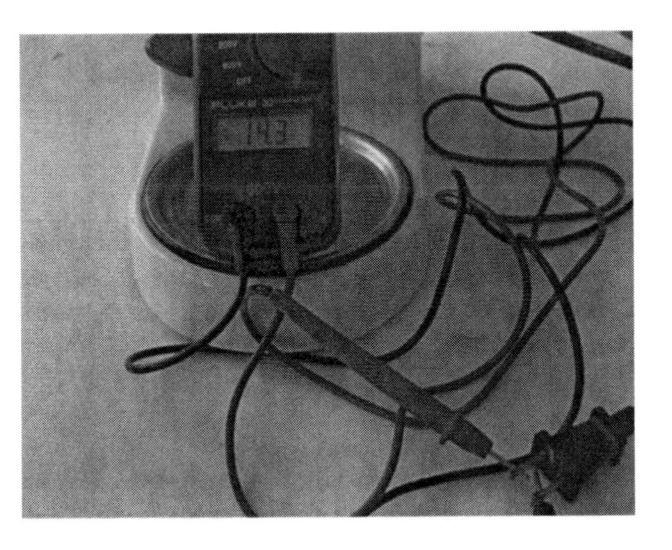

EXHIBIT 22.1
Checking for Electrical Continuity of Coffeemaker with an Ohm Meter

relation to the remainder of the fire scene. If the appliance shows more severe damage than surrounding items, this could indicate that the fire originated at the appliance. All other causes for the fire damage that occurred at the appliance must be considered, including fall down. Appliances with plastic housings can suffer severe fire damage as a result of exposure to fire. The type of housing on the appliance must be carefully considered in determining the origin area. The investigator should verify that the appliance was connected to an electrical power supply at the time of the fire. It is important for the investigator to secure and document the power supply.

The reconstruction of the fire scene and replacement of the appliance in its prefire position may be necessary to document the fire patterns and indicators that the fire originated at the appliance. After the fire scene has been reconstructed, the appliance and scene should be documented as previously discussed.

DETERMINING THE CAUSE OF A FIRE INVOLVING AN APPLIANCE

This section applies and expands on the material in Chapter 18, "Cause Determination," of NFPA 921.

To determine how the appliance could generate sufficient heat energy to start a fire, the investigator should ask the following questions:

- Could the heat be generated under normal operating conditions?
- Could the heat be generated under abnormal operating conditions?

The goal of these questions is to determine what material the heat ignited and how the material was ignited by this heat.

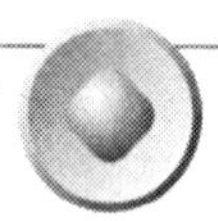

Read NFPA 921, Section 24.4, "Cause Analysis Involving Appliances," through Section 24.4.6, "Testing Exemplar Appliances." ◀◀◀

Appliance Operation

The investigator should thoroughly understand how the appliance operated and its safeguards. A good reference source for the operation and design of appliances can be found in home appliance repair books. The investigator should document any modifications made by the manufacturer, end user, installer, or service personnel.

Electricity

Evaluate and determine whether the electricity was on and supplied to the appliance before the fire. Specific conditions must be met for an electrical fault or overload to have occurred and sufficient heat must have been generated to cause ignition. (See Chapter 6, "Electricity and Fire," of this *User's Manual*.)

Disassembly

Before beginning disassembly, a specific reason for carrying out the process should be determined. Each step of the disassembly process should be documented. (Caution: The investigator should disassemble only if he or she has proper expertise and has given notification to interested parties, such as the appliance manufacturer and the insured.) The investigator should take notes and photographs or videos documenting the entire disassembly process. X-rays might be considered if disassembly is not possible.

Exemplar Appliance

An exemplar is an exact duplicate of the appliance in question. An exemplar can be used to help understand how an appliance operates and can be used in testing a proposed ignition scenario. The model and serial numbers on the appliance in question can help the investigator obtain an exact duplicate. If the exemplar is not exactly the same, the investigator must determine whether it is suitable to use. Testing should verify that the appliance could not only generate the heat, but could also ignite the fuel load.

APPLIANCE COMPONENTS

ACTIVITY

Read NFPA 921, Section 24.5, "Appliance Components," through Paragraph 24.5.1.8, "Ceramics." ◀◀◀

Housings

The housing of an appliance is the outer shell of the appliance that contains the working components. Although most housings are made of metal or plastic, they may also be made of wood, glass, or ceramic.

Metal Housings

Most metal housings are fabricated from steel or stainless steel. Steel housings may be coated with an enamel or plastic coating that may be a factor when the appliance is exposed to fire.

Steel Properties

Steel melts only at high temperatures not normally found in structure fires. Ordinary unfinished steel oxidizes in fire and turns dull blue-gray. If the steel has been deeply oxidized, it might be possible to flake off the oxidation, or oxidation may have been severe enough to penetrate completely through the steel.

Steel Colors (Postfire)

Postfire patterns and colors depend on many factors. Ordinary steel housings can have a mottled appearance with colors that include blue-gray, white, black, and reddish brown. Bare galvanized steel could have a whitish coating. Protective coatings on the steel can cause many varied colors.

Aluminum

Aluminum housings are generally made from formed sheets or castings. Pure aluminum has a low melting temperature of 1220°F (660°C). Aluminum alloys have slightly lower melting temperatures.

Other Metals

Other metals, such as zinc or brass, are used for appliance housings, often for decorative purposes. Zinc melts at 786°F (419°C) and is found as a lump of gray metal. Brass is an alloy and softens over a range of temperatures; however, it generally has a melting temperature of 1740°F (950°C). Although brass is sometimes used as a housing material, its primary use is for electric terminals.

Plastic housings, or housings made from carbon and other elements, are used in appliances that do not normally operate at high temperatures.

Damage

Following a brief fire, the plastic housing could be melted and partially charred. The fire investigator must determine whether the heat source was inside the appliance or external. X-rays provide a view of the internal metal components. In a severe fire, the entire plastic housing might be consumed. This does not necessarily indicate that the fire started with the appliance.

Phenolic Plastics

Phenolic plastics are highly resistant to heat and are often used to make coffeepot handles and circuit breaker cases. These plastics do not melt and do not support combustion. Phenolic plastics form a thin, gray ash layer when moderately heated and turn to a gray ash in a sustained fire. A thin, gray ash layer on the inside, and not the outside, of a phenolic plastic component may indicate that the heating was internal.

Wood Housings

Wood is occasionally used for an appliance housing. It might be completely consumed in a fire or may have indicative fire patterns.

Glass Housings

Glass is most commonly used for covers and doors and for decorative purposes. It readily cracks and may soften and drip under fire conditions.

Ceramic Housings

Ceramic is generally employed for use as a novelty housing. It does not melt in fire. However, the decorative glaze could possibly melt. Ceramic is also used to support or house the electrical components.

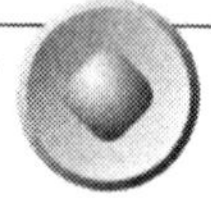

Conduct a survey of appliances in your kitchen and laundry room, noting the housing construction. ◀◀◀

Power Sources

Read NFPA 921, Paragraph 24.5.2, "Power Sources," through Paragraph 24.5.2.4, "Overcurrent Protection." ◀◀◀

The power source for common appliances is generally the alternating current (ac) supplied by the electrical utility company. Information in this chapter is limited to single-phase power that is 240 V or less. Power in the United States is supplied at 60 Hz, 120/240 V ac. Most appliances operate on 120 V ac. However, appliances that require 240 V (e.g., dryers and ranges) can also work on the same system.

Electric Cords

Electric cords can comprise two or three conductors. These conductors are stranded to provide flexibility. Cords with two conductors are found on appliances made before 1962 or on some newer double-insulated appliances. Cords with three conductors are found on newer appliances on which the third conductor is used as a ground for the appliance. These cords can be used for 120-V and 240-V appliances.

Stranded conductors usually survive fires and may be brittle following exposure to the fire.

Plugs

Plugs made prior to 1987 and rated for 20 A or less have two straight prongs of equal width. Plugs made after 1987 for 20 A or less have the neutral prong wider than the "hot" prong. This is known as a polarized plug. Some newer plugs have a third prong for grounding. Plugs may have the conductors attached to the prongs inside a molded plastic housing.

Step-Down Transformer (Adapters)

Some appliances operate at lower voltages such as 6 V, 12 V, or 24 V. A step-down transformer reduces the 120 V provided at the receptacle to the required voltage. The transformer might be a part of the appliance or might be separate from the appliance and plug into the wall receptacle (adapter). Leading from the adapter to the appliance is a thinner, two-conductor wire. Shorting of the lower-voltage wire is not likely to cause a fire.

Batteries

Batteries are used for portable devices and some security devices and can include car batteries, common dry cells, and button batteries of the type used in hearing aids. The remains of batteries are usually found after the fire, and it is important to determine what they were connected to. Under normal conditions, battery-powered circuits generally do not allow sufficient heat buildup to cause ignition. There are, however, certain conditions in which one battery can provide sufficient power to ignite some materials.

Protective Devices

Overcurrent protection devices often employed in appliances and electrical service panel are fuses and circuit breakers.

Fuses. If the current is moderate (less than twice the rating), the fuse metal simply melts, thus breaking the circuit. If the current is excessive, the metal vaporizes, leaving an opaque deposit on the glass tube or window.

Circuit Breakers. Circuit breakers operate thermally or magnetically. A circuit breaker in a fire can trip as it is heated from an external fire or when the current flow exceeds the rating of the circuit breaker. The higher the overcurrent, the faster the breaker trips. Circuit breakers in appliances may have a reset button.

ACTIVITY

Read NFPA 921, Paragraph 24.5.3, "Switches," through Paragraph 24.5.3.2.2, "Temperature Switches." ◀◀◀

Switches

Many different switch designs are used in common appliances. Switches are used to turn the appliance on or off or to change operating conditions. (See Exhibit 22.2.) Postfire examination of a switch can sometimes determine its state and that of the appliance at the time of the fire. The investigator should not operate the switch. The remains of the switch can be very delicate. Careful documentation of the position of the knobs, levers, or shafts while the switch is in place should be performed. The electrical continuity of the switch should be checked while it is in place. It is important to make sure that the power to the appliance is off or that the appliance has been unplugged before testing. The investigator should not disassemble the switch unless he or she is qualified to do so.

Table 22.1 illustrates the functions of different switch designs.

Fluid Pressure (Capillary Tube). A fluid pressure switch operates by fluid in a sensing bulb that is located in a hot area. The fluid expands in the bulb when overheated and applies pressure to a bellows device, which in turns opens a set of contacts, breaking the flow of electricity.

Bimetal. Bimetal switches are much more common. They are composed of two dissimilar pieces of metal that are joined together to form a flat piece. One type of metal expands at a faster rate that the other type, causing the flat piece to bend. This bending motion can open a set of electrical contacts halting the flow of electricity. After a severe fire, the bimetal device can be distorted far beyond its operating position. This can be a result of heat from the fire and does not necessarily indicate a defective component.

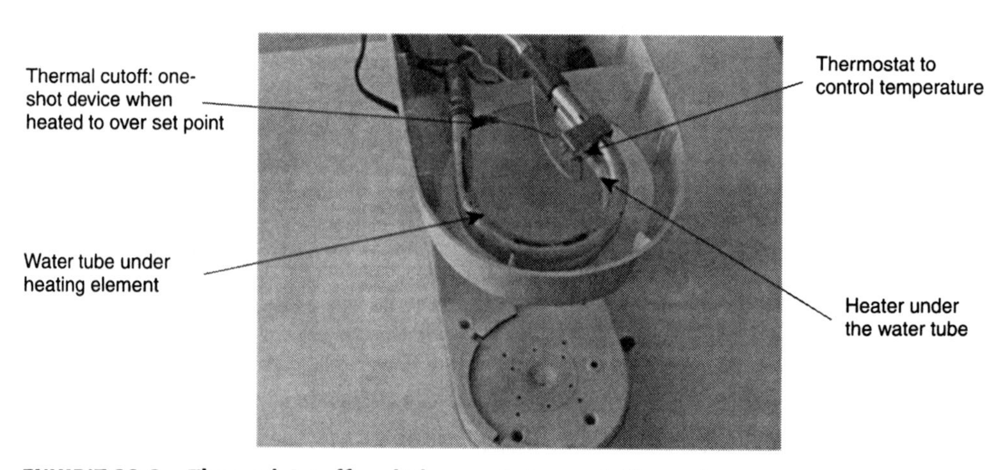

EXHIBIT 22.2 Thermal Cutoff and Thermostat on a Coffeemaker

TABLE 21.1 Switch Design

Switch	*Intended Function*	*Activation*	*Mode of Operation*
Manual	For user to operate	User	Push button Lever Knob Toggle
Automatic	Keep unit operating as intended	Overcurrent	Fuses Circuit breakers
	Prevent unsafe operation	Temperature	Thermostats: • Bimetal • Expanding metal • Fluid pressure (capillary tube) High-temperature limits: • Thermal fuse • Cutoff
		Motion	Tip-over switches Door switches

Expanding Metal. The expanding rod switch employs a long rod that is exposed in the heated area. As the temperature rises, the rod expands, opening a set of electrical contacts.

Melting. Some cutoff devices operate by means of a material that melts when the normal operating temperature is exceeded. When the material melts, it opens a set of contacts that stops the flow of electricity. These devices can be deliberately bypassed to allow the appliance to be operated without any thermal protection.

Motion Switch. Appliances such as a portable electric heater employ a motion (tip-over) switch to ensure that the appliance is operated in its correct designed position. A typical motion switch uses a weighted arm that hangs down and would open the contacts should the heater be tipped over.

Damage. Mechanical switches can fail due to an overload that can overheat internal parts or weld the contacts. Poor internal connections can cause destructive heating and failure. Switches normally show damage created by a parting arc on their metal contact surfaces. Contacts, such as in a thermostat, are normally pitted because of frequent opening and closing. Safety cutoff switches should not have pitting on the contact faces. If they do, this is a sign of frequent operation and possibly evidence of a defect. Most switches are designed to snap open or closed to avoid surface pitting, erosion, or possible welding. Welded contacts do not necessarily prove that the switch caused the fire. Electrically welded contacts have normal shapes and have their faces stuck together. Contacts that are melted together in one lump were probably exposed to external heating.

Solenoids and Relays

Read NFPA 921, Paragraph 24.5.4, "Solenoids and Relays," through Paragraph 24.5.9, "Miscellaneous Components." ◀◀◀

Solenoids and relays are used to control high-power circuits with a low-power circuit. The remains of these devices normally exist after a fire, and their contacts should be inspected to determine whether they were stuck together.

Transformers

Transformers are devices that reduce voltage from 120 V or 240 V to a lower voltage and isolate the appliance from its power source. The windings in the transformer may deteriorate after long-term use or use in an area with inadequate ventilation. As the windings deteriorate, impedance drops, and more current flows, which in turn generates more heat. This can lead to severe heating, which can cause the windings to fail by melting or create a ground fault. The heat that is generated may ignite the insulation of the transformer or combustibles in the vicinity of the transformer. Components of a transformer often survive a fire. Internal damage of the windings will be shown by a clear pattern of internal heating, arcing from turn to turn, or a pattern of fire travel out from the source.

Motors

Motors range from ⅓ to ¼ horsepower in major appliances to small motors in smaller appliances. If the rotor stops while the motor is energized or at start, the impedance falls and current increases. This can cause the windings to heat sufficiently to ignite the insulation and plastic materials. Protection for the motor is often a fuse link or a thermal cutoff switch. Small motors that drive cooling fans are generally not sources of ignition. Small motors do not have sufficient torque to generate heat by friction to cause ignition.

Heating Elements

Heating elements can ignite combustibles that are in contact with the element itself. Appliances with heating elements—with the exception of cooking appliances—are designed to maintain a distance between the element and surrounding combustibles. Sheathed elements are found in ovens and ranges. They are made of a resistance wire surrounded by an insulator and encased in a metal sheath. The sheath may be made of steel. Baseboard and space heaters have sheaths made of aluminum. Aluminum sheaths generally melt from external fire exposure. Open elements are composed of wires or ribbons constructed from nickel-chromium-iron alloy. Some appliances use a fan to remove heat from the element to stop it from glowing and to disperse the heat. The element can be tested for continuity and resistance with the use of a volt/ohm meter.

Lighting

Appliances often employ lighting to illuminate work areas, light dials, or illuminate internal cavities. This lighting is normally of low wattage and not prone to ignite combustibles. The lighting types normally found are incandescent and fluorescent. Some types of higher-wattage incandescent lighting may ignite combustibles if they are in contact. Fluorescent lighting operates at a higher voltage. However, the tubes normally do not get hot enough to ignite combustibles.

Fluorescent Lighting Systems

Fluorescent lighting systems are commonly employed in office settings. They utilize one or more glass tubes filled with a starting gas and low-pressure mercury. An electrical discharge is sent down the length of the glass tube, exciting the mercury gas. When the mercury gas is excited, it creates UV light, which is converted to visible light by the coating on the inside of the tube, known as the phosphor or fluorescent powder.

Fluorescent Light Ballasts

There are two main types of fluorescent light ballasts: magnetic and electronic ballasts. Magnetic ballasts incorporate either a reactor or a transformer. Interior fluorescent light fixtures manufactured after 1968 are required to have thermal protection in the ballast. A "P" on most metal housings indicates that thermal protection exists. All fixtures, indoor and outdoor, manufactured after 1990 are required to have thermal protection in the ballast. However, thermal protection does not ensure that a failure cannot occur. Both the electronic and magnetic fluorescent light ballasts contain pitch or potting compound within the ballast. This pitch can ooze out because of either internal heating or fire exposure. This pitch will not ignite other materials unless it is already burning. Electronic ballasts employ a circuit board and smaller magnetic components to operate fluorescent lamps. Some electronic ballasts employ resetting thermal protectors, while others use fuses for thermal protection.

Some common failures of ballasts that initiate fires are arc penetrations into combustible ceiling materials or nearby combustibles and extreme coil overheating that conduct heat into nearby combustibles. If the investigator suspects a ballast failure as the cause of a fire, the ballast along with the fixture and the wiring should be preserved for examination by qualified experts. The fixture itself should be examined for failures, such as arcing inside the fixture and failure of the lampholders.

High-Intensity Discharge Lighting Systems

High-intensity discharge (HID) lighting systems are often employed in warehouses, manufacturing facilities, or "big box" retail stores. They utilize a lamp that has a short tube filled with various metal vapors, such as sodium, mercury vapor, or metal halide. An electrical discharge is created along the length of the tube exciting the metal vapors in the tube that in turn creates light. HID lights operate at higher pressures than common fluorescent lighting systems. HID systems also employ a ballast and a capacitor for starting voltage and to limit the current flowing through the lamp. These ballasts may be electronic or magnetic and some of them are protected by fuses or thermal protectors. The ballasts are typically mounted inside the fixture enclosure above the lampholder and reflector assembly. If the investigator suspects an HID lighting fixture as the cause of the fire, the whole fixture must be preserved for laboratory examination. (See NFPA 921, Paragraph 24.5.8.2.5.)

Miscellaneous Components

Dimmers and speed controllers are examples of miscellaneous components. Older appliances may contain rheostats or wire resistors. Components in newer appliances are solid state and are usually destroyed in a fire unless the fire is of brief duration.

Timers can be built into the appliance or stand alone as separate devices. They can be driven by small motors and be badly damaged after a fire. Failure of the timer results from gears wearing out or losing teeth and are generally not the cause of the fire. Thermocouples measure temperature differences. A thermopile is a series of thermocouples. They are used in gas appliances to keep a valve open when the pilot is burning. Newer appliances use electric igniters instead of standing pilots.

COMMON RESIDENTIAL APPLIANCES

Read NFPA 921, Section 24.6, "Common Residential Appliances," through Paragraph 24.6.15, "Lighting."

Ranges and Ovens

In the common household range or oven, heat is provided by electricity passing through resistance heating coils or by the burning of natural gas or propane. The temperature in the oven is controlled by a thermostat and valve or switch on the fuel or power supply. In a gas range, a burner fuel supply valve that is manually operated controls fuel flow rate. Ignition is by a standing pilot flame or by an electric igniter. A timing device (temperature control) cycles the flow of electricity through the heating coil to maintain the selected temperature of the electric range. Exhibit 22.3 shows the elements of a common household range.

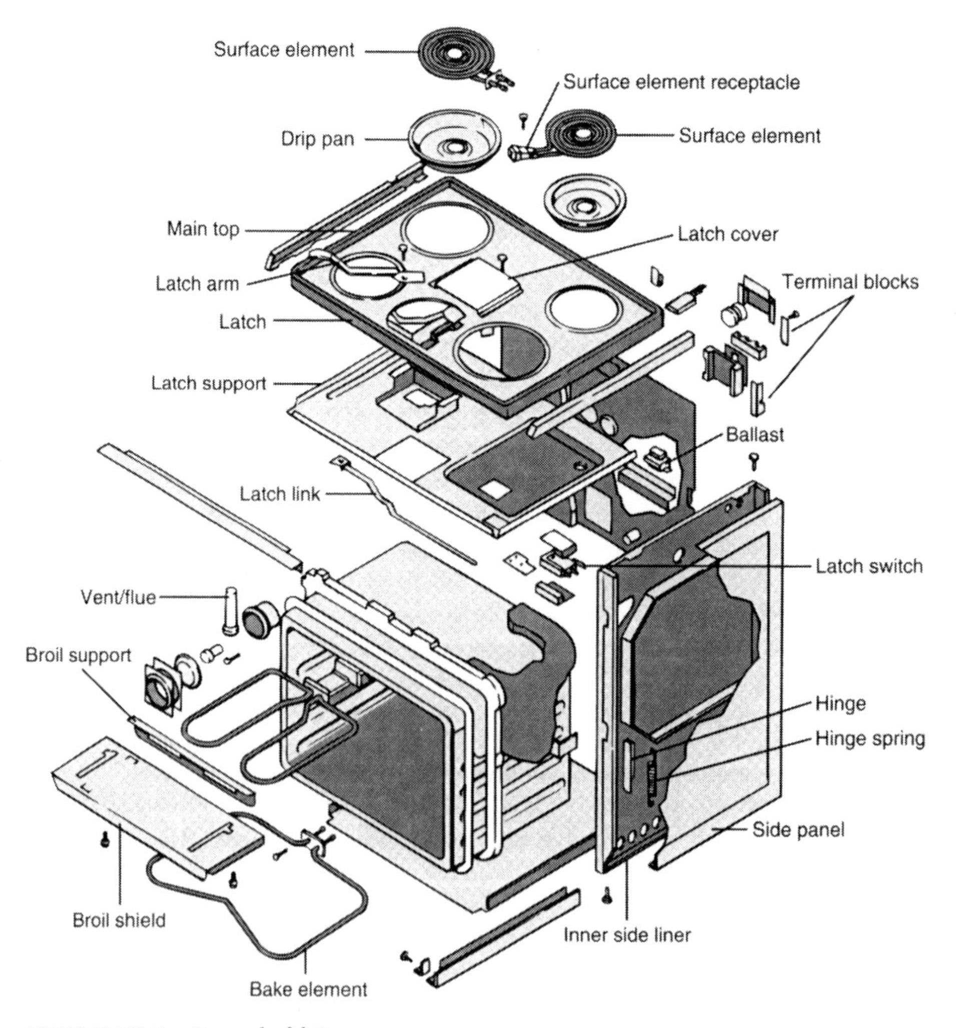

EXHIBIT 22.3 Household Range

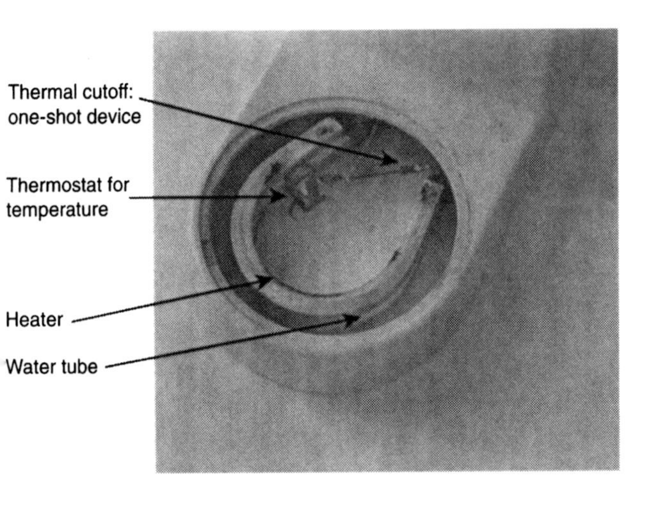

EXHIBIT 22.4
Coffeemaker Components

Coffeemakers

Coffeemaker components consist of a water reservoir, heating tube, carafe, and housing. The heating tube boils the water flowing through it from the water reservoir and forces the water over the ground coffee. The coffee then drips into the carafe. The carafe sits on a warming plate that is heated by the same resistance heating element that heats the heating tube. The resistance heater is controlled by a thermostat that cycles the flow of electricity as needed. A thermal cutoff may be employed as a protection device. Some coffeemakers include automatic timers that turn off the coffeemaker after a selected period of time and turn the appliance on at a predetermined time. Exhibit 22.4 shows a coffeemaker with the cover burner removed.

Toasters

The common toaster uses resistance heaters with a sensor that controls the toasting time. The sensor is a bimetal strip. Newer designs employ an electronic timer to set the toasting time.

Electric Can Openers

The electric can opener uses an electric motor and generally can only operate when the lever is depressed.

Refrigerators

Components contained in a household refrigerator (Exhibit 22.5) include the following:

- Evaporator
- Condenser
- Compressor
- Tubing to connect the components

Warm air inside older appliances evaporates the heat exchange medium (Freon®), turning it to a vapor and transferring the heat to the Freon from the

EXHIBIT 22.5
Refrigerator

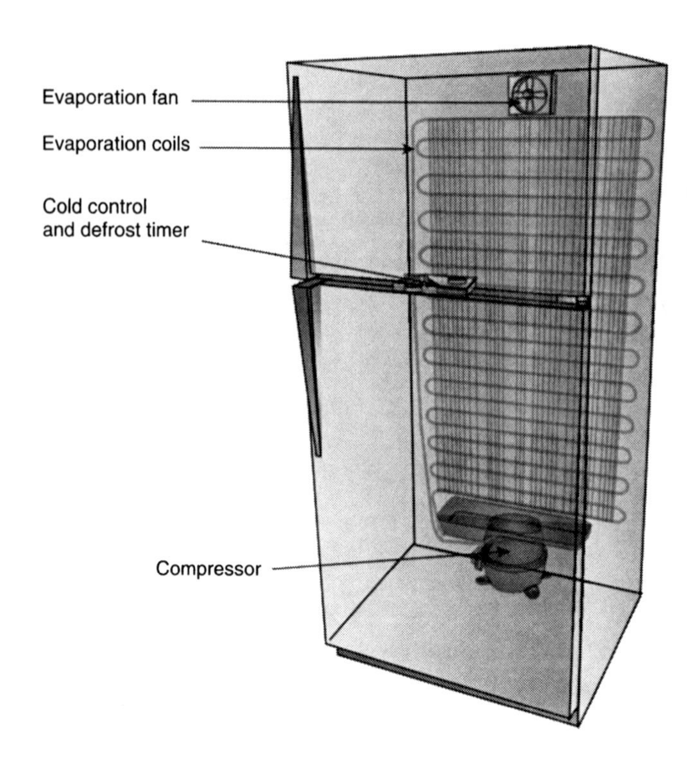

enclosure. The coolant vapor moves to the compressor, and the compressor compresses the vapor. As the coolant condenses to a liquid in the condensing coil, it gives off the heat picked up in the enclosure. The air around the evaporator is cooled, and the air around the condenser is heated. The cool air is circulated in the refrigerator, and the warm air is dissipated into the room. The cycle is controlled by a thermostat or a timing device. The compressor is typically powered by an electric motor equipped with a thermal cutoff switch contained in a sealed container that functions as a heat sink. Possible additional systems include the following:

- Lighting
- Ice maker/water fill valve
- Ice and water dispenser/water fill valve
- Fan for the condenser and possibly one for the evaporator
- May have heating coils for
  - Automatic defrosters
  - Prevention of water condensation
  - Drain pans and ice makers
- Antisweat heaters

Dishwashers

The dishwasher is a household appliance that employs a pump and a motor. The pump motor may or may not have thermal protection. An electric resistance heater may be used to dry the dishes after the water has been drained. Some fires have been caused by faults in the push-button controls that then ignite the plastic housing.

Microwave Ovens

A microwave oven uses a magnetron to generate microwave energy that heats the food. Microwaves are usually equipped with a thermal cutoff switch. The appliance is usually equipped with internal lighting and possibly a food rotation tray, and it may have thermal cutoff switches above the enclosure.

Portable Space Heaters

There are generally two kinds of portable space heaters. A convective heater employs a fan to move air across a hot surface, heating the air and dispersing it throughout the room. A radiative heater transfers heat by radiation only. These appliances often are equipped with tip-over switches or other protective devices.

Electric Blankets

An electric blanket employs an electric heating element inside of the blanket. The controls are typically located on the blanket power cord. Thermal cutoffs are located near the elements in the blanket. There could be as many as 12 to 15 thermal cutoffs. The blanket is designed to be laid flat but could char and ignite if operated while folded up.

Window Air Conditioner Units

Window air conditioner units function and operate similarly to refrigerators, and use similar components and principles for cooling. These appliances may be powered by 120 V or 240 V.

Hair Dryers

Hair dryers employ a high-speed fan to force air across a heating element to heat air for drying hair. They usually have one or more thermal cutoff switches that are resettable.

Hair Curlers (Curling Irons)

Hair curling irons use an electric resistance heater inside of a wand that is equipped with a thermal cutoff switch. Some models allow water to be added to generate steam.

Clothes Irons

Clothes irons use an electric resistance heater that can heat in both the vertical and horizontal positions and are provided with one or more thermostats or thermal cutoff switches. Water can be added to many models for generating steam.

Clothes Dryers

Clothes dryers dry clothing by circulating heated air through a rotating drum that contains the wet clothing. The air is discharged from the drum through a filter (lint trap) and into a duct that is typically discharged to the exterior of the house. The dryer uses either electricity in 120 V or 240 V or the combustion of fuel gas to heat the air. The components in a clothes dryer are timing controls, humidity sensors, heat source selectors, intensity selectors, thermal cutoffs, blower motor, and heating elements. Exhibit 22.6 shows the internal components of a clothes dryer.

Common fire causes in dryers are lint that may collect in other parts of the dryer and be ignited, frictional heating from a piece of clothing caught between

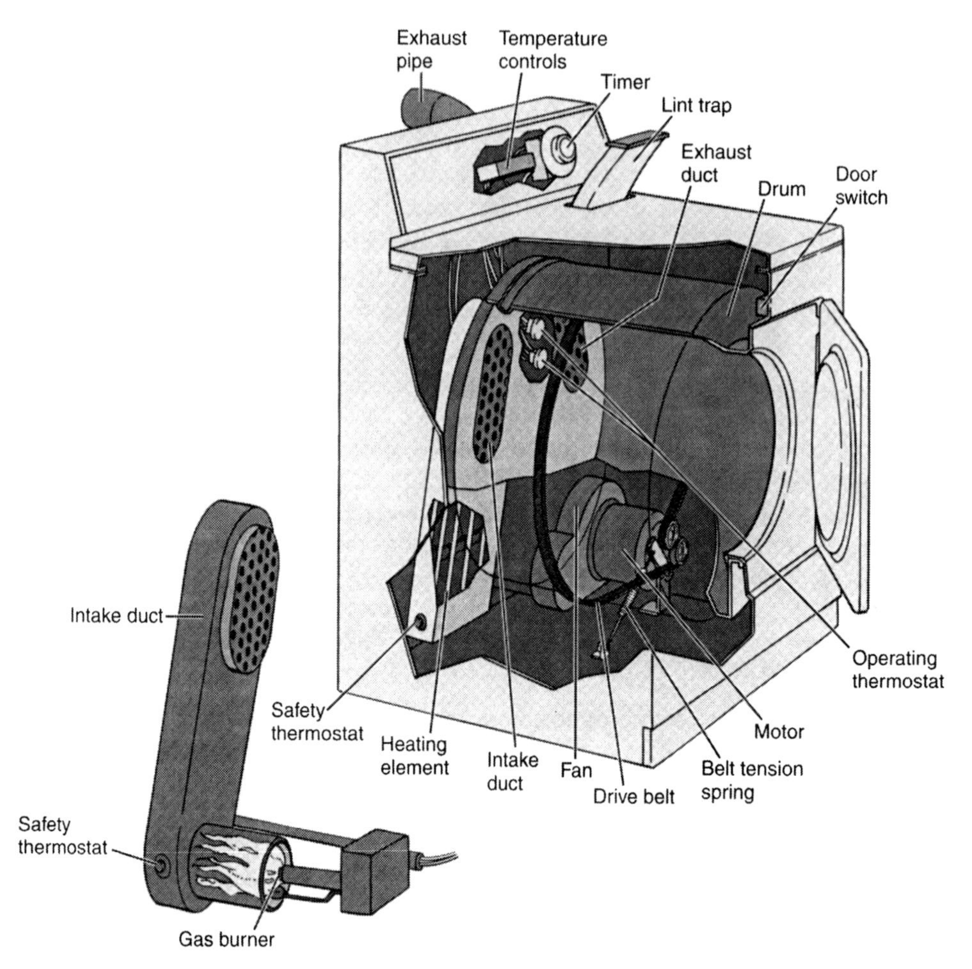

EXHIBIT 22.6 Clothes Dryer

the moving parts, and the heating of contents such as vegetable oil–soaked rags and plastic bags.

Consumer Electronics

Consumer electronics are very common household appliances. Some examples are videocassette recorders (VCRs), radios, compact disc (CD) players, DVD players, video cameras, and personal computers. All of these appliances have a power supply, circuit boards, and a housing.

ACTIVITY 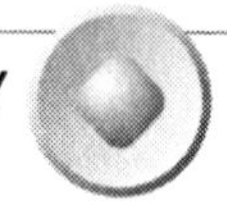

Locate and disassemble as many appliances as you can to identify the protection devices and to observe the operation of the appliance. Good sources for appliances are flea markets, garage sales, and used appliance stores. Use a volt/ohm meter to check the continuity and resistance of the appliance in different operational settings. (Refer to Chapter 6 of this *User's Manual* for a review of Ohm's Law.)

QUESTIONS FOR CHAPTER 22

1. What items should be documented if an appliance is determined to be in the area of fire origin?
   - **A.** The appliance in its original location
   - **B.** The location of the combustibles and materials found while disassembling the appliance
   - **C.** The position of all controls and other components
   - **D.** The power source for the first material ignited
2. How should an investigator obtain the manufacturer name, model number, serial number, date of manufacture, and any warnings that may be located on an appliance that was determined to be in the area of fire origin?
   - **A.** Move the appliance after all parties have been notified.
   - **B.** Attempt to find an exemplary unit to obtain the manufacturer and model.
   - **C.** Move the appliance only after it has been properly photographed and diagrammed, including the position of all the controls, power settings, and movable parts.
   - **D.** Locate the information on the front panel of the appliance or similar model.
3. True or false: The condition of a plastic appliance in and of itself is an adequate indicator of the point of origin.
   - **A.** True
   - **B.** False
4. What information can be gained by examination of damage patterns on an appliance?
   - **A.** The rate of heat release
   - **B.** The point of flashover
   - **C.** The source of ignition energy
   - **D.** The direction of fire spread
5. Once an investigator determines that an appliance is in the area of origin and is the likely cause, what is the next step?
   - **A.** Determine the first material ignited and how ignition took place.
   - **B.** Check for recalls or other safety information on the appliance.
   - **C.** Disassemble the appliance in an attempt to identify the manufacturing defect in the appliance.
   - **D.** Determine whether the appliance malfunctioned.
6. What types of connectors are found on newer model electrical cords that carry power to the appliance?
   - **A.** Three-conductor cords equipped with a ground
   - **B.** A ground conductor if the appliance has over an 18 horsepower motor or is over 2000 watts
   - **C.** One neutral conductor, one power conductor, and one ground conductor, except if they are double insulated, in which case they have only two conductors
   - **D.** Polarized cords that may have only two conductors

7. What is the function of the wider prongs on 20-amp cords made after 1987?
   A. Used on appliances with diodes to allow for ground fault protection
   B. Allow for plug-ins so that the wider blade can be used in the hot or positive receptacle
   C. Can be plugged into the neutral position on the receptacle
   D. For universal use so that the appliance can be used in homes with new receptacles
8. What can an investigator determine by examining the brass prongs (or male ends) of a cord?
   A. Intensity of temperature around the appliance
   B. If the prongs were plugged into a receptacle
   C. If the polarized plug was improperly utilized, causing the fire
   D. The amount of current flow through the appliance
9. Which statement most closely relates to the protection of appliances?
   A. Appliances with magnetic circuit breakers that experience a dead short fail to operate effectively.
   B. Small appliances generally do not have thermal protection devices.
   C. Remains of the protected device may indicate whether it was operating at the time of the incident.
   D. The fusing element is always melted following a fire and provides little information.
10. How is temperature controlled in many heat-producing appliances?
   A. Resistant-operated switches
   B. Bimetallic thermostats
   C. TCO heating element
   D. Low-voltage transformer
11. What function do thermal cutoffs provide for heat-producing appliances?
   A. Reset thermostats that are designed to reset when cooled
   B. Measure the current flow to the appliance and cut off the thermal energy when it exceeds a current demand
   C. Used in coffeemakers to divert the heat from the water reservoir tank to the warming plate
   D. Operate by melting of an internal pellet when the appliance overheats turning the unit off
12. Which statement best describes a transformer?
   A. A mechanical switch used to open contacts in order to shut the appliance off if the temperature rises above a certain level
   B. A protection device that prevents an appliance from overheating
   C. A mechanism used to reduce voltages from the normal 120 V and to isolate the rest of an appliance from the supply circuit
   D. A copper winding that can detect a defect in an appliance and cut off the circuit supply

13. What specific type of protective device are motors generally equipped with in countertop appliances?

A. A thermal protector that may reset when the motor cools
B. Copper motor windings to reduce circuit supply
C. Internal overcurrent protection
D. Aluminum fan propellers that keep the motor cool

14. With what are frost-free refrigerators equipped to keep them from "sweating"?

A. Dehumidifiers that monitor the humidity in the unit
B. Interior surfaces constructed of polyethylene that do not allow the accumulation of ice or moisture
C. Air ventilation systems that cause humidity to condense at the bottom part of the refrigerator
D. Heaters located just under the interior faces

15. What is the function of a ballast on a fluorescent light fixture?

A. It transforms the voltage from normal household voltage to a higher voltage.
B. It acts as a thermal cutoff device that cannot be reset.
C. It reduces the amount of current flow when it senses higher temperatures.
D. It is an electronic device that does not have the failure potential of transformers.

Motor Vehicle Fires

▶▶▶ OBJECTIVES

Upon completion of Chapter 23, the user will be able to

- Define the investigative techniques used to analyze a vehicle fire
- Discuss the various vehicle systems and components with relation to fire cause
- Outline the methodology used in the examination of a vehicle fire

The investigation of vehicle fires is often complex and difficult. The fire investigator should be able to recognize the burn and char indicators, which vary from vehicle to vehicle. The investigator should also have a good understanding of various fuel characteristics within the vehicle, fire behavior associated with those fuels, and a general knowledge of the working mechanical components. This chapter discusses the various aspects of vehicle systems and methodologies that assist the investigator in arriving at a successful conclusion.

INVESTIGATIVE TECHNIQUES

Before conducting any vehicle fire investigation, the investigator should prepare himself or herself by researching the various aspects of the vehicle. The investigator should be aware of the vehicle's safety features; have a general knowledge of the type and use of the vehicle, including its motor and components; and have a plan for conducting the investigation. Table 23.1 provides a suggested method in planning the investigation.

VEHICLE FIRE FUELS

Fuels in vehicle fires fall into three categories: liquid fuels, gaseous fuels, and solid fuels.

TABLE 23.1 Planning the Investigation of a Motor Vehicle Fire

Safety Concerns	*Qualifications*	*Vehicle Examination*
Before viewing the vehicle, in any position: • Ensure its stability • Check that the electrical system on the vehicle is disconnected • Be aware of undeployed air bags and bumpers. (Sodium azide, the expelling agent for the air bag, is a serious safety hazard.) • Check for fuel and other fluid leaks that may pose a fire hazard • Take the necessary steps to neutralize hazards • Use proper protective clothing and equipment	Perform an analysis of the vehicle systems: • Know the specific function of each system • Determine whether a system malfunctioned • Determine whether a system has been altered • Determine whether a malfunction or alteration could be responsible for the fire • Preserve evidence for future laboratory examination If you are not qualified to perform an analysis of the vehicle systems, seek guidance from a qualified individual.	Differentiate three major compartments: • Engine compartment • Passenger compartment or interior • Cargo compartment Differentiate ignition scenarios: • Electrical • Mechanical • Human intervention (actions or inactions) • Determine an area of fire origin

Liquid Fuels

Liquid fuels include engine fuels (gasoline, diesel), transmission fluid, power steering fluid, and brake fluid. One or more of these fuels are present in most vehicles. The ignition potential of these fuels depends upon the properties of each fuel, the physical state of the fuel (liquid, atomized, or spray form), and the nature of the ignition source.

Gaseous Fuels

Gaseous fuels for motor vehicles are commonly propane or natural gas. These gaseous fuels not only represent alternative motor fuels, but they are also used in recreational vehicles for heating, cooking, and refrigeration. Other gaseous fuels that the investigator might encounter are hydrogen and oxygen, which are used in newer hybrid vehicles. Larger quantities of the various gaseous fuels can be found in larger vehicles or as cargo.

Solid Fuels

Solid fuels encompass any fuel within or on a vehicle that is not liquid or gaseous in its normal state. Although this is less common in accidental fires, solid fuels may be the first material ignited. Examples include wiring insulation reaching its ignition temperature due to resistive heating of the conductors, seat materials becoming ignited by improperly discarded or misused smoking materials, or heat from friction causing drive belts, bearings, and tires to ignite.

Solid fuels may contribute significantly to the fire spread. Solid fuels in vehicles also include plastics that have heat release rates similar to those of ignitible liquids. Plastics can sag and drop flaming pieces. Aluminum and magnesium or their alloys found in engine and vehicle components can ignite and provide an additional fuel load. Most metals need to be in powder or melted form to burn.

However, solid magnesium, which is present in some vehicles, burns vigorously once it is ignited by a competent external heat source.

The presence of melted metals in a vehicle is not necessarily indicative of the presence of an ignitible liquid accelerant. Many of the alloys or metals found in vehicles are not in a pure form or state and therefore have melting temperatures lower than those given on charts and graphs that show such metal in their pure form. When one type of metal falls or melts onto another, the process of "alloying" can occur. For instance, melted aluminum could run or drip onto another metal, causing it to melt at a lower-than-normal temperature.

Investigators should not interpret the presence of melted metals to be an indicator of the use of an ignitible liquid in the belief that only an ignitible liquid can produce temperatures high enough to melt metals.

IGNITION SOURCES

In most cases, the sources of ignition energy in vehicles are the same as those associated with structure fires. Open flames, mechanical and electrical failures, and discarded smoking materials can often lead to a vehicle fire.

Open Flames

Open flames were a common source of fires in a carbureted vehicle when the engine backfired through the carburetor. It is important to note that if the air cleaner is in place, propagation is unlikely. It is important to realize that fuel injection systems have replaced carburetors in most of today's modern vehicles.

Open flames can occur in ashtrays from a match and could ignite combustibles. Open flames in recreational vehicles include appliance pilot lights, burners, and ovens.

Electrical Sources

The primary source of electrical energy in a vehicle that is not running is the battery. During the investigation, it is important to remember that some vehicles have dual batteries. Some components that can remain energized when the vehicle is not operating include headlights, power seats, and interior lights.

The investigator should determine whether the vehicle was running at the time of the fire to determine what potential electrical ignition sources were energized.

Fuses, circuit breakers, or fusible links (Exhibit 23.1) are designed to provide protection for electrical circuits in most vehicles. Vehicle electrical systems are often direct current (dc) with the negative (ground) side, of the system connected to the body, frame, and engine. The positive side of the electrical system provides current via the electrical wiring to devices such as starters, alternators, and radios. Any time a positive lead goes to ground, there is the potential for a fire.

The inside cover of most circuit breaker panels lists the equipment protected and the size of the fuse (Exhibit 23.2). It is important to view the fuse panel to determine whether a problem exists with an electrical component. The vehicle owner might have elected to overload the circuit by using a larger-rated circuit breaker.

Recreational vehicles (RVs) may have alternating current (ac) circuits as well as dc circuits (see Exhibit 23.3). They may have a converter to convert ac to dc.

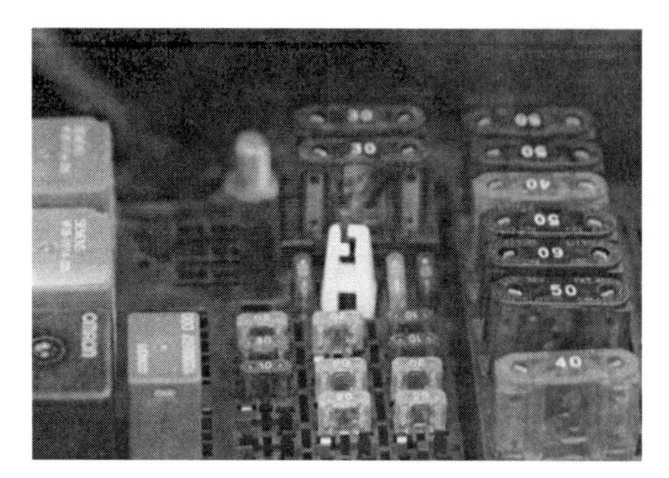

EXHIBIT 23.1 Circuit Breaker/Fuse Relay Panel

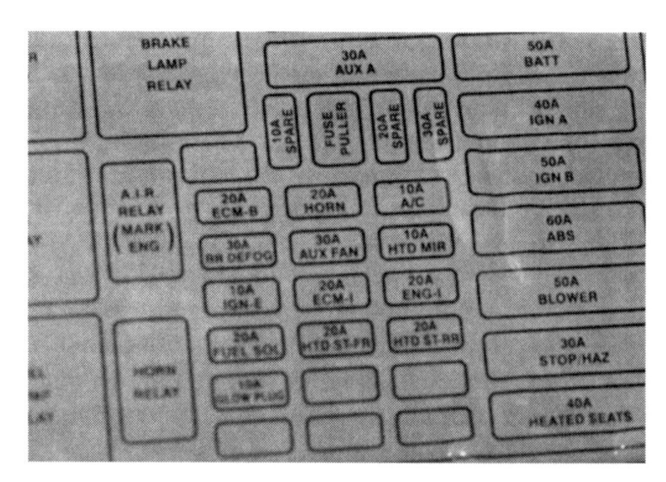

EXHIBIT 23.2 Inside Cover of a Circuit Breaker

Two or more batteries or an auxiliary battery as a secondary power source are common as is an onboard generator.

Another means by which electrical energy might be an ignition source is through resistance heating, which might be associated with overloaded wiring. Aftermarket equipment can overload the vehicle's electrical systems. Aftermarket items include stereo systems and associated hardware as well as different or enhanced vehicle lighting (running lights, spotlights, fog lights) or even radar detectors. On occasion, aftermarket equipment might lack overcurrent protection. A high-resistance fault in the electrical wiring can raise the temperature of the wiring insulation to its ignition point without activating the circuit protection. Faults in high-current devices can ignite readily available combustible materials. (See Exhibit 23.4.)

Deteriorated or worn insulation can produce electrical arcing when a charged wire comes into contact with a grounded surface. Crushed or cut wires and crushed batteries can be a source of electrical arcing. It is important to remember that the battery and starter cables might not be electrically protected (by fuses, or overcurrent protection) and that they carry a large amount of current. If these

EXHIBIT 23.3 RV with More Than One Battery for Auxiliary System

EXHIBIT 23.4 Fire Damage from Overloaded Circuit

lines are severed during a crash, arcing can occur and result in a fire. Wiring that is subject to chafing may become devoid of insulation, allowing unintended contact with other metals, thereby creating sparks or resistive heating.

Lamp filaments of broken bulbs are a potential ignition source if the fuel is in a gas, vapor, or liquid spray form. Most lamp filaments are designed to function in a vacuum. The operating temperature of the exposed filament can be high as 2550°F (1399°C), but when a filament is broken and exposed to air, it operates for only a few seconds and then burns away, removing the ignition source. Exhibit 23.5 illustrates a brake lamp with the filament intact.

During the investigation of a vehicle fire, the investigator should look for evidence of external electrical ignition sources. This could include ports for recreational connections, engine heaters, and battery chargers. Battery chargers can provide an additional electrical source to consider or eliminate during the investigation.

EXHIBIT 23.5 Brake Lamp with Filament Intact

Hot Surfaces

Exhaust system components can provide sufficient temperature to ignite diesel spray and to vaporize gasoline. If a transmission is overloaded, fluid can be forced out the dipstick tube and can ignite on contact with heated exhaust system components.

Engine oil and brake fluid can also ignite when in contact with heated exhaust system components. These liquids usually ignite soon after the engine is shut off, when the temperature of exhaust system components can increase. The ignition temperatures for fluids on hot surfaces is, in general terms, approximately 360°F (182°C) above their given ignition temperatures. The following factors all affect ignition of liquids by a hot surface:

- Ventilation
- Liquid flash point
- Liquid boiling point
- Liquid vapor pressure
- Liquid vaporization rate
- Misting of the liquid, roughness of the hot surface
- Length of exposure of the liquid to the heated surface

Other combustible materials, such as plastics, papers, and vegetation, can reach their ignition temperature when in contact with heated vehicle components.

Mechanical Sparks

Mechanical sparks occur from metal-to-metal contact and can be associated with broken drive pulleys or bearings. Metal-to-pavement contact (e.g., a broken drive shaft, tire rim, or exhaust pipe) can also produce sparks. The vehicle must be running or moving to produce such an ignition source. Vehicle speeds of 5 mph (8 kmh) have been determined to create orange sparks with a temperature of 1470°F (800°C), and greater speeds have created white sparks with a temperature of 2190°F (1200°C). Aluminum-to-pavement sparks are not considered a competent ignition source for most materials. Usually, these types of sparks will not ignite solids. Consideration has to be given to the adjacent fuels around the area from which the sparks are emanating and the amount of time the sparks are in contact with these fuels. If gasoline liquids or atomized liquids are present when the sparks are produced, ignition of these fuels can occur.

Many types of farm equipment have mechanical attachments that can also ignite crops and grass if they are not properly maintained. Highway departments and cities have weed abatement programs to mow and cut potential grass hazards around roads and freeways. These programs often use flail mowers, such as the one shown in Exhibit 23.6. Flail mowers are metal blades attached to a drive unit and pulled by a tractor. Flail mowers have been responsible for a number of fires under the right conditions. When the humidity is low and fuels are dry, mechanical sparks can occur when the metal blades hit a rock. The mechanically produced sparks have the potential energy to ignite surrounding fuels under these conditions. Exhibit 23.7 shows a close view of one type of flail mower. Centrifugal force of the rotating blades in this flail mower can strike a rock or object, causing a shower of sparks.

EXHIBIT 23.6 Flail Mower

EXHIBIT 23.7 Close View of Flail Mower

Smoking Materials

Another ignition source identified in vehicle fires is improperly discarded or misused smoking materials. In the hypothesis for this ignition scenario, the ignition sequence for discarded smoking materials centers around the vehicle's ability to confine and retain the heat of the smoking material, and the ventilation aspects of the vehicle. Ignition of nearby fuels such as papers or clothing that are in the vehicle should be considered. Once ignition of these materials occurs, the fire often spreads to other fuels, including the foam materials in the seat.

SYSTEMS AND THEIR FUNCTIONS IN A VEHICLE

The importance of knowing how the various systems in vehicles operate is critical for the fire investigator and can be enhanced by viewing various technical manuals available at the library, car dealerships, and parts suppliers. Quite often, familiarity with a particular vehicle system can be gained by visiting a dealership or used car lot and asking for specific information about the vehicle from the service manager or mechanic. The parts department might also have schematics and views of particular systems.

Fuel Systems

Two basic fuel systems are used in modern vehicles: vacuum/low-pressure carbureted systems and high-pressure fuel-injection systems. Both take fuel from the gas tank through the use of a fuel pump and deliver the fuel through a fuel line and filter system to the engine, where the fuel is atomized and burned.

Vacuum/Low-Pressure Carbureted Systems. Vacuum/low-pressure carbureted systems draw fuel from the fuel tank, usually by means of a mechanical pump attached to the engine, and operate only when the engine is operating. Electric fuel pumps are sometimes used. The fuel is then pumped into the carburetor at a pressure of 3 to 5 psi (20.7 to 34.5 kPa), where it is mixed with air in the carburetor (usually at a ratio of 15:1) and drawn into the engine via the intake manifold and

EXHIBIT 23.8
Vehicle Fuel System and Associated Components

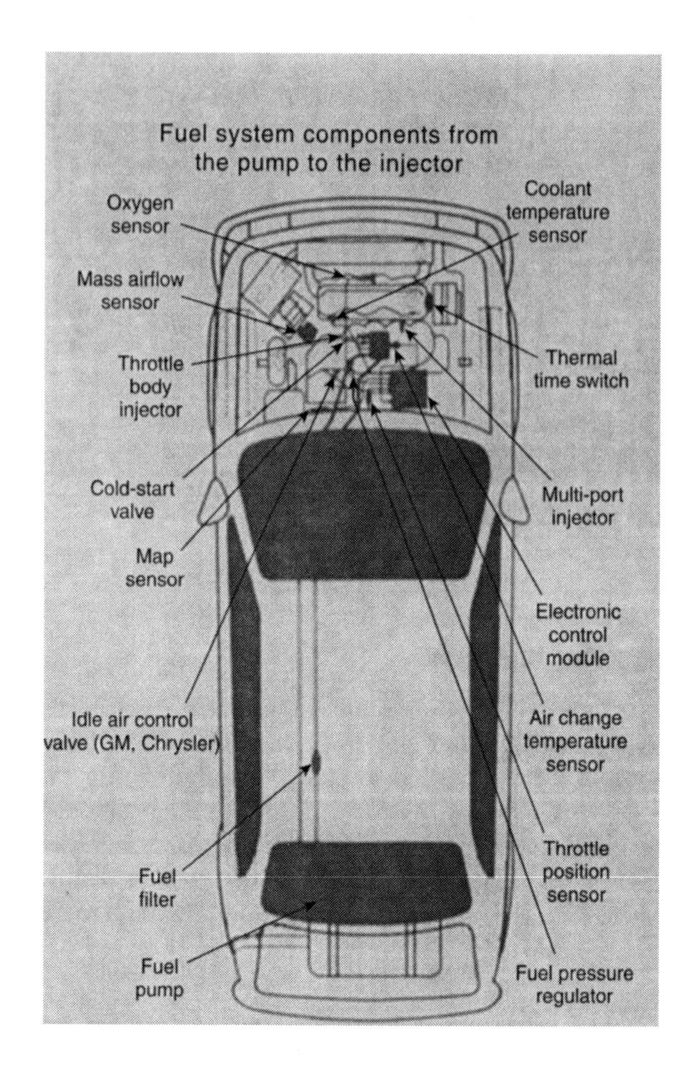

into the cylinder, where combustion takes place. Exhibit 23.8 shows a generic view of a vehicle fuel system and its associated components.

Potential problems with vacuum/low-pressure carbureted systems are outlined in Table 23.2.

High-Pressure Fuel-Injected Systems. High-pressure fuel-injected systems also retrieve the fuel from the fuel tank and pump the fuel to the engine via an electric fuel pump, which is energized whenever there is a key in the ignition in the "on" or "run" position.

TABLE 23.2 Vacuum/Low-Pressure Carbureted System Problems

Location of a Leak	*Problem*
Vacuum side	Air is drawn into the system, and the engine will not run.
Pressure side	Fluids can leak as a fine mist or a heavy stream. Ignition is possible if here is an ignition source. If the fuel line leaks as a result of a fire, the leaking fuel vapors can be ignited.

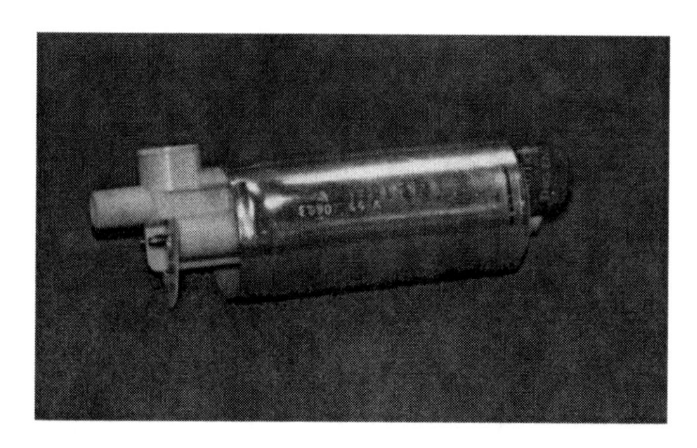

EXHIBIT 23.9
Electric Fuel Pump

Some of the fuel pumps on modern cars are integrated with the on-board computer, which can trigger an inertia switch in a crash that could disable the fuel pump. The fuel is pumped at 35 to 70 psi (241.3 to 482.6 kPa) in most vehicles with fuel injection systems. Exhibit 23.9 depicts an electric fuel pump used in most new vehicles with fuel injection systems. The fuel is pumped to either a single venturi-mounted fuel injector (throttle body) or a fuel rail assembly on the engine. In both systems, the excess fuel is pumped back to the fuel tank.

Potential problems with high-pressure fuel-injected systems may include leaks under pressure at various fittings. In some vehicles, the pressure developed from a leak may be sufficient to propel the fuel several feet.

The horizontal aluminum piping in Exhibit 23.10 depicts the fuel rail on this vehicle. The fuel injector shown in Exhibit 23.11 is attached to the fuel rail. Failures can and do occur due to age and type of materials used for the connection between the rail and the supply line. (See Exhibit 23.12.)

When being interviewed, the vehicle owner might report fuel-related problems that may assist the investigator in determining an event that led to a fire. A leak on the supply side of the system or problems with the operation of the vehicle should be noted by the operator. Starting difficulty, erratic operation, and stalling all might be problems that were noticed. A leak on the return side of the system can go undetected with no operational problems noted. Fuel lines that enter the tank on the bottom can feed fuel to a fire by gravity once they are

EXHIBIT 23.10 Fuel Rail

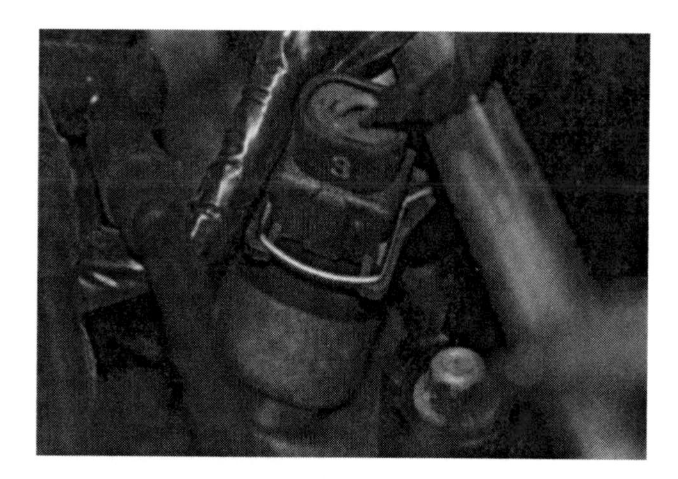

EXHIBIT 23.11 Fuel Injector

EXHIBIT 23.12
Connection Between Rail and Supply Line

compromised. Residual pressure in the tank or pressure that is created by the fire could force fuel out of a compromised fuel line. Table 23.3 can assist in testing a hypothesis in a suspected fuel-related fire.

Diesel Fuel Systems. Diesel fuel systems typically use two pumps to deliver the fuel from the tank to the engine with lift pumps that operate under high volume but with low pressure. The diesel fuel is pumped from the tank to a fuel injector pump near the engine, generally with a high-volume, low-pressure pump. The diesel fuel injector pump then pumps the fuel into the cylinders. Combustion air is provided by natural aspiration or through a turbocharger.

Fuel leaks can occur over time due to the vibrations of a diesel engine. If any diesel fuel comes into contact with a hot surface, the diesel fuel may ignite. Another potential, although rare, problem associated with diesel engines, is a

TABLE 23.3 Fuel-Related Problems

Location of a Leak	*Problem Noted*
Supply side	Operating problems will be noted. Unintended ignition may occur.
Return side	No operating problems will be noted. Unintended ignition may occur.

"runaway" engine in which combustible air is drawn in through the air intake system, causing the engine to maintain a very high rpm level. This high rpm level causes the engine to tear itself apart or possibly explode in a fireball.

Natural and Propane Fuel Systems. Natural and propane fuel systems are stored as liquids under very high pressures, which are reduced by a regulator at or near the engine. The fuel flows through a regulator and into the carburetor, where the fuel is mixed with air before entering the engine. Leaks found postfire might not be indicative of leaks prefire partly because of the different coefficients of expansion between the fittings and piping materials. If a leak does occur, fugitive gases can find an ignition source and flash back to that source. In some cases, the burning fuel can cause the fuel tank to rupture during a fire. Recreational vehicles (RVs) often have natural gas or LPG systems for heating, refrigeration, and cooking.

Turbochargers. Turbochargers can be found on gasoline and diesel-powered engines. The turbocharger increases the power to the engine by forcing air into the cylinder. The turbocharger rotates at up to 100,000 rpm. It also uses exhaust gases for propulsion. Turbochargers provide one of the hottest points on the engine (1500°F/815°C and up). If a fuel leak or oil leak occurs, the turbocharger can ignite readily.

The turbocharger shown in Exhibit 23.13 is from a 52-ft (15.8-m) offshore power boat. A fuel line was compromised, sending atomized fuel in the direction of this turbocharger and causing a flash fire that was extinguished by the vessel's halon extinguishing system.

Emission Control Systems

Emission control systems are found on most vehicles today. These systems are designed to reduce or control gases and collect gasoline vapors while the engine is operating. The emission control systems regulate the fuel input, timing adjustments, and the recirculation of exhaust gases through the EGR valve. A charcoal canister collects excess fuel vapors.

Problems that can occur in the emission control system include vapor leakage from the hoses or charcoal canister or overfilling of the fuel tank, which forces liquid into the charcoal canister. This could result in rough idling, stalls, backfires, and overheating of the catalytic converter. (Refer back to Exhibit 23.8.)

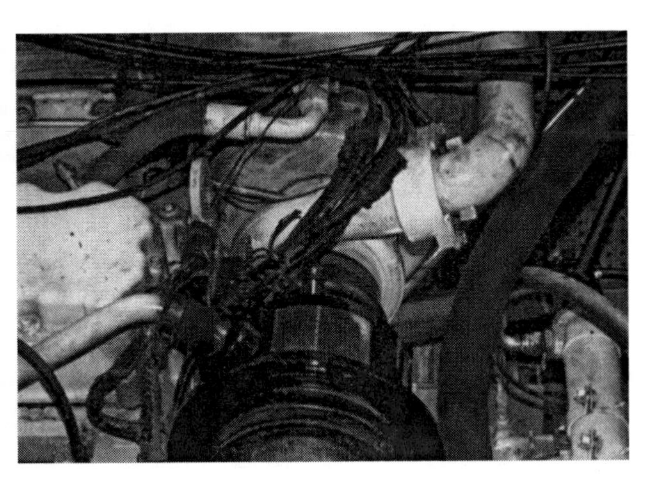

EXHIBIT 23.13
Turbocharger

Exhaust Systems

Starting at the engine, exhaust system components include an exhaust manifold that is connected directly to the engine, exhaust piping, a catalytic converter, and a muffler. During normal operation of a properly maintained and operating vehicle, the temperatures that enter the catalytic converter can measure 650°F (343°C), and it is normally the hottest part of the exhaust system. An improperly operating engine can raise the temperature of the exhaust system enough to ignite undercoating and interior carpeting. A hot exhaust system can ignite grass and other external combustibles.

Many states require construction tractors and agriculture vehicles to be equipped with spark arrestors (Exhibit 23.14). Carbon particles from the exhaust system of farm and construction vehicles have sufficient energy to ignite dry grass and other light fuels.

Motor Vehicle Electrical Systems

The storage battery is the primary energy source in a vehicle. Once the ignition key is turned on, the starter draws energy from the battery to start the engine. Once the engine is running, the alternator provides electrical power to the vehicle. The vehicle battery is normally a 12-V, dc negative ground system. The conductor size determines the amount of electrical energy the electrical system can safely carry. Simply put, the bigger the conductor, the more electrical energy the system can carry. The largest conductors extend from the battery to the alternator

EXHIBIT 23.14
Tractor Spark with Spark Arrestors

and to a grounding location and are rarely protected by overcurrent protection equipment. Arcing on this conductor is commonly found following a fire in the vehicle. The battery can also be a fuel source. The battery can generate hydrogen gas and ignite easily if an ignition source is present. Small amounts of hydrogen and oxygen gases might be found in sealed no maintenance batteries, and these gases could be released if the battery is damaged.

12-Volt Electrical Systems. Electrical energy throughout the vehicle comes through the positive leg of the electrical system and is routed through a busbar with overcurrent protection in the form of circuit breakers or fuses. The electrical system for the vehicle is grounded through the frame of the vehicle. Most of the electrical devices in a vehicle have only one conductor (positive). The rest of the circuit is completed through the car frame or body.

Only a few of the electrical circuits have power when the engine is shut off. Those with power may include circuits from the battery to the starter, from the battery to the ignition switch, and to power seats. Circuits from the ignition switch to the accessories, such as the clock and cigarette lighter, on-board computers, and aftermarket accessories, may also be powered with the ignition off. To determine whether the vehicle was running at the time of the fire, the engine control fuses should be inspected. If none of the fuses have opened, the engine might not have been operating.

Wiring diagrams typically show the color of conductor insulation, the device location, and its function. Most vehicle wiring is stranded and can become brittle when heated. The investigator should document the wiring before moving it.

Other Electrical Systems. Electrical systems on some older vehicles have +6-V systems. Some can have positive grounding systems. Large trucks, buses, and heavy equipment may have +24-V systems.

Recreational vehicles can have electrical systems that have a combination +12 V dc and 120 V ac. Converters are used to change voltage from 120 V ac to 12 V dc. Inverters change power from 12 V dc to 120 V ac. Many motor homes may have on-board generators that produce 120 V ac.

Mechanical Power Systems

Lubrication Systems. Most vehicle engines use hydrocarbon-based oils for lubrication. The oil is pumped through the engine at 35 to 60 psi (241 to 413 kPa). An oil leak can occur in a number of locations within the engine compartment. Although rare, if an oil leak occurs, the oil vapors are susceptible to ignite given the proper circumstances. Lack of oil in the engine can also cause a fire if the engine fails.

Liquid Cooling Systems. Most vehicle engines contain a liquid coolant, which is circulated through the engine by a water pump. Generally, the coolant is 50 percent ethylene glycol and 50 percent water. Operating temperatures of the coolant generally range from 180°F to 195°F (82°C to 91°C) with an operating pressure of 16 psi (110 kPa).

Air-Cooled Systems. Air-cooled systems are uncommon in most vehicles. An air-cooled system circulates the air through the engine by a fan and ductwork. Normally, the fans are belt driven, and if the belt breaks, engine failure can occur.

Mechanical Power Distribution

Mechanical power produced by the engine must be delivered to the drive wheels for the vehicle to be put in motion. This power distribution or transfer is accomplished through a transmission that requires lubrication.

Mechanically Geared Transmissions. Mechanically geared transmissions are the means of power transfer from the engine through the clutch assembly (which is located between the engine and the transmission) to the vehicle wheels. Most transmissions contain a transmission fluid for lubrication and the movement of internal valves. If a leak occurs in the vicinity of heated exhaust system components or another suitable ignition source, a fire can occur.

Hydraulically Geared Transmissions. Hydraulically geared or automatic transmissions have a transmission fluid that is cooled by routing through a cooler in the radiator. The fluid can be added to the transmission through the tube provided for the transmission fluid level indicator (dipstick). If the transmission is overfilled or overtaxed while the vehicle is being driven or a load is being pulled that is beyond what the vehicle was designed for, fluid can be expelled from the transmission through the fluid level indicator opening or tube and subsequently result in a fire. If a transmission is filled with the incorrect type of transmission fluid, the fluid can expand and escape through the fluid level indicator opening and onto the exhaust system.

Accessories, Braking, and Windshield Washer Systems

Other accessories attached to the engine that can fail include air conditioning compressors, power steering pumps, air pumps, and vacuum pumps.

Hydraulic brake systems operate under high pressure, and if a leak occurs, it can generate a spray of brake fluid in the form of ignitible vapors.

Windshield washer systems contain a fluid that typically has a high alcohol content. If the spray is focused on a hot surface, an ignitible vapor can form. The investigator should determine the condition of the reservoir to ascertain whether fluid leaked. Some windshield systems use nozzle heaters, which can cause a fire to occur.

BODY SYSTEMS

Many panels in modern vehicles are made of combustible materials. These panels can contribute to the fuel load once a fire has started. Combustible metals are used in some panels, and these can burn intensely. The bulkhead can have a number of penetrations for plastic ductwork that burn in a fire and provide an avenue of fire spread from one compartment to another.

Interior Finishes and Accessories

Most of the interior seats and padding provide a significant fuel load in a vehicle fire. During your investigation of a vehicle fire the investigator should document any existing or missing accessories within the vehicle passenger compartment.

Cargo Areas

It is important to determine whether the fire started in a cargo area or spread to it from another location.

The contents of the cargo area should be inventoried. The spare tire should be examined and the depth of the tire tread noted. The investigator should compare this information with the owner statements.

RECORDING MOTOR VEHICLE FIRE SCENES

It is advantageous to document the vehicle at the fire scene using the same procedures used for documenting a structural fire scene. The investigator should start with the documentation of the area where the fire occurred and then focus on the vehicle.

Vehicle Identification

Vehicle identification begins with the make, model, and year as well as other identifying features, including the vehicle identification number (VIN). (See Exhibit 23.15.) The VIN is commonly found on the dashboard in front of the driver's position but can be stamped on the engine side of the bulkhead (also known as the firewall). Sometimes the VIN will be located on a label affixed to the driver's door pillar or on the engine block. With a VIN in hand, the investigator can obtain information about the manufacturer, country of origin, body style, engine type, model year, assembly plant, and production number.

Vehicle Fire Scene History

It is advisable to determine the actions and events that led up to the fire by conducting interviews with the person(s) who first observed the fire. The owner, passengers, and operator of the vehicle at the time of the fire or the last person to drive the vehicle; fire department personnel; and police officers might also provide pertinent information. A list of information that can be determined through interviews is contained in Paragraphs 13.4.1 through 13.4.3 of NFPA 921.

Vehicle Particulars

To ensure that the investigator is not missing any information, it could be worthwhile to inspect a similar (exemplar) vehicle. The checklist in Paragraph 25.7.3 of NFPA 921 can be used for assistance in gathering information.

Information about fire causes in vehicles of the same make, model, and year could be important to the investigation. This information can be obtained from

EXHIBIT 23.15 Vehicle Identification Number

the National Highway Traffic Safety Administration, the Insurance Institute for Highway Safety, and the Center for Auto Safety.

Recording at the Scene

The investigator should also record the fire scene by making a scene diagram. The scene should be photographed, as should surrounding buildings, highway structures, vegetation, other vehicles, tire and foot impressions, fire damage, signs of fuel discharge, and any parts or debris.

The vehicle should be photographed in a systematic manner similar to that for a structure fire. The investigator should begin on the outside, documenting all surfaces (including top and underside, if possible), damaged and undamaged areas, tires and tire tread depth. Document the engine compartment, taking overview photographs of all sides and focusing on specific engine areas and components. The passenger area should be photographed from driver's side to passenger's side, then reversing the documentation with different angles, front to rear, overhead and floor. The investigator should document the paths of fire spread into or out of compartments and cargo spaces. Photos of the cargo space should include photographs of the spare tire and any special equipment such as stereo gear or add-on devices when necessary. The investigator should also document the vehicle removal process when possible.

Recording Away from the Scene

It is always advisable to visit the scene even if the vehicle has been removed. The investigator should secure any background information, including the date and time of loss, the location of loss, the name of the operator, names of any passengers, witness statements, police and fire department reports, the vehicle's current location, and the method of transportation. Documentation should be made of any parts that might be missing or damaged when the vehicle is inspected after it has been moved from the scene. If possible, the vehicle should be protected from the elements.

MOTOR VEHICLE EXAMINATIONS

The examination of a vehicle fire is a tedious and often complex task that can take longer than investigation of a simple house fire. The first step in the investigation is to determine an area of fire origin. As with structural fire investigations, the investigator should work from the area of least damage to the area of greatest damage.

The investigation process of vehicles can be divided into three compartments: the engine compartment, passenger and driver area, and the cargo space.

One of the most significant indicators of what compartment was first involved is to observe how the windshield reacted to the fire. Table 23.4 provides the investigator with an overview of windshield indicators.

Examination of Vehicle Systems

The inspection of gas-fueled vehicle systems should be systematic, from outside to inside, starting with the gas tank and working forward to the filler pipe and fuel cap. The fuel supply lines and vapor return lines should be documented in the process. The fuel lines should be inspected if they are in the immediate area of the catalytic converter or exhaust manifold. The investigator should note all signs of fire-related damage or rupture if they are present. He or she should complete

TABLE 23.4 Windshield Indicators of First Involvement

Compartment	*Indicators*
Passenger	Top of the windshield fails Radial burn patterns on the hood
Engine	Bottom of the windshield fails Radial patterns on the doors

the inspection and documentation of the fuel lines as they enter the fuel rail or carburetor system and, if applicable, the fuel lines to the individual injectors for each cylinder.

Switches, Handles, and Levers

During inspection of the interior compartment, the investigator should inspect the position of switches, handles, and levers if possible. The ignition switch should also be checked to see whether the key is present or absent. If this area is severely damaged, the investigator can check the floor below for any signs of keys or the ignition switch. He or she should ascertain the position of the windows prior to the fire, and determine what position the transmission gearshift lever was.

Exhibit 23.16 shows an electrical failure in the steering column. Ignition switch fires are also known to occur.

TOTAL BURNS

Total burns are challenging fires to investigate. The investigator should determine the condition of the vehicle prior to the fire and whether any components were missing. He or she should be sure to look at the floorboards for the presence of ignitible liquids. It is also advisable to analyze the fluid level in the various systems

EXHIBIT 23.16 Results of Electrical Failure in Steering Column

and take samples. An engine oil analysis can indicate the condition of the engine prior to the fire. Laboratory analysis of the engine oil filter can provide information regarding the condition of the vehicle engine at the time of the fire. Similarly, the prefire condition of the transmission can frequently be determined by analysis of the transmission fluid. Exhibit 23.17 illustrates the results of a total burn.

STOLEN VEHICLES

History has shown that the chances of an accidental fire after a vehicle is stolen are low. Often, vehicles are stolen for parts and are burned to cover the crime. Parts that may have been removed include the wheels, major body panels, engines, transmissions, air bags, stereos, and passenger seats. Sometimes the vehicle was stolen to conceal another crime.

Through debris sifting and inspecting, it could be possible to recover the ignition switch locking tumblers. Forensic laboratory inspection of the tumblers can indicate whether the ignition switch was defeated or whether a key was used to last operate the vehicle.

VEHICLES IN STRUCTURES

When a vehicle is located in a structure where a fire has occurred, the vehicle should be considered a potential ignition source until it is properly eliminated. The vehicle might be under structural debris that has to be removed before inspection. An external fire could have caused the fuel in the fuel tank to be released and to become involved in the fire. This possibility should be considered during the investigation.

RECREATIONAL VEHICLES

Recreational vehicles are similar to houses and mobile homes in terms of contents and materials. Some have plywood flooring, carpets with foam padding, wall paneling, and polyurethane foam furniture. As part of an investigation, the investigator should determine the type and condition of the appliances. Care should also be

EXHIBIT 23.17 Results of Total Burn

used in removing components because of close installation tolerances. Assistance from an electrical or mechanical engineer could be indicated in the evaluation of a recreational vehicle's heating and electrical systems if a particular system is suspected of having been the ignition source or there is a need to eliminate a system in a suspected arson.

Exhibit 23.17 depicts a total loss to a large motor home. On completion of the fire scene investigation, two separate and distinct points of fire origin were determined. One area of fire origin was in the front driver's position (Exhibit 23.18), and the second area was located at the rear of the motor home (Exhibit 23.19).

EXHIBIT 23.18 Fire Origin in Front Driver's Position

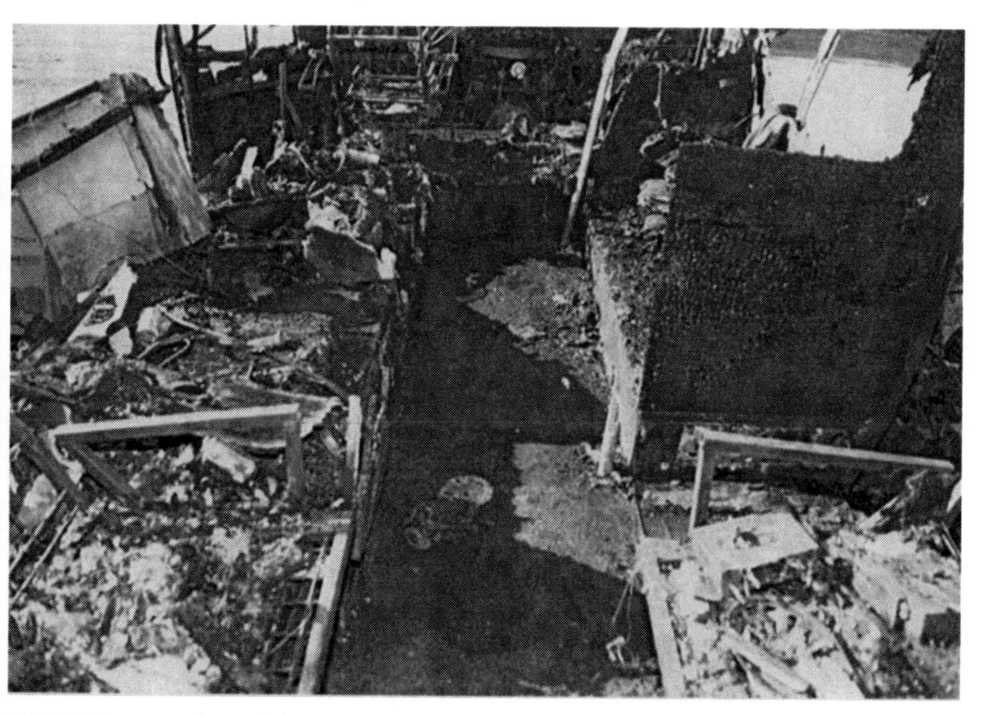

EXHIBIT 23.19 Fire Origin in Rear of Vehicle

HEAVY EQUIPMENT

Heavy equipment includes earth-moving equipment and mining, forestry, landfill, and agricultural equipment. These types of equipment normally serve one or more specific materials-handling functions. Large equipment is often diesel powered with a hydraulic transmission. These systems are sometimes susceptible to failure due to overloading the engine or transmission, a failure of the hydraulic systems, bearing failures, or the unintended ignition of the materials being handled.

◗◗◗ QUESTIONS FOR CHAPTER 23

1. True or false: Once a fuel is ignited in a motor vehicle fire, any additional fuels contained in the vehicle have no effect on the fire growth rate and ultimate damage sustained.

   A. True

   B. False

2. True or false: Flash point is of little or no significance when a fuel is released in spray form.

   A. True

   B. False

3. Which is most likely to cause an electrical overload to factory installed wiring?

   A. Bundled cables such as the wiring harness

   B. Accessory wiring under the dash

   C. Self-regulating wires in heated seats

   D. Electrical wiring to aftermarket stereo or alarm equipment

4. Which would not be considered an ignition source?

   A. Metal-to-metal sparks

   B. Aluminum-to-pavement sparks

   C. Metal-to-pavement sparks

   D. Motor bearings sparks

5. In high-pressure fuel-injected systems, what pressure is typically used when pumping fuel from a fuel storage tank?

   A. 15–35 psi (103–241 kPa)

   B. 35–70 psi (241–482 kPa)

   C. 70–120 psi (482–827 kPa)

   D. 120–150 psi (827–1034 kPa)

6. True or false: A turbocharger uses exhaust gases for propulsion and is the hottest point on an engine.

   A. True

   B. False

7. From what does the starter motor draw its energy?

   A. Alternator

   B. Battery

   C. Turbocharger

   D. Starter

8. Where do most engine fires typically penetrate the interior of a vehicle?

   A. Driver's side near the steering column

   B. Passenger's side at the bottom of the windshield

   C. Heater ducts under the front dashboard

   D. Passenger's side through the glove compartment

9. True or false: Engine oil analysis can be performed to determine the pre-fire condition of the engine, but it does not establish possible motive for the destruction of the vehicle.

   A. True

   B. False

10. True or false: Recreational vehicles and motor home fires are similar in many ways to fires in houses and mobile homes.

    A. True

    B. False

Wildfire Investigations

OBJECTIVES

Upon completion of Chapter 24, the user will be able to

- Identify additional factors that influence wildfire investigations
- Classify the physical characteristics of combustible wild land materials
- List major factors that affect fire spread

Wildfire investigations involve special techniques, practices, equipment, and terminology. This chapter describes wildfire fuels, factors that affect fire spread, indicators of directional patterns, conducting an origin investigation, and special safety considerations.

WILDFIRE FUELS

Fuels are classified as ground fuels, which include all flammable materials lying on or immediately above the ground or in the ground, and aerial fuels, which include all green and dead materials located in the upper forest canopy.

Flammability Analysis

Prior to flammability analysis, the physical characteristics of the fuel must be classified by the burning characteristics of individual materials and by the combined effects of the various types of materials present.

Ground fuels characteristics are as follows:

- Duff is not a major influence on the fire spread rate because it is typically moist and tightly compressed with little surface exposure.
- Dead leaves and coniferous litter are a highly flammable material and should be considered separately in evaluating ground fuels. Examples include needles dropped from coniferous trees.

- Grass, weeds, and other small plants are ground fuels that influence the rate of fire spread, based primarily on their degree of curing.
- Fine, dead wood, which consists of twigs, small limbs, bark, and rotting material with a diameter of less than 2 in. (50.8 mm), ignites easily and often carries fire from one area to another.
- Downed logs, stumps, and large limbs require long periods of hot, dry weather before they become highly flammable.
- Low brush and reproduction vegetation may either accelerate or slow down the spread rate of a fire because understory vegetation prevents other ground fuels from drying out rapidly.

Aerial fuels characteristics are as follows:

- Tree branches and crowns are a highly flammable fuel with arrangements that allow free circulation of air.
- Snags, or tree stumps, are important aerial fuels that influence fire behavior. Fires start in snags because they are drier and much easier to ignite.
- Moss hanging on trees is the lightest and the most apt to flash of all aerial fuels and provides a means of spreading fires from ground fuels to other aerial fuels. Moss reacts quickly to changes in relative humidity.
- Crowns of high brush are aerial fuels because they are separated by distance from ground fuels.

FACTORS AFFECTING FIRE SPREAD

Factors to be considered in determining the possible area of fire origin are the wind speed and direction as they relate to the movement of the fire.

Lateral Confinement

When the fire cannot spread laterally due to terrain, the radiation feedback can cause the fuel to burn more rapidly.

Wind Influence

Wind influences a fire by pushing the flames ahead and preheating fuel, drying out vegetation, creating airborne firebrands, and blowing embers and sparks ahead of the fire.

There are three classifications of winds:

1. *Meteorological winds* are caused by atmospheric pressure differentials in upper-level air masses that generate regional weather patterns, forming wind and pressure belts.
2. *Diurnal winds* are formed by solar heating and nighttime cooling, creating rising air. When air cools after sunset, this air sinks and causes downslope winds.
3. *Fire winds* are the result of the entrainment of air by the rising fire plume and influence the spread of the fire.

The direction in which the local wind is blowing primarily determines the route of the fire's advance, led by the fire head or the area of greatest fire intensity. The fire heel will generally be backing or burning slowly against the wind.

Definition

fire head

The portion of a fire that is moving most rapidly, subject to influences of slope and other topographic features. Large fires burning in more than one drainage or fuel type can develop additional heads. (Source: NFPA 921, 2004 edition, Paragraph 26.3.3)

fire heel

Located at the opposite side of the fire from the head. The fire at the heel is less intense and is easier to control. (Source: NFPA 921, 2004 edition, Paragraph 26.3.4)

uel Influence

gnitibility, rate of burning, and fire spread are influenced by the following fuel haracteristics:

- *Species of vegetation:* Moisture content, shape, and density
- *Fuel size:* The smaller the ratio of surface area to mass, the easier the fuel will be to ignite, and the faster it will be consumed. (See Table 24.1.)
- *Moisture present in the fuel:* Depends on the type and condition of the vegetation, solar exposure, weather, and geographic location
- *Oil content within vegetation*
- *Fuel types:* Ground, surface, and crown can burn along their entire length underground and can ignite a surface fire in a different location.

opography

opography relates to the form of natural and human-made earth surfaces. It ffects the intensity and spread of the fire and greatly influences winds.

lope. Slope is the change in elevation over a given distance, measured by the ise over the run (Exhibit 24.1). It allows the fuel on the uphill side to be preeated more rapidly than if it were on level ground, and is an important factor in ire spread. Fuel that is uphill of the fire can be preheated more rapidly, and wind urrents moving uphill during the day accelerate the fire spread. Burning fuels an roll downhill, igniting other fuels.

ABLE 24.1 Fuel Size

Size	Burning Characteristics	Examples
mall (fine) fuels	Easiest to ignite Rapidly consumed	Seedlings and small trees Twigs Dry grass Brush Dry field crops Pine needles and cones
Large (heavy) fuels	Harder to ignite Burns slower	Large-diameter trees and brush Large limbs Logs Stumps

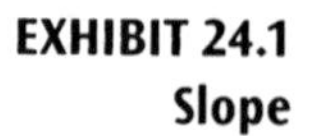

EXHIBIT 24.1
Slope

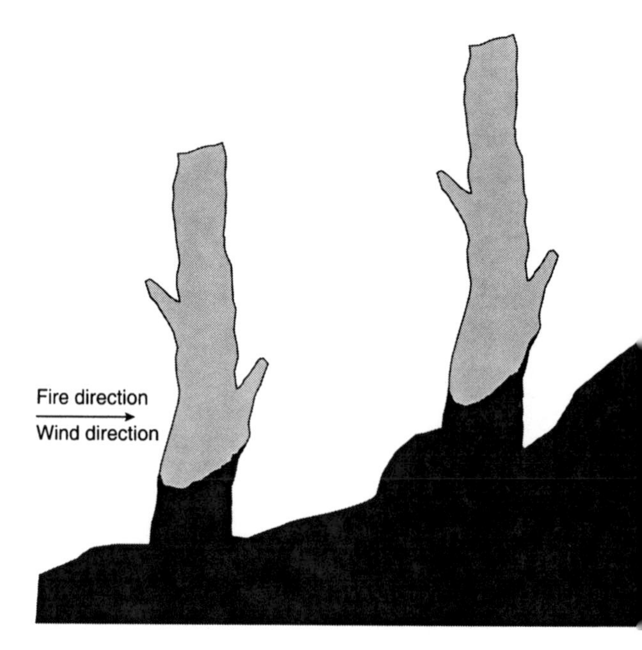

Aspect. Aspect is the direction the slope faces. If a slope faces the sun, it is typ cally drier and may have a more combustible fuel type or character of vegetatio than slopes without this solar heating. This results in a greater ease of ignition an faster spread. Table 24.2 summarizes these relationships.

Weather

Weather plays an important role in the behavior of wildfires because it influenc the atmosphere with respect to atmospheric stability, temperature, relative humi ity, wind velocity, cloud cover, and precipitation.

Weather history is a description of atmospheric conditions over the precedin few days or several weeks and should be analyzed to determine what influences might have had on the fire's ignition and burning characteristics. Descriptions c conditions over the past days and weeks can be analyzed to determine what rol weather played in the fire.

The ambient temperature influences the temperature of the fuel. The sun one factor that affects temperature. As the radiant solar energy of the sun hea the ground and vegetation, the fuel becomes more susceptible to ignition.

- Ambient air temperature directly affects the fuel temperature.
- Altitude affects air temperature.

Higher altitudes = lower air pressure = cooler air temperatures

TABLE 24.2 Effect of Slope Aspect on Fire

Direction	*Conditions*	*Effect*
Facing toward the sun	Has more vegetation Dryer	Greater ease of ignition and fire spread
Facing away from the sun	Has less vegetation More moist	Less ease of ignition and fire spread

Relative humidity is the measure of water vapor suspended in air. Humidity is usually expressed as relative humidity. The moisture in the air directly affects the amount of fuel moisture. Warm air can hold more water than can cool air.

- The moisture content of the air and the moisture content of the fuel are directly related to each other.
- Fine fuels are more susceptible to atmospheric moisture content.
- Warm air can hold more moisture.

relative humidity
The amount of moisture in a given volume of air, compared to how much the air could hold at that same temperature.

Fire Suppression

Fire suppression is all activities that lead to the extinguishment of the fire. Protection by fire crews of potential areas of origin is of extreme importance in establishing the fire origin and cause. The investigator and fire suppression crews protect probable areas of origin. Common fire suppression activities include the following:

- *Firebreaks, fire lines, or control lines:* Any natural or human-made barriers that are used to stop the spread or reroute the direction of the fire by separating the fuel from the fire
- *Air drop:* The aerial application of water, slurry, or retardant mixture directly onto the fire or threatened area or along a strategic position ahead of the fire
- *Firing out:* The process of burning the fuel between a firebreak and the approaching fire to extend the width of the fire barrier
- *Class A foam:* A mechanically generated aggregation of bubbles that have a lower density than water

Class A foam is made by introducing air into a mixture of water and foam concentrate. The bubbles adhere to the wildfire fuels or other Class A fuels and gradually release the moisture they contain. Bubbly water adsorbs heat more efficiently than plain water does, and the bubble mass provides a barrier to oxygen, necessary to sustain combustion. The reduced rate of water release results in more efficient conversion of water to steam, providing enhanced cooling effects. The surfactant that is contained in the solution allows the water to penetrate the fuels and reach deep-seated fires. The bubble mass also provides a protective barrier for unburned and exposed fuels.

firebreak
A natural or human-made barrier to stop the spread of the fire

air drop
Aerial application of suppressant either on the fire or on unburned fuels

firing out
Deliberate burning of unburned fuel between the fire and a firebreak

Other Natural Mechanisms of Fire Spread

Wind-borne firebrands are hot embers picked up by wind and blown into unburne fuel great distances from the original fire. They can start spot fires remote fron the main body that could be mistakenly interpreted as deliberately set fires.

A fire storm is an intense and violent convection fire, fed by self-induce winds. Fire storms are characterized by powerful indrafts that have the ability t propel objects.

Animals and birds can spread fire via flaming fur or feathers. A bird's feath ers or an animal's fur can be ignited by contact with power lines and can star a wildfire when the electrocuted body falls to the ground. Animals can becom ignited by the fire and travel to unburned areas.

INDICATORS OF DIRECTIONAL PATTERN

Analysis of the directional pattern shown by multiple indicators in a specific are can identify the path of fire spread through the area. By a systematic approach t backtracking the progress of the fire, the investigator can retrace the path of th fire to the point of origin. This procedure is the accepted and standard techniqu in wildfire investigation. Visual indicators include differential damage, char pat terns, discoloration, carbon staining, and the shape, location, and condition o residual, unburned fuel.

Wildfire V-Shaped Patterns

Wildfire V-shaped patterns are horizontal ground surface burn patterns generate by the fire spread. When viewed from above, they are generally shaped like the let ter V. These are not to be confused with the traditional plume-generated vertica V-patterns associated with structural fire investigations.

Wildfire V-shaped patterns have the following characteristics:

- Horizontal, not vertical patterns
- Affected by wind direction and slope
- Legs widen up the slope in the direction of the wind
- Origin is near the base of the V

The degree and type of damage to fuels indicate the intensity and the direc tion of fire passage. Leaves, branches, and limbs sustain greater damage on the side from which the fire approached, as shown in Exhibit 24.2 and Table 24.3.

Grass Stems

As a low-intensity fire burns the bottoms of grass stems, the stems fall over int the fire. If they fall into a burned area ahead of the fire head, they might stay unburned. Unburned stalks of grass on the ground generally point in the direc tion of the fire approach.

Trees

Trees are significant indicators of fire direction, particularly in areas of fronta fire damage (Exhibit 24.3). Fire movement is recorded at ground level around the root base and tree trunk and at flame height by the lower foliage and crown

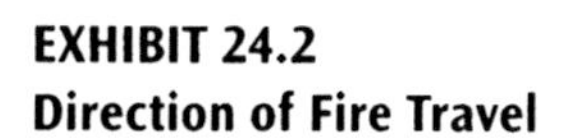

EXHIBIT 24.2
Direction of Fire Travel

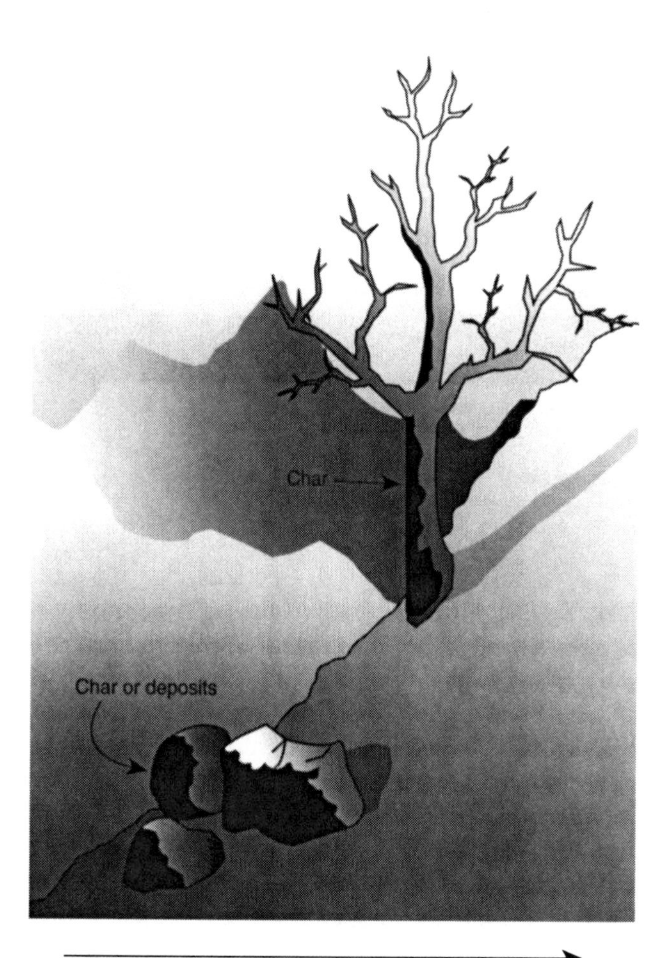

TABLE 24.3 Damage Indicators of Fire Passage

Fuel	*Appearance*	*Interpretation*
General	Fuel is more damaged on one side than another	Fire approached from the side with more damage
Grass stems	Unburned grass stalks or seed heads lying on the ground	Point in the direction the fire approached
Brush	More upper foliage is burned	Exit side
	Some upper branch tips fall unburned to the ground	Entry side
	Ash deposits not found on fuels	Fuels still burning when the ash was deposited
	Brush cupping Tips of burnt stubs are blunt or cupped	Upwind side of fire
	Brush cupping Tips of burnt stubs are sharp	Downwind side of fire
	Brush die-out pattern Decreasing fire intensity, charring, and burned branch size	Fire died out after entering brush growth

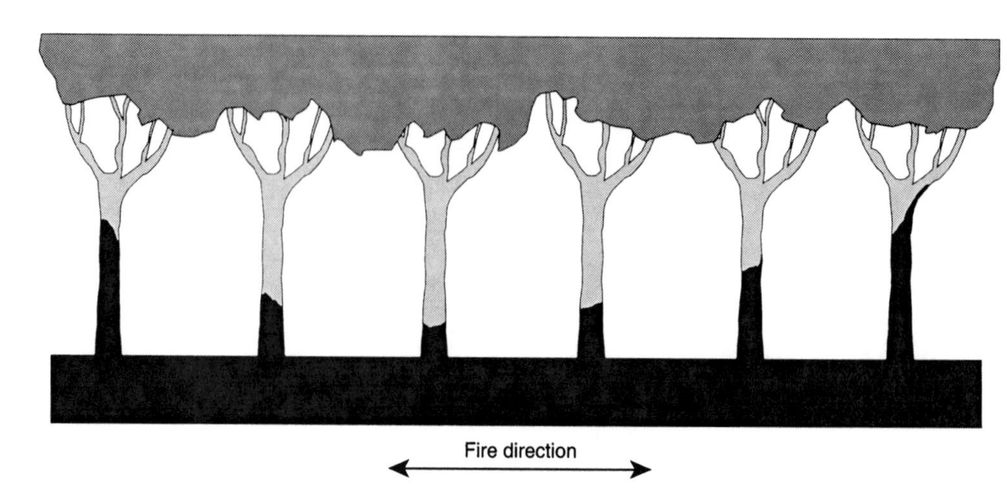

EXHIBIT 24.3 Trees Indicate Fire Direction

canopy. The char to the trunk surface of a tree is affected by the topography of the land surface. A fire burning uphill or with the wind creates a char pattern that slopes to a greater degree than a ground slope.

Crown damage can also be used to interpret fire direction. Convection and radiant heat travel ignite lower limbs and then spread upward into the rest of the tree. This action progresses in intensity as the wind action drives the fire away from the windward foliage and branches. (See Tables 24.4 and 24.5.)

Cupping can be observed on the side of a tree stump or brush. Cupping is a charred surface on the fuel that is caused by the exposure of the surface to the windward side of the fire. This side is charred deeper than the opposite side, which is protected from exposure.

TABLE 24.4 Trunk Char

Fire Direction	*Wind Direction*	*Observation*
Upslope	Upslope	Angle of char on the tree will be greater than the slope angle of the hill.
Downhill	Upslope	Char line on the tree is nearly parallel to the slope of the hill.
Down slope	Down slope	Angle of tree char will be higher on the downhill side.
Upslope	Down slope	Char angle is level with the slope of the hill with only light upslope damage to the tree trunk.

TABLE 24.5 Crown Damage

Observation	*Interpretation*
Triangular unburned area is observed on the side of the crown.	The unburned area is on the windward, or approaching side of the fire.

Exposed and Protected Fuels

A protected area in fuels immediately adjacent to a noncombustible object indicates that the fire direction was from the opposite side of the protected area. Staining and sooting on one side of the object indicate that the fire direction was from the stained side.

CONDUCTING AN ORIGIN INVESTIGATION

In identifying the area of origin, considering the factors of wind, topography, and fuels, the origin is normally located close to the heel or rear of the fire.

Interviews with various parties can help the investigator to narrow down the area where the fire first started. Observations of reporting parties can help, as they have observed the fire at a relatively small stage.

Observations of the initial attack crew can also be useful. Crew members might have observed people or vehicles, can report weather conditions, and can detail initial fire conditions and location.

Airborne personnel can make observations similar to those of ground-based crews and have a broader perspective on the location of the fire.

Witnesses can often provide vital information in the investigation of a wildfire and often are familiar with the area and can give information about the area of origin and a possible cause by describing the condition of the fire (e.g., smoke conditions, intensity, rate of spread, weather).

Satellite or imaging tools that are used primarily to establish fire suppression tactics can also be used to assist in the establishment of the area of origin. Based on the direction of fire spread and data indicating the fire location when first detected, these tools can provide a broader perspective on the fire and fire spread.

The fire scene must be secured from entry by civilians and fire suppression personnel in order to protect evidence such as tracks and potential ignition sources.

Identifying Evidence. The investigator should use metal flagging stakes or labeled flags to identify evidence. Labeled flags can also be used to mark the original position of an item of evidence that has been removed from the scene.

Analyzing fire spread may involve many steps to establish and plot the location of the head, flanks, and heel of the fire when it was discovered as well as at the time of arrival. Wind direction and available weather information should be documented and plotted, based upon the observations of those who discovered the fire. Weather information involves plotting fire head, fire flank, fire heel, and wind direction and speed.

Origin Area. Care must be continually exercised not to destroy evidence or other signs, such as footprints or vehicle tracks.

Search Techniques

If the identified area of origin is small, the area can be searched in its entirety from its perimeter. However, if the area is large, the site should be broken into segments for systematic close-up examination. (See Exhibit 24.4.)

The loop technique is also referred to as the spiral method. This method is effective in a small area, but as the loop or circle widens, evidence can be more

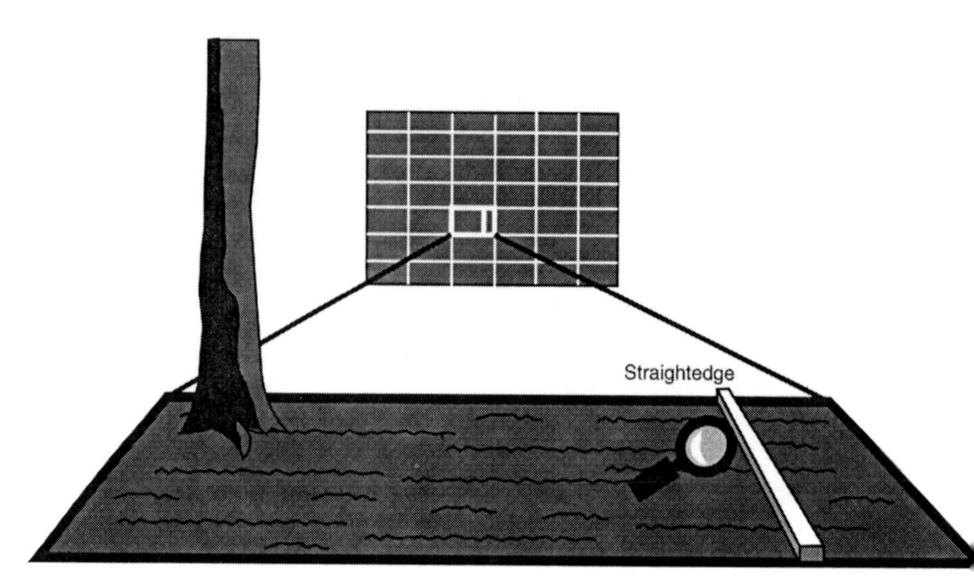

EXHIBIT 24.4 Searching Each Segment

easily overlooked or even damaged as the investigator moves into the area o
origin.

The grid technique is one of the best procedures for covering a large are
with more than one searcher. The lane technique is also referred to as the stri
method. This method can be used effectively if the area to be covered is larg
and open. It is relatively quick and simple to implement and can even be pe
formed by a single investigator in small areas.

Search Equipment

The following is a list of some of the equipment that can be used on a wildfir
investigation:

- *Magnifying glass or reading glasses:* To enhance small details
- *Magnets:* To locate ferrous metal fragments or particles
- *Straightedge:* To segment the origin area
- *Probe:* To uncover small pieces of evidence from the surrounding veget tion
- *Comb:* To separate evidence from debris
- *Hand held lights:* To locate items in low-light areas
- *Air blower:* To separate light ash from items of interest
- *Metal detector:* To locate metals that may be of evidentiary value and th could not be easily located by a magnet
- *Sifting screen:* To separate a suspected item of evidence from the surrounc ing dirt and vegetation
- *Global positioning satellite (GPS) recorder:* To obtain the accurate longitudin and latitudinal position of the fire origin

GPS data can be cross-referenced against on-site survey information, ligh ning strike data, aerial photography, or satellite imagery for fire cause determ nation. (See NFPA 921, Section 26.7, "Cause Determination.")

Security of the Area or Point of Origin

The area of origin should be secured and should not be unhampered. If the area has been tampered with, it could mislead the investigation and affect evidence credibility. The area of origin must remain undisturbed by investigators and others.

Investigation of the Area of Origin

The investigator should conduct the examination of the origin area with minimum disturbance, seeking to expose the evidence of fire spread at the site of the origin rather than destroying or removing it during the investigation. A photographic record should be maintained throughout the investigative process.

FIRE CAUSE DETERMINATION

The objective of every origin and cause investigation is to establish and confirm the cause of the fire. Wildfires are not always started by the actions of people. Many are ignited by natural causes, such as lightning.

Natural Causes

Lightning. The action of a lightning strike causes fulgurites by melting sand in the soil in the root area. It might be possible to verify lightning strikes through a lightning detection service. Physical evidence at the origin is usually noticeable.

Spontaneous Heating. Spontaneous heating can be caused by biological or chemical action and is more likely on warm, humid days.

Human Fire Cause

Humans can cause fires by action or by failing to act. Fires caused by humans can be classified as accidental or incendiary.

Campsite. Circles of rocks, or pits with a large amount of ash, or a pile of wood are good indicators of a campfire. Even camp areas that have burned completely leave evidence of their prior existence. Discarded food containers, metal tent stakes, or metal grommets from a tent might be found, indicating the possibility of a campfire.

Smoking. Discarded ignited smoking materials, such as cigarettes, cigars, pipe tobacco, and matches, can start wildfires. Conditions that must exist for a fire to be started by smoking materials include the following:

- The fine fuel moisture must be dry, below about 15 percent
- The fuel source must be fine or powdery (litter or punky wood)
- Approximately 30 percent of the side of a cigarette must be in contact with the fuel bed

Debris Burning. Debris burning occurs at dumpsites as well as at residences from garbage and other debris set on fire. Witnesses often are the best way to confirm whether debris burning was the cause of the fire.

Sunlight and Glass Refraction. The sun's rays can be focused to a point of intense heat if they are concentrated by certain glass or shiny objects. This refraction process bends light rays, similar to that which occurs through a magnifying glass. The shiny, concave end of a metal can may focus sunlight, but its short focal distance makes the potential as a possible cause unlikely.

Incendiary. These fires are often set in more than one location and in areas that are frequently traveled. Matches, fuses, and other ignition devices might be found in the area of origin. The firesetter might decide to use a time-delay ignition device. Items to look for include cigarettes, rope, rubber bands, candles, and wire.

Juveniles. The carelessness of a child, coupled with curiosity about fire can lead to wildfires around residences, schools, playgrounds, campsites, and wooded areas. Matches, lighters, or other ignition devices might be found in the area of origin.

Fireworks. Fireworks provide means of ignition through sparks and flaming debris. Sparklers are a smaller hazard but may ignite dry grass or other fuels. Most sparklers include a metal (wire) or wood core that might be found at or near the point of fire origin. The remains of fireworks or their packaging might be found near the area of origin. Also, some fireworks have the potential to create small indentations in the ground due to their explosive force.

Prescribed Fire (Controlled Burn)

A prescribed fire is a fire that resulted from intentional ignition by a person or a naturally caused fire that is allowed to continue to burn according to approved plans to achieve resource management objectives.

prescribed fire (controlled burn)

A fire resulting from intentional ignition by a person or a naturally caused fire that is allowed to continue to burn according to approved plans to achieve resource management objectives. (Source: NFPA 921, 2004 edition, Paragraph 26.7.2.6)

Machinery and Vehicles

Any power or motorized equipment that uses electricity or flammable products in its operation is capable of starting a fire when in proximity to combustible vegetation. Fires can be started via exhaust systems, faulty spark arresters, vehicle fires, and friction. Defective or failed parts add to the fire potential through friction heating of bearing-worn brakes, "frozen" shafts, and abrasion.

Railroad

Wildfires are sometimes started along railroads. Occasionally, fire used to clear a right-of-way escapes. Locomotives and rolling stock can also cause trackside fires from exhaust carbon, external buildup of lubricants, fuel line failure, hot brake metal, or overheated wheel bearings (hot box). Fires can also result from derailments, cutting or grinding actions, or warning flares.

Utilities

Electric. Overhead power lines can cause wildfires when trees contact a conductor and ignite the branch or foliage involved. This contact usually leaves a brand on the portion of the tree that made contact and creates a pit or flash mark on the power conductor. After ignition, burning portions of the tree may fall to the ground and ignite ground litter. In addition to tree contact, conductors can be blown against each other (phase to phase) during a windstorm, creating a hot metal globule that falls to the ground. Conductors and transformers can fail, starting the pole or other equipment on fire, and can drop flaming or hot material onto the ground. Underground conductors can be damaged by heavy equipment or digging operations, resulting in fire. Electric fences are likewise a source of energy resulting in ignition of combustible material.

Oil and Gas. Oil- and gas-drilling activities in wildfire areas can cause a fire. Most of the hazardous activities associated with fire take place during the drilling operations. A number of these hazards have been discussed in the smoking, equipment use, and electricity sections of this chapter. A well blowout and subsequent ignition can cause a wildfire to ensue. Depending on the minerals to be extracted, gas-fired equipment such as separators might be present at the well site after the drilling process has been completed. Likewise, the proximity of pipelines carrying gas or liquid fuel can provide sources of the initial fuel or may be a contributing factor to wildfire spread.

EVIDENCE

Evidence collection procedures for wildfire investigations are the same as those outlined in Chapter 14 of this *User's Manual,* "Physical Evidence."

SPECIAL SAFETY CONSIDERATIONS

Hazards

Safety is a major concern on all fires. Investigating wildfires carries its own specific hazards. The following are some considerations that must be addressed while on the fire scene:

- Maintenance of personnel accountability
- Escape routes
- Underground burning hazards
- Falling or rolling debris
- Weather

Personal Protective Equipment

Protective clothing and safety equipment vary depending on the circumstances. If investigating while the fire is still in progress in the investigator's immediate vicinity, the investigator must comply with applicable federal, state, provincial, or local requirements for personal protective equipment applicable to wildfire fighting.

Information on the proper type of protective equipment can be found in NFPA 1977, *Standard on Protective Clothing and Equipment for Wildland Fire Fighting,* and NFPA 1500, *Standard on Fire Department Occupational Safety and Health Program.*

QUESTIONS FOR CHAPTER 24

1. Why is it important to consider weather history when investigating wildfires?
2. True or false: If a wildfire is not attended to quickly, it can reach flashover in a short period of time.
   - A. True
   - B. False
3. What does the degree of damage to a wildland fuel indicate?
   - A. Moisture content
   - B. Wind direction
   - C. Fire intensity
   - D. Oil content
4. What is the difference between wildfire V-shaped patterns and compartment fire V-pattern?
   - A. Heat transfer
   - B. Ignition source
   - C. Heat release rate
   - D. Orientation of the surface
5. True or false: The collection and documentation of wildfire evidence is different from that for structural fires.
   - A. True
   - B. False

CHAPTER 25

Management of Major Investigations

OBJECTIVES

Upon completion of Chapter 25, the user will be able to

- List issues that should be covered in an agreement between interested parties
- Identify the importance of a team leader committee
- List the responsibilities and duties of an investigation team

This chapter is a resource to assist the investigator in the management of investigations in which more than one (usually several) parties are interested in conducting an investigation. The definition of *major investigation* is broad because the need to use these management functions can vary with the size of the incident, the complexity of the investigation, and the needs or demands of the other interested parties. Investigators often conduct investigations in which more than one party is present and the management functions outlined in this chapter might not be needed. This chapter provides tools for management or coordinating scene management for multiple interested parties, complex investigations that usually require several days to complete. This often occurs in high-rise buildings, large complexes, or facilities that have multiple occupancies. However, it can occur with much smaller fires in which several parties have an interest.

This chapter not only provides tools for private-sector investigations, but also can assist with public-sector investigations. It is recognized that the public or law enforcement investigation may have concerns for access of private investigators during certain investigative activities. However, with proper coordination and management, private-sector interests may be preserved so that private investigators have the opportunity to conduct separate independent investigations and to view or examine the evidence before it is preserved or altered.

The issue of spoliation of evidence, which is discussed in Chapter 9, "Legal Considerations," continues to evolve through the courts. The courts have generally recognized that other parties might have the right to participate in the scene investigation before it is altered. The tools provided in this chapter help the investigator in charge of the scene to coordinate the scene activities. This chapter does not describe how the investigation should be conducted, nor does it describe how the evidence should be handled. These issues are addressed in other chapters of this *User's Manual*. The management structure outlined in this chapter is needed to coordinate the operations of the different parties involved. The manager of the investigation should keep in mind that the goal in conducting a major investigation is usually the coordination so that parties can conduct their investigation to determine the cause and origin and the analysis of factors that contributed to the damage, injuries, or loss of life.

This chapter covers two distinct subjects. One part describes coordinating the actions of the various interested parties. The second part describes the management function of developing an individual team or resources as one of the interested parties participating in a major investigation.

COORDINATING ACTIONS OF DIFFERENT PARTIES

Understanding Between Parties

In coordinating the actions of the different parties, often the manager of the major investigation not only needs to coordinate the parties so that they can conduct their activities, but also needs to identify and manage functions of personal necessities. These might include providing portable toilets, providing enough parking, providing for personal comfort of workers and interested parties, ensuring that safety issues have been identified and managed, and ensuring continuity and coordination of the requests.

The first functions of the manager of the major investigation are to ensure that a site evaluation is conducted to identify potential safety issues and then to address those issues. Information relating to the handling of those issues is found in Chapter 10, "Safety." The investigator who manages the scene also needs to coordinate access to the site to ensure that only those that have been authorized gain access to the site. This access might depend on whether certain safety training has occurred or environmental issues are maintained.

Often, one of the most difficult responsibilities borne by the manager of the major investigation is to identify the interested parties who should have access to the site. This is usually coordinated with some legal advice. There might be parties who have a financial or legal interest in participating in the investigation and others with an interest, such as the media. The manager could also attempt to identify other parties that the team want present, such as representatives from parties that are thought to have contributed to the cause or spread of the incident.

During the course of an investigation, the investigator in charge could identify other potential interested parties, such as building component manufacturers, appliance systems providers, and installers of protection and detection systems. These groups should be notified to participate in the investigation. Interested parties that have become aware of the incident and wish to participate in the investigation can also approach the investigator in charge.

Interested parties should be allowed to participate in the investigation (joint investigation) provided that an agreeable plan (protocol) is established. Joint investigations help to avoid accusations of evidence altering or other indiscretions. With proper planning and communication, criminal investigations can be conducted parallel with civil investigations. Private investigators can provide various types of support to public investigators.

It is important that there be an understanding between the parties of how the investigation will be conducted, who has control of the investigation, and the rules or parameters for participating in the investigation.

The understanding between the parties should include an agreement regarding how individual parties access the scene or how evidence will be handled. This can be a complex issue during the investigation when there are multiple parties and each party has multiple investigators who want to have access to certain areas of the site. It may not be feasible to allow access to all parties, because there would be too many people for the allowable area, safety issues, or lack of space to properly view the activities. A consensus or compromise needs to address these issues, which may include allowing access to only one person per interested party or setting up audio and video feeds so that the investigation can be viewed from a safe location.

Disseminating information between the parties needs to be addressed early or lack of information could lead to difficulties between the parties or even legal action. The issue of interviews should also be addressed early, along with how the information should be shared. There could be a need for confidentiality agreements before interviews can be conducted or information exchanged to ensure that proprietary or secret information is held in confidentiality. As has been described in other chapters of this *User's Manual,* interviews are an important part of an investigation so that a final hypothesis can be developed.

Read NFPA 921, Section 27.2, "Understanding Between the Parties." ◀◀◀

Agreement Between Parties

All of the parties involved in an investigation should reach a consensus on site access, sharing of information, custody and examination of evidence, acquiring information, release of information, and scene examination protocol.

This consensus may be accomplished by establishing a "Memorandum of Understanding" or other written agreement through which all interested parties have input and obtain direction. NFPA 921 provides examples of useful tools to assist in the formulation and implementation of these types of multi–interested-party agreements. For an example of a Memorandum of Understanding, see Figure 27.3.2(a) in NFPA 921.

The Memorandum of Understanding creates an agreement, in either written or verbal form, that outlines the parameters and sets the rules by which the investigation is to be conducted. The Memorandum of Understanding encompasses various aspects of the investigation, including who is responsible for the overall investigation and who is responsible for the costs related to equipment and other concerns involved in conducting the investigation, the documentation of the scene, and the securing of the evidence.

Read NFPA 921, Section 27.3, "Agreement Between Parties." ◀◀◀

ORGANIZATION OF THE INVESTIGATION

Each interested party should form an investigation team that can include more than one person. This team should have one person appointed as team leader or spokesperson. The team spokesperson represents the interested party as a member of the team leader committee.

Each interested party's investigative team might consist of various special experts including engineers, lawyers, metallurgists, and others who are deemed necessary. Each team is considered one person under the organization of the investigation and should provide one team leader to participate on the team leader committee.

ACTIVITY  Read NFPA 921, Section 27.4, "Organization of the Investigation." ◀◀◀

Team Leader Committee

The team leader committee should comprise one representative from each interested party. Only one person from each interested party should be the spokesperson for that group. Someone should be appointed to chair the committee, and another person should be appointed to record the minutes of each meeting. The committee's function is to coordinate all of the interested parties' activities. A planning meeting should be held before starting the on-scene investigation to discuss safety considerations, to identify the concerns of the various interested parties, and to set forth the ground rules for conducting the investigation.

The team leader committee will most likely prepare the Memorandum of Understanding and the flow chart that will be used during the investigation. The team leader committee will work through a consensus process. If other representatives from the team attend the team leader committee meeting, it is usually agreed that they voice their concerns or votes through their respective team leaders.

The following are some of the concerns that should be addressed by the team leader committee:

- Safety
- Planning
- Occupant access
- Organizing the investigative team
- Regular meetings
- Resources
- Preliminary information
- Lighting
- Access for investigators
- Securing the scene
- Sanitary and comfort needs
- Communications
- Interviews
- Plans and drawings
- Search patterns
- Evidence
- Release of information

ACTIVITY  Read NFPA 921, Section 27.5, "Team Leader Committee." ◀◀◀

Planning

A plan should be developed, reviewed, and modified as needed during the course of the investigation. This plan should begin with scene safety concerns and recommendations.

Read NFPA 921, Section 27.6, "Planning."

Occupant Access

The occupants should be allowed access to the scene if it is safe to do so as indicated by the team leader committee and if such access will not compromise the investigation.

Read NFPA 921, Section 27.7, "Occupant Access."

Regular Meetings

The team leader committee should have short, daily meetings with participants to keep those involved up to date regarding the status of the investigation. The daily briefing meetings need to be conducted to discuss activities that have occurred, safety issues, and the plan for the day. The management committee should meet at least weekly to discuss the scheduling, safety issues, and other needs for the scene management issues.

Read NFPA 921, Section 27.9, "Regular Meetings."

Resources

Necessary resources can be obtained from outside jurisdictions, organizations, or other agencies beyond the authority having jurisdiction. Using the team or task force approach may enable access to these resources.

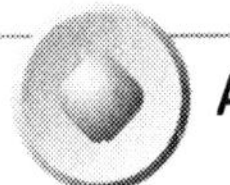

Read NFPA 921, Section 27.10, "Resources."

Preliminary Information

Preliminary information should be gathered before developing the investigation flow chart and plan. This should include the condition of the scene, the magnitude of the incident, incident-specific information, the type of structure and use of the structure, and the extent of various interested parties' concerns.

ACTIVITY

Read NFPA 921, Section 27.11, "Preliminary Information."

Safety

A safety plan should be developed as explained in the section on the team leader committee. In some situations, the appointment of a safety officer is warranted.

The safety plan should address concerns for both the investigators and other building occupants. It might be necessary to monitor the air quality during the investigation. Investigator fatigue should be considered a safety issue during major investigations.

ACTIVITY 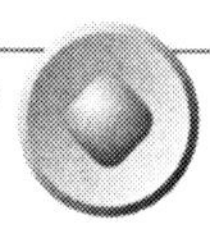 Read NFPA 921, Section 27.12, "Safety."

Lighting

It may be advantageous to have temporary lighting installed to illuminate areas of concern. This work should be performed by an electrician for safety and code concerns. Local code requirements should be addressed before and during lighting design and installation.

ACTIVITY 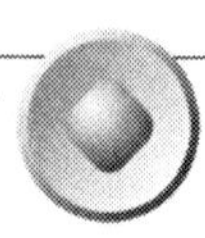 Read NFPA 921, Section 27.13, "Lighting."

Access for Investigator

Transportation concerns to, from, and within the scene should be addressed by the team leader committee. Means of access to elevated areas should be determined.

ACTIVITY 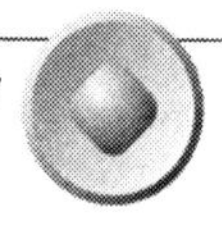 Read NFPA 921, Section 27.14, "Access for Investigator."

Securing the Scene

Scene security should be initiated with the first responders and should be maintained throughout the investigation. Access to the investigation site should be limited to those necessary and authorized to conduct the investigation. Temporary barriers and security guards might be necessary to secure the scene adequately.

ACTIVITY 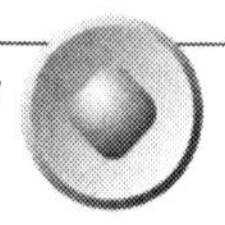 Read NFPA 921, Section 27.15, "Securing the Scene."

Sanitary and Comfort Needs

The team leader committee ensures that provisions are made for uncontaminated areas to be used for meetings and meals. Sanitary facilities and drinking water stations should be provided. A first-aid station should be considered, based on the potential for injury during the investigation.

ACTIVITY 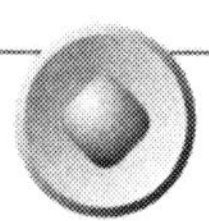 Read NFPA 921, Section 27.16, "Sanitary and Comfort Needs."

Communications

The team leader committee should establish a central communication point, often referred to as a command post. Team communications are important to the continuity and coordination of an investigation and may be provided through portable radios, temporary hard-wired telephone systems, and cellular telephones.

ACTIVITY

Read NFPA 921, Section 27.17, "Communications." ◀◀◀

Interviews

One or more members from each investigation team should conduct preliminary interviews with the fire chief, fire prevention personnel, suppression personnel, police officers, passersby, neighbors, property owner(s), employees, tenants, and any others that could have information.

Examples of the kinds of information to be gleaned from interviews include fire discovery; events before the fire department arrived; fire suppression efforts; movement of the fire; building construction; contents; actions of occupants prior to, during, and after fire discovery; processes conducted within the structure; information regarding utilities; information regarding fuel loads; and information regarding any hazardous materials.

Interviews can be conducted in conjunction with other fire scene investigation activities, document collection, scene photography, safety plan formulation, and so forth. Whenever possible, joint interviews should be conducted involving other interested parties in the investigation. Joint interviews are not always feasible, such as in a suspect interview, due to legal considerations.

ACTIVITY

Read NFPA 921, Section 27.18, "Interviews." ◀◀◀

Plans and Drawings

Plans and drawings of the structure can be instrumental in identifying the location of equipment or materials and can also assist in developing fire scene plans. Plans and drawings can be obtained from the building owner, architect, contractor, and the public building department.

ACTIVITY

Read NFPA 921, Section 27.19, "Plans and Drawings." ◀◀◀

Search Patterns

The following methods can be used to search the scene:

- Grid (usually the most effective for fire scenes)
- Spiral
- Strip
- Area

Regardless of the search pattern method that is implemented, each section of the search area should overlap each adjacent section to ensure that a thorough search is conducted.

ACTIVITY 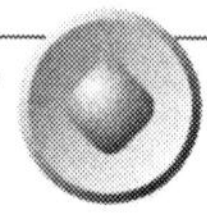

Read NFPA 921, Section 27.21, "Search Patterns." ◀◀◀

Evidence

Procedures and protocol should be identified at the beginning of the investigation through the team leader committee. One evidence technician should be responsible for evidence control and documentation. A secure location in which to keep the evidence should be established. All interested parties should be notified prior to any movement, alteration, or destructive testing of any of the evidence.

ACTIVITY 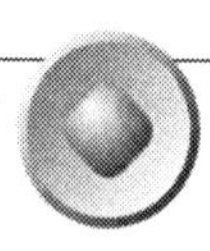

Read NFPA 921, Section 27.22, "Evidence." ◀◀◀

Release of Information

One person or agency should be responsible for releasing information regarding the investigation. Prior to the actual release interested parties should agree on information to be released.

ACTIVITY 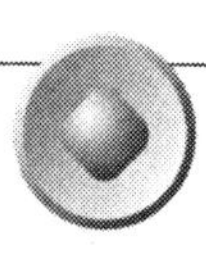

Read NFPA 921, Section 27.23, "Release of Information." ◀◀◀

ORGANIZATION OF AN INVESTIGATION TEAM

Interested Party Investigative Team

This section discusses the coordination and development of the functions of an interested parties investigative team. This discussion is not about management of the scene, but about management functions of the individual investigative team.

Each interested party might form an investigation team to take part in the investigation. The investigation of smaller fire scenes or losses could involve investigation teams representing various interested parties similar to the teams that are involved in larger scenes or losses. A team structure might include the following positions:

- Team leader
- Secretary
- Evidence technician/custodian
- Photographer
- Team member responsible for diagrams and sketches
- Team member(s) responsible for conducting interviews
- Safety officer

The interested party teams should meet daily to share information that has been developed by the individual team members. It is likely that the members will be doing different activities concurrently. Each interested party team will be conducting its own investigation activities to represent their needs.

Read NFPA 921, Section 27.8, "Organization of an Investigation Team." ◀◀◀

Interested Party Teams

Each of the interested parties might have its own team that has several of the same type of functions outlined in the referenced team leader committee information. It is also common that a responsible party's entire team consists of one individual who participates physically in the investigation. There is only one scene, and all parties should have access to the scene and all of the evidence contained therein. Even though an investigation can be conducted with one or more other interested parties, each investigator is analyzing the data in an effort to develop his or her own hypothesis and conclusions. The Memorandum of Understanding does not provide for a consensus regarding findings. Moreover, determinations and findings can vary and are the responsibility of each of the interested parties.

Chapter 12, "Planning the Investigation," provides information on conducting the investigation and the use of resources to assist in conducting the investigation.

Using a previous fire incident, prepare or preplan a fire investigation involving a multi-use occupancy that brings together several interested parties. This activity should include the preparation of a Memorandum of Understanding and an investigation flow chart that includes the components discussed in Chapter 27 of NFPA 921. ◀◀◀

▶▶▶ QUESTIONS FOR CHAPTER 25

1. What are some of the typical scenarios that would classify a fire as a major incident?
2. What should an on-site investigator do during a major investigation if he or she determines that a particular item might be evidence?
   - A. Document the evidence in place and remove it.
   - B. Provide the information to a law enforcement evidence technician, who will then document and remove the item.
   - C. Follow an established and agreed-upon protocol for the collection of evidence.
   - D. Photograph and note its position for later retrieval.
3. Which statement is true regarding the completion of a major investigation in which several parties are involved?
   - A. The investigators will concur at the completion of the investigation.
   - B. Joint investigations can occur simultaneously and involve separate and independent examinations through which each party can develop an independent hypothesis.
   - C. Parties conducting the investigation cooperate and most likely come to the same conclusions.
   - D. Major investigations seldom result in an investigation in which conclusions are drawn and findings are substantiated.
4. On a joint investigation, what course of action should be taken by fire officials to best serve the needs of all concerned?
   - A. Through proper planning and communication, allow other investigators to assist or observe.
   - B. Conduct the scene investigation first before allowing other investigators to participate.
   - C. Do not conduct an investigation but instead allow access to the scene to anyone.
   - D. Dismantle the scene with heavy equipment during the scene investigation without regard for the interests of others.
5. Prior to a joint investigation, what agreement between interested parties should be developed?
   - A. A legal document stating that all investigators should have liability insurance coverage
   - B. A consensus developed by all interested parties that prohibits any one party from bringing suit against any other party
   - C. A legal document stating which party or parties will be responsible for the cost of the investigation
   - D. A consensus agreement between all of the interested parties
6. Name three key items that should be included on a Memorandum of Understanding.

7. When should occupant access be allowed during a major investigation in which multiple tenants are involved?
   A. When the investigation is complete
   B. After the entire area has been examined and safety concerns have been addressed
   C. Supervised access before conducting the investigation
   D. After the property management personnel have completed their evaluation
8. Who should be included as members of a team leader committee?
   A. Each of the interested parties and any of their representatives
   B. One representative from the parties affected by the incident
   C. One representative from each of the investigative parties
   D. Only those approved by management of the building
9. Who has the responsibility for monitoring safety conditions at the fire or explosion scene?
   A. The local police department
   B. The evidence technician
   C. All team members
   D. The building owner
10. Give three examples of preliminary witnesses who should be interviewed at the scene.

Chapter Organization of NFPA 921

The content of NFPA 921, *Guide for Fire and Explosion Investigations,* was reorganized for the 2004 edition to conform to the *NFPA Manual of Style.* To aid the *User's Manual* reader, Table A.1 compares the table of contents of the 2001 and 2004 editions of NFPA 921 by chapter number. Table A.2 presents the same information, sorted alphabetically by chapter subject.

TABLE A.1 Comparison of 2001 and 2004 Editions of NFPA 921, by Chapter Number

	2001 Edition	*2004 Edition*
Chapter 1	Administration	Administration
Chapter 2	Basic Methodology	Mandatory References
Chapter 3	Basic Fire Science	Definitions
Chapter 4	Fire Patterns	Basic Methodology
Chapter 5	Building Systems	Basic Fire Science
Chapter 6	Electricity and Fire	Fire Patterns
Chapter 7	Building Fuel Gas Systems	Building Systems
Chapter 8	Fire-Related Human Behavior	Electricity and Fire
Chapter 9	Legal Considerations	Building Fuel Gas Systems
Chapter 10	Safety	Fire-Related Human Behavior
Chapter 11	Sources of Information	Legal Considerations
Chapter 12	Planning the Investigation	Safety
Chapter 13	Recording the Scene	Sources of Information
Chapter 14	Physical Evidence	Planning the Investigation
Chapter 15	Origin Determination	Documenting the Investigation
Chapter 16	Cause Determination	Physical Evidence

(continues)

TABLE A.1 Comparison of 2001 and 2004 Editions of NFPA 921, by Chapter Number *(continued)*

	2001 Edition	2004 Edition
Chapter 17	Failure Analysis and Analytical Tools	Origin Determination
Chapter 18	Explosions	Cause Determination
Chapter 19	Incendiary Fires	Analyzing the Incident for Cause and Responsibility
Chapter 20	Fire and Explosion Deaths and Injuries	Failure Analysis and Analytical Tools
Chapter 21	Appliances	Explosions
Chapter 22	Motor Vehicle Fires	Incendiary Fires
Chapter 23	Wildfire Investigations	Fire and Explosion Deaths and Injuries
Chapter 24	Management of Major Investigations	Appliances
Chapter 25	Referenced Publications	Motor Vehicle Fires
Chapter 26	N/A	Wildfire Investigations
Chapter 27	N/A	Management of Major Investigations
Annex A	Appendix A	Explanatory Material
Annex B	Appendix B	Bibliography
Annex C	Appendix C	Referenced Publications

[Source: NFPA's *Field Guide for Fire Investigators*, 2004 edition, Table A.1]

TABLE A.2 Comparison of 2001 and 2004 Editions of NFPA 921, by Chapter Subject

Chapter Title	2001 Edition	2004 Edition
Administration	Chapter 1	Chapter 1
Analyzing the Incident for Cause and Responsibility	N/A	Chapter 19
Appliances	Chapter 21	Chapter 24
Basic Fire Science	Chapter 3	Chapter 5
Basic Methodology	Chapter 2	Chapter 4
Bibliography	Appendix B	Annex B
Building Fuel Gas Systems	Chapter 7	Chapter 9
Building Systems	Chapter 5	Chapter 7
Cause Determination	Chapter 16	Chapter 18
Definitions	Chapter 1	Chapter 3
Documenting the Investigation	Chapter 13	Chapter 15
Electricity and Fire	Chapter 6	Chapter 8

TABLE A.2 Comparison of 2001 and 2004 Editions of NFPA 921, by Chapter Subject *(continued)*

Chapter Title	*2001 Edition*	*2004 Edition*
Explanatory Material	Appendix A	Annex A
Explosions	Chapter 18	Chapter 21
Failure Analysis and Analytical Tools	Chapter 17	Chapter 20
Fire-Related Human Behavior	Chapter 8	Chapter 10
Fire and Explosion Deaths and Injuries	Chapter 20	Chapter 23
Fire Patterns	Chapter 4	Chapter 6
Incendiary Fires	Chapter 19	Chapter 22
Legal Considerations	Chapter 9	Chapter 11
Management of Major Investigations	Chapter 24	Chapter 27
Mandatory References	Appendix C	Chapter 2
Motor Vehicle Fires	Chapter 22	Chapter 25
Origin Determination	Chapter 15	Chapter 17
Physical Evidence	Chapter 14	Chapter 16
Planning the Investigation	Chapter 12	Chapter 14
Referenced Publications	Appendix C	Annex C
Safety	Chapter 10	Chapter 12
Sources of Information	Chapter 11	Chapter 13
Wildfire Investigations	Chapter 23	Chapter 26

[Source: NFPA's *Field Guide for Fire Investigators*, 2004 edition, Table A.2]

Text of NFPA 1033

This appendix provides the reader with the full text of NFPA 1033, *Standard for Professional Qualifications for Fire Investigator,* 2003 edition. All material within this appendix is extracted from that publication.

NFPA 1033

Standard for

Professional Qualifications for Fire Investigator

2003 Edition

This edition of NFPA 1033, *Standard for Professional Qualifications for Fire Investigator*, was prepared by the Technical Committee on Fire Investigator Professional Qualifications, released by the Technical Correlating Committee on Professional Qualifications, and acted on by NFPA at its May Association Technical Meeting held May 18–21, 2003, in Dallas, TX. It was issued by the Standards Council on July 18, 2003, with an effective date of August 7, 2003, and supersedes all previous editions.

This edition of NFPA 1033 was approved as an American National Standard on July 18, 2003.

Origin and Development of NFPA 1033

In 1972, the Joint Council of National Fire Service Organizations (JCNFSO) created the National Professional Qualifications Board (NPQB) for the fire service to facilitate the development of nationally applicable performance standards for uniformed fire service personnel. On December 14, 1972, the board established four technical committees to develop those standards using the National Fire Protection Association (NFPA) standards-making system. The initial committees addressed the following career areas: fire fighter, fire officer, fire service instructor, and fire inspector and investigator.

The original concept of the professional qualification standards as directed by the JCNFSO and the NPQB was to develop an interrelated set of performance standards specifically for the uniformed fire service. The various levels of achievement in the standards were to build upon each other within a strictly defined career ladder. In the late 1980s, revisions of the standards recognized that the documents should stand on their own merit in terms of job performance requirements for a given field. Accordingly, the strict career ladder concept was revised, except for the progression from fire fighter to fire officer, in order to allow civilian entry into many of the fields. These revisions facilitated the use of the documents by other than the uniformed fire services.

The Committee on Fire Inspector and Investigator Professional Qualifications met and produced the first edition of NFPA 1031, *Professional Qualifications for Fire Inspector, Fire Investigator, and Fire Prevention Education Officer.* This document was adopted by the NFPA in May of 1977.

In 1986, the joint council directed the committee to develop separate documents for each of the job functions the original document addressed. This direction was coupled with the decision to remove the job of fire investigator from the strict career path previously followed and allow for civilian entry. The first edition of this new document, NFPA 1033, *Standard for Professional Qualifications for Fire Investigator*, was adopted by the NFPA in June of 1987.

In 1990, responsibility for the appointment of professional qualifications committees and the development of the professional qualifications standards was assumed by the NFPA. The Professional Qualifications Correlating Committee was appointed by the NFPA Standards Council and assumed the responsibility for coordinating the requirements of all of the documents in the professional qualifications system.

The NFPA Standards Council established the Technical Committee on Fire Investigator Professional Qualifications in 1990 to address the need for specific expertise in the area of fire investigation to review and revise the existing document. This committee completed a job task analysis and developed specific job performance requirements for the job of fire investigator.

The intent of the Technical Committee on Fire Investigator Professional Qualifications was to develop clear and concise job performance requirements that can be used to determine that an individual, when measured to the standard, possesses the skills and knowledge to perform as a fire investigator. These job performance requirements are applicable to fire investigators both public and private.

In the 2003 edition of the document the Technical Committee made changes to bring it into conformance with the new NFPA *Manual of Style.*

In Memoriam, September 11, 2001

We pay tribute to the 343 members of FDNY who gave their lives to save civilian victims on September 11, 2001, at the World Trade Center. They are true American heroes in death, but they were also American heroes in life. We will keep them in our memory and in our hearts. They are the embodiment of courage, bravery, and dedication. May they rest in peace.

Technical Correlating Committee on Professional Qualifications

Douglas P. Forsman, *Chair*
Union Colony Fire & Rescue Authority, CO [E]

Fred G. Allinson, Seattle, WA [L]
Rep. National Volunteer Fire Council
Stephen P. Austin, State Farm Insurance Company, DE [I]
Rep. International Association of Arson Investigators Inc.
Timothy L. Bradley, North Carolina Fire Commission, NC [E]
Rep. TC on Fire Service Instructor Professional Qualifications
(Vote Limited to Professional Qualifications System Management)
Boyd F. Cole, SunnyCor Incorporated/SmartCoat Inc., CA [M]
Rep. TC on Emergency Vehicle Mechanic Technicians Professional Qualifications
(Vote Limited to Professional Qualifications System Management)
Yves Desjardins, Ecole nationale des pompiers du Quebec, Canada [U]
David T. Endicott, Stevensville, MD [U]
Rep. TC on Fire Fighter Professional Qualifications
(Vote Limited to Professional Qualifications System Management)
Gerald C. Evans, Salt Lake City Fire Department, UT [L]
Rep. TC on Public Safety Telecommunicator Professional Qualifications
(Vote Limited to Professional Qualifications System Management)
Kelly Fox, Washington State Council of Fire Fighters, WA [L]
Rep. International Association of Fire Fighters
Jon C. Jones, Jon Jones & Associates, MA [SE]
Rep. TC on Industrial Fire Brigades Professional Qualifications
(Vote Limited to Professional Qualifications System Management)
Alan E. Joos, Utah Fire and Rescue Academy, UT [E]
Rep. International Fire Service Accreditation Congress
Charles E. Kirtley, City of Guymon, Oklahoma, Fire Department, OK [U]
Rep. TC on Public Fire Educator Professional Qualifications
(Vote Limited to Professional Qualifications System Management)
Barbara L. Koffron, Phoenix Fire Department, AZ [U]
Rep. TC on Fire Inspector Professional Qualifications
(Vote Limited to Professional Qualifications System Management)
Michael J. McGovern, Lakewood Fire Department, WA [U]
Gerard J. Naylis, U.S. Consumer Product Safety Commission, NY [C]
Rep. TC on Fire Investigator Professional Qualifications
(Vote Limited to Professional Qualifications System Management)
Chris Neal, Fire Protection Publications, OK [M]
Rep. TC on Fire Officer Professional Qualifications
(Vote Limited to Professional Qualifications System Management)
David K. Nelson, David K. Nelson Consultants, CA [SE]
Rep. TC on Wildfire Suppression Professional Qualifications
(Vote Limited to Professional Qualifications System Management)
William E. Peterson, Plano Fire Department, TX [M]
Rep. International Fire Service Training Association
Hugh A. Pike, U.S. Air Force Fire Protection, FL [E]
Rep. TC on Rescue Technician Professional Qualifications
(Vote Limited to Professional Qualifications System Management)
Richard Powell, Saginaw Township Fire Department, MI [L]
Rep. TC on Accreditation and Certification
(Vote Limited to Professional Qualifications System Management)
Johnny G. Wilson, Georgia Firefighter Standards and Training Council, GA [E]
Rep. National Board on Fire Service Professional Qualification

Alternates

Jack R. Reed, Iowa Professional Fire Fighters, IA [L]
(Alt. to K. Fox)
Michael W. Robinson, Baltimore County Fire Department, MD [E]
(Alt. to J. G. Wilson)

Frank E. Florence, NFPA Staff Liaison

This list represents the membership at the time the Committee was balloted on the final text of this edition. Since that time, changes in the membership may have occurred. A key to classifications is found at the back of the document.

NOTE: Membership on a committee shall not in and of itself constitute an endorsement of the Association or any document developed by the committee on which the member serves.

Committee Scope: This Committee shall have primary responsibility for the management of the NFPA Professional Qualifications Project and documents related to professional qualifications for fire service, public safety, and related personnel.

Technical Committee on Fire Investigator Professional Qualifications

Gerard J. Naylis, *Chair*
U.S. Consumer Product Safety Commission, NY [C]

Stephen P. Austin, State Farm Insurance Company, DE [I]
Rep. International Association of Arson Investigators Inc.
Joseph Bertoni, U.S. Department of the Treasury, VA [U]
Bill L. Buckley, S E A, Inc., FL [SE]
David R. Fischer, State Fire Marshal, NV [SE]
Mary M. Galvin, State of Connecticut, CT [C]
John F. Goetz, James Valentine Incorporated, PA [SE]
Douglas Lee Holmes, Introspect, TX [E]
Rep. City of Nassau Bay
David B. Hooton, Spectrum Solutions, TN [SE]
David W. Kircher, Essex County Courthouse Prosecutor's Office, NJ [U]
Hunter B. Lacy, Royal & SunAlliance, NC [I]
Hal Lyson, Robins, Kaplan, Miller & Ciresi, MN [L]
Russell K. Mason, Central County Fire and Rescue, MO [U]
John L. McMahon, Grinnell Mutual Reinsurance Company, IA [I]
Frank J. Molina, Salt River Pima-Maricopa Indian Community, AZ [E]
G. Terry Smith, University of Illinois, IA [SE]
Dennis W. Smith, Kodiak Enterprises, Inc., IN [SE]
Paul T. Steensland, U.S. Forest Service, CA [E]
Rep. National Wildfire Coordinating Group
George Wendt, Morris County Prosecutor's Office, NJ [U]

Alternates

Daniel L. Churchward, Kodiak Enterprises, Inc., IN [SE]
(Alt. to D. W. Smith)
David S. Evinger, Robins, Kaplan, Miller & Ciresi, MN [L]
(Alt. to H. Lyson)
Andrew M. Giglio, U.S. Fire Administration, MD [SE]
(Voting Alt. to U.S. Fire Administration Rep.)
Gerald Haynes, U.S. Department of the Treasury, Washington, DC [U]
(Alt. to J. Bertoni)
Jack A. Ward, Universal Fire Specialists, Inc., FL [SE]
(Alt. to S. P. Austin)

Frank E. Florence, NFPA Staff Liaison

This list represents the membership at the time the Committee was balloted on the final text of this edition. Since that time, changes in the membership may have occurred. A key to classifications is found at the back of the document.

NOTE: Membership on a committee shall not in and of itself constitute an endorsement of the Association or any document developed by the committee on which the member serves.

Committee Scope: This Committee shall have primary responsibility for documents on professional competence required of fire investigators.

NFPA 1033

Standard for

Professional Qualifications for Fire Investigator

2003 Edition

IMPORTANT NOTE: This NFPA document is made available for use subject to important notices and legal disclaimers. These notices and disclaimers appear in all publications containing this document and may be found under the heading "Important Notices and Disclaimers Concerning NFPA Documents." They can also be obtained on request from NFPA or viewed at www.nfpa.org/disclaimers.

NOTICE: An asterisk (*) following the number or letter designating a paragraph indicates that explanatory material on the paragraph can be found in Annex A.

Changes other than editorial are indicated by a vertical rule beside the paragraph, table, or figure in which the change occurred. These rules are included as an aid to the user in identifying changes from the previous edition. Where one or more complete paragraphs have been deleted, the deletion is indicated by a bullet (•) between the paragraphs that remain.

Information on referenced publications can be found in Chapter 2 and Annex C.

Chapter 1 Administration

1.1 Scope. This standard shall identify the professional level of job performance requirements for fire investigators.

1.2* Purpose. The purpose of this standard shall be to specify the minimum job performance requirements for service as a fire investigator in both the private and public sectors.

1.2.1 The intent of this standard is not to restrict any jurisdiction from exceeding the minimum requirements.

1.2.2 Job performance requirements for each duty are the tasks an individual must be able to perform in order to successfully carry out that duty; however, they are not intended to measure a level of knowledge. Together, the duties and job performance requirements define the parameters of the job of fire investigator.

1.3 General.

1.3.1 The fire investigator shall be at least age 18.

1.3.2 The fire investigator shall have a high school diploma or equivalent.

1.3.3 The authority having jurisdiction shall conduct a thorough background and character investigation prior to accepting an individual as a candidate for certification as a fire investigator.

1.3.4 The job performance requirements for fire investigator shall be completed in accordance with established practices and procedures or as they are defined by law or by the authority having jurisdiction.

1.3.5* The job performance requirements found in this standard are not required to be mastered in the order they appear. Training agencies or authorities shall establish instructional priority and the training program content to prepare individuals to meet the job performance requirements of this standard.

1.3.6* Evaluation of job performance requirements shall be b individuals who are qualified and approved by the authorit having jurisdiction.

1.3.7* The fire investigator shall remain current with investi gation methodology, fire protection technology, and code re quirements by attending workshops and seminars and/o through professional publications and journals.

Chapter 2 Referenced Publications

2.1 General. The documents or portions thereof listed in thi chapter are referenced within this standard and shall be con sidered part of the requirements of this document.

2.2 NFPA Publication. National Fire Protection Associa tion, 1 Batterymarch Park, P.O. Box 9101, Quincy, MA 02269-9101.

NFPA 472, *Standard for Professional Competence of Responders t Hazardous Materials Incidents,* 2002 edition.

2.3 Other Publications. (Reserved)

Chapter 3 Definitions

3.1 General. The definitions contained in this chapter shal apply to the terms used in this standard. Where terms are no included, common usage of the terms shall apply.

3.2 NFPA Official Definitions.

3.2.1* Approved. Acceptable to the authority having jurisdic tion.

3.2.2* Authority Having Jurisdiction (AHJ). An organization office, or individual responsible for enforcing the require ments of a code or standard, or for approving equipment materials, an installation, or a procedure.

3.2.3 Labeled. Equipment or materials to which has been at tached a label, symbol, or other identifying mark of an organiza tion that is acceptable to the authority having jurisdiction an concerned with product evaluation, that maintains periodic in spection of production of labeled equipment or materials, an by whose labeling the manufacturer indicates compliance wit appropriate standards or performance in a specified manner.

3.2.4* Listed. Equipment, materials, or services included in a list published by an organization that is acceptable to the author ity having jurisdiction and concerned with evaluation of product or services, that maintains periodic inspection of production o listed equipment or materials or periodic evaluation of services and whose listing states that either the equipment, material, o service meets appropriate designated standards or has bee tested and found suitable for a specified purpose.

3.2.5 Shall. Indicates a mandatory requirement.

3.2.6 Should. Indicates a recommendation or that which i advised but not required.

3.2.7 Standard. A document, the main text of which contain only mandatory provisions using the word "shall" to indicat requirements and which is in a form generally suitable fo mandatory reference by another standard or code or for adop tion into law. Nonmandatory provisions shall be located in a appendix or annex, footnote, or fine-print note and are not t be considered a part of the requirements of a standard.

3.3 General Definitions.

3.3.1 Due Process. The compliance with the criminal and civil laws and procedures within the jurisdiction where the incident occurred.

3.3.2 Fire Department. An organization providing rescue, fire suppression, and related activities. For the purposes of this standard, the term "fire department" includes any public, private, or military organization engaging in this type of activity.

3.3.3 Fire Investigator. An individual who has demonstrated the skills and knowledge necessary to conduct, coordinate, and complete an investigation.

3.3.4 Investigation. A systematic inquiry or examination.

3.3.5 Job Performance Requirement. A statement that describes a specific job task, lists the items necessary to complete the task, and defines measurable or observable outcomes and evaluation areas for the specific task.

3.3.6 Requisite Knowledge. Fundamental knowledge one must have in order to perform a specific task.

3.3.7 Requisite Skills. The essential skills one must have in order to perform a specific task.

3.3.8 Task. A specific job behavior or activity.

3.3.9 Tools.

3.3.9.1* ***Investigator's Special Tools.*** Tools of a specialized or unique nature that might not be required for every fire investigation.

3.3.9.2* ***Standard Equipment and Tools.*** Investigator's tools and equipment that every investigator must carry.

Chapter 4 Fire Investigator

4.1 General.

4.1.1* The fire investigator shall meet the job performance requirements defined in Sections 4.2 through 4.7.

4.1.2* The fire investigator shall meet the requirements of 4.2.1 through 4.2.3 of NFPA 472.

4.1.3* The fire investigator shall employ all elements of the scientific method as the operating analytical process throughout the investigation and for the drawing of conclusions.

4.1.4* Because fire investigators are required to perform activities in adverse conditions, site safety assessments shall be completed on all scenes and regional and national safety standards shall be followed and included in organizational policies and procedures.

4.1.5* The fire investigator shall maintain necessary liaison with other interested professionals and entities.

4.1.6* The fire investigator shall adhere to all applicable legal and regulatory requirements.

4.1.7 The fire investigator shall understand the organization and operation of the investigative team within an incident management system.

4.2 Scene Examination. Duties shall include inspecting and evaluating the scene so as to determine the area or point of origin, source of ignition, material(s) ignited, and act or activity that brought ignition source and materials together and to assess the subsequent progression, extinguishment, and containment of the fire.

4.2.1 Secure the fire ground, given marking devices, sufficient personnel, and special tools and equipment, so that unauthorized persons can recognize the perimeters of the investigative scene and are kept from restricted areas and all evidence or potential evidence is protected from damage or destruction.

(A) Requisite Knowledge. Fire ground hazards, types of evidence, and the importance of fire scene security, evidence preservation, and issues relating to spoliation.

(B) Requisite Skills. Use of marking devices.

4.2.2 Conduct an exterior survey, given standard equipment and tools, so that evidence is preserved, fire damage is interpreted, hazards are identified to avoid injuries, accessibility to the property is determined, and all potential means of ingress and egress are discovered.

(A) Requisite Knowledge. The types of building construction and the effects of fire upon construction materials, types of evidence commonly found in the perimeter, evidence preservation methods, the effects of fire suppression, fire behavior and spread, and burn patterns.

(B) Requisite Skills. Assess fire ground and structural condition, observe the damage and effects of the fire, and interpret burn patterns.

4.2.3 Conduct an interior survey, given standard equipment and tools, so that areas of potential evidentiary value requiring further examination are identified and preserved, the evidentiary value of contents is determined, and hazards are identified in order to avoid injuries.

(A) Requisite Knowledge. The types of building construction and interior finish and the effects of fire upon those materials, the effects of fire suppression, fire behavior and spread, evidence preservation methods, burn patterns, effects of building contents on fire growth, and the relationship of building contents to the overall investigation.

(B) Requisite Skills. Assess structural conditions, observe the damage and effects of the fire, discover the impact of fire suppression efforts on fire flow and heat propagation, and evaluate protected areas to determine the presence and/or absence of contents.

4.2.4 Interpret burn patterns, given standard equipment and tools and some structural or content remains, so that each individual pattern is evaluated with respect to the burning characteristics of the material involved.

(A) Requisite Knowledge. Fire development and the interrelationship of heat release rate, form, and ignitibility of materials.

(B) Requisite Skills. Interpret the effects of burning characteristics on different types of materials.

4.2.5 Correlate burn patterns, given standard equipment and tools and some structural or content remains, so that fire development is determined, methods and effects of suppression are evaluated, false origin area patterns are recognized, and all areas of origin are correctly identified.

(A) Requisite Knowledge. Fire behavior and spread based on fire chemistry and physics, fire suppression effects, and building construction.

(B) Requisite Skills. Interpret variations of burn patterns on different materials with consideration given to heat release rate, form, and ignitibility; distinguish impact of different types of fuel loads; evaluate fuel trails; and analyze and synthesize information.

4.2.6 Examine and remove fire debris, given standard equipment and tools, so that all debris is checked for fire cause evidence, potential ignition source(s) is identified, and evidence is preserved without investigator-inflicted damage or contamination.

(A) Requisite Knowledge. Basic understanding of ignition processes, characteristics of ignition sources, and ease of ignition of fuels; debris-layering techniques; use of tools and equipment during the debris search; types of fire cause evidence commonly found in various degrees of damage; and evidence-gathering methods and documentation.

(B) Requisite Skills. Employ search techniques that further the discovery of fire cause evidence and ignition sources, use search techniques that incorporate documentation, and collect and preserve evidence.

4.2.7 Reconstruct the area of origin, given standard and, if needed, special equipment and tools as well as sufficient personnel, so that all protected areas and burn patterns are identified and correlated to contents or structural remains, items potentially critical to cause determination and photo documentation are returned to their prefire location, and the area(s) or point(s) of origin is discovered.

(A) Requisite Knowledge. The effects of fire on different types of material and the importance and uses of reconstruction.

(B) Requisite Skills. Examine all materials to determine the effects of fire, identify and distinguish among different types of fire-damaged contents, and return materials to their original position using protected areas and burn patterns.

4.2.8* Inspect the performance of building systems, including detection, suppression, HVAC, utilities, and building compartmentation, given standard and special equipment and tools, so that a determination can be made as to the need for expert resources, an operating system's impact on fire growth and spread is considered in identifying origin areas, defeated and/or failed systems are identified, and the system's potential as a fire cause is recognized.

(A) Requisite Knowledge. Different types of detection, suppression, HVAC, utility, and building compartmentation such as fire walls and fire doors; types of expert resources for building systems; the impact of fire on various systems; common methods used to defeat a system's functional capability; and types of failures.

(B) Requisite Skills. Determine the system's operation and its effect on the fire; identify alterations to, and failure indicators of, building systems; and evaluate the impact of suppression efforts on building systems.

4.2.9 Discriminate the effects of explosions from other types of damage, given standard equipment and tools, so that an explosion is identified and its evidence is preserved.

(A) Requisite Knowledge. Different types of explosions and their causes, characteristics of an explosion, and the difference between low- and high-order explosions.

(B) Requisite Skills. Identify explosive effects on glass, walls, foundations, and other building materials; distinguish between low- and high-order explosion effects; and analyze damage to document the blast zone and origin.

4.3 Documenting the Scene. Duties shall include diagramming the scene, photographing, and taking field notes to be used to compile a final report.

4.3.1 Diagram the scene, given standard tools and equipment, so that the scene is accurately represented and evidence, pertinent contents, significant patterns, and area(s) or point(s) of origin are identified.

(A) Requisite Knowledge. Commonly used symbols and legends that clarify the diagram, types of evidence and patterns that need to be documented, and formats for diagramming the scene.

(B) Requisite Skills. Ability to sketch the scene, basic drafting skills, and evidence recognition and observational skills.

4.3.2* Photographically document the scene, given standard tools and equipment, so that the scene is accurately depicted and the photographs support scene findings.

(A) Requisite Knowledge. Working knowledge of high-resolution camera and flash, the types of film, media, and flash available, and the strengths and limitations of each.

(B) Requisite Skills. Ability to use a high-resolution camera, flash, and accessories.

4.3.3 Construct investigative notes, given a fire scene, available documents (e.g., prefire plans and inspection reports), and interview information, so that the notes are accurate, provide further documentation of the scene, and represent complete documentation of the scene findings.

(A) Requisite Knowledge. Relationship between notes, diagrams, and photos, how to reduce scene information into concise notes, and the use of notes during report writing and legal proceedings.

(B) Requisite Skills. Data-reduction skills, note-taking skills, and observational and correlating skills.

4.4 Evidence Collection/Preservation. Duties shall include using proper physical and legal procedures to retain evidence required within the investigation.

4.4.1 Utilize proper procedures for managing victims and fatalities, given a protocol and appropriate personnel, so that all evidence is discovered and preserved and the protocol procedures are followed.

(A) Requisite Knowledge. Types of evidence associated with fire victims and fatalities and evidence preservation methods.

(B) Requisite Skills. Observational skills and the ability to apply protocols to given situations.

4.4.2* Locate, collect, and package evidence, given standard or special tools and equipment and evidence collection materials, so that evidence is identified, preserved, collected, and packaged to avoid contamination and investigator-inflicted damage and the chain of custody is established.

(A) Requisite Knowledge. Types of evidence, authority requirements, impact of removing evidentiary items on civil or criminal proceedings (exclusionary or fire-cause supportive evidence), types, capabilities, and limitations of standard and special tools used to locate evidence, types of laboratory tests available, packaging techniques and materials, and impact of evidence collection on the investigation.

(B) Requisite Skills. Ability to recognize different types of evidence and determine whether evidence is critical to the investigation.

4.4.3 Select evidence for analysis given all information from the investigation, so that items for analysis support specific investigation needs.

(A) Requisite Knowledge. Purposes for submitting items for analysis, types of analytical services available, and capabilities and limitations of the services performing the analysis.

(B) Requisite Skills. Evaluate the fire incident to determine forensic, engineering, or laboratory needs.

4.4.4 Maintain a chain of custody, given standard investigative tools, marking tools, and evidence tags or logs, so that written documentation exists for each piece of evidence and evidence is secured.

(A) Requisite Knowledge. Rules of custody and transfer procedures, types of evidence (e.g., physical evidence obtained at the scene, photos, and documents), and methods of recording the chain of custody.

(B) Requisite Skills. Ability to execute the chain of custody procedures and accurately complete necessary documents.

4.4.5 Dispose of evidence, given jurisdictional or agency regulations and file information, so that the disposal is timely, safely conducted, and in compliance with jurisdictional or agency requirements.

(A) Requisite Knowledge. Disposal services available and common disposal procedures and problems.

(B) Requisite Skills. Documentation skills.

4.5 Interview. Duties shall include obtaining information regarding the overall fire investigation from others through verbal communication.

4.5.1 Develop an interview plan, given no special tools or equipment, so that the plan reflects a strategy to further determine the fire cause and affix responsibility and includes a relevant questioning strategy for each individual to be interviewed that promotes the efficient use of the investigator's time.

(A) Requisite Knowledge. Persons who can provide information that furthers the fire cause determination or the affixing of responsibility, types of questions that are pertinent and efficient to ask of different information sources (first responders, neighbors, witnesses, suspects, and so forth), and pros and cons of interviews versus document gathering.

(B) Requisite Skills. Planning skills, development of focused questions for specific individuals, and evaluation of existing file data to help develop questions and fill investigative gaps.

4.5.2 Conduct interviews, given incident information, so that pertinent information is obtained, follow-up questions are asked, responses to all questions are elicited, and the response to each question is documented accurately.

(A) Requisite Knowledge. Types of interviews, personal information needed for proper documentation or follow-up, documenting methods and tools, and types of nonverbal communications and their meaning.

(B) Requisite Skills. Adjust interviewing strategies based on deductive reasoning, interpret verbal and nonverbal communications, apply legal requirements applicable, and exhibit strong listening skills.

4.5.3 Evaluate interview information, given interview transcripts or notes and incident data, so that all interview data is individually analyzed and correlated with all other interviews, corroborative and conflictive information is documented, and new leads are developed.

(A) Requisite Knowledge. Types of interviews, report evaluation methods, and data correlation methods.

(B) Requisite Skills. Data correlation skills and the ability to evaluate source information (e.g., first responders and other witnesses).

4.6 Post-Incident Investigation. Duties shall include the investigation of all factors beyond the fire scene at the time of the origin and cause determination.

4.6.1 Gather reports and records, given no special tools, equipment, or materials, so that all gathered documents are applicable to the investigation, complete, and authentic; the chain of custody is maintained; and the material is admissible in a legal proceeding.

(A) Requisite Knowledge. Types of reports needed that facilitate determining responsibility for the fire (e.g., police reports, fire reports, insurance policies, financial records, deeds, private investigator reports, outside photos, and videos) and location of these reports.

(B) Requisite Skills. Identify the reports and documents necessary for the investigation, implement the chain of custody, and organizational skills.

4.6.2 Evaluate the investigative file, given all available file information, so that areas for further investigation are identified, the relationship between gathered documents and information is interpreted, and corroborative evidence and information discrepancies are discovered.

(A) Requisite Knowledge. File assessment and/or evaluation methods, including accurate documentation practices, and requisite investigative elements.

(B) Requisite Skills. Information assessment, correlation, and organizational skills.

4.6.3 Coordinate expert resources, given the investigative file, reports, and documents, so that the expert's competencies are matched to the specific investigation needs, financial expenditures are justified, and utilization clearly furthers the investigative goals of determining cause or affixing responsibility.

(A) Requisite Knowledge. How to assess one's own expertise, qualification to be called for expert testimony, types of expert resources (e.g., forensic, CPA, polygraph, financial, human behavior disorders, and engineering), and methods to identify expert resources.

(B) Requisite Skills. Apply expert resources to further the investigation by networking with other investigators to identify experts, questioning experts relative to their qualifications, and developing a utilization plan for use of expert resources.

4.6.4 Establish evidence as to motive and/or opportunity, given an incendiary fire, so that the evidence is supported by documentation and meets the evidentiary requirements of the jurisdiction.

(A) Requisite Knowledge. Types of motives common to incendiary fires, methods used to discover opportunity, and human behavioral patterns relative to fire-setting.

(B) Requisite Skills. Financial analysis, records gathering and analysis, interviewing, and interpreting fire scene information and evidence for relationship to motive and/or opportunity.

4.6.5 Formulate an opinion of the person(s) and/or product(s) responsible for the fire, given all investigative findings, so that the opinion regarding responsibility for a fire is supported by the records, reports, documents, and evidence.

(A) Requisite Knowledge. Analytical methods and procedures (e.g., hypothesis development and testing, systems analysis, time lines, link analysis, fault tree analysis, and data reduction matrixing).

(B) Requisite Skills. Analytical and assimilation skills.

4.7 Presentations. Duties shall include the presentation of findings to those individuals not involved in the actual investigations.

4.7.1 Prepare a written investigation report, given investigative findings, documentation, and a specific audience, so that the report accurately reflects the investigative findings, is concise, expresses the investigator's opinion, and meets the needs or requirements of the intended audience(s).

(A) Requisite Knowledge. Elements of writing, typical components of a written report, and types of audiences and their respective needs or requirements.

(B) Requisite Skills. Writing skills, ability to analyze information and determine the reader's needs or requirements.

4.7.2 Express investigative findings verbally, given investigative findings, notes, a time allotment, and a specific audience, so that the information is accurate, the presentation is completed within the allotted time, and the presentation includes only need-to-know information for the intended audience.

(A) Requisite Knowledge. Types of investigative findings, the informational needs of various types of audiences, and the impact of releasing information.

(B) Requisite Skills. Communication skills and ability to determine audience needs and correlate findings.

4.7.3 Testify during legal proceedings, given investigative findings, contents of reports, and consultation with legal counsel, so that all pertinent investigative information and evidence is presented clearly and accurately and the investigator's demeanor and attire are appropriate to the proceedings.

(A) Requisite Knowledge. Types of investigative findings, types of legal proceedings, professional demeanor requirements, and an understanding of due process and legal proceedings.

(B) Requisite Skills. Communication and listening skills and ability to differentiate facts from opinion and determine accepted procedures, practices, and etiquette during legal proceedings.

4.7.4 Conduct public informational presentations, given relative data, so that information is accurate, appropriate to the audience, and clearly supports the information needs of the audience.

(A) Requisite Knowledge. Types of data available regarding the fire loss problem and the issues about which the community must know.

(B) Requisite Skills. Ability to assemble, organize, and present information.

Annex A Explanatory Material

Annex A is not a part of the requirements of this NFPA document but is included for informational purposes only. This annex contains explanatory material, numbered to correspond with the applicable text paragraphs.

A.1.2 See Annex B.

A.1.3.5 See Annex B.

A.1.3.6 Those responsible for conducting evaluations should have experience levels or qualifications exceeding those being evaluated or be certified as a fire investigator by an accredited agency, and be trained or qualified to conduct performance evaluations. The latter, for authorities having jurisdiction, can be based on instructor certification. Many agencies select specialists to evaluate various sections of Chapter 4, which the committee accepts as best practice.

A.1.3.7 Fire investigation technology and practices are changing rapidly. It is essential for an investigator's performance and knowledge to remain current. It is recommended that investigators be familiar with the technical information and procedural guidance presented in materials such as NFPA 921 and *Fire Protection Handbook.*

A.3.2.1 Approved. The National Fire Protection Association does not approve, inspect, or certify any installations, procedures, equipment, or materials; nor does it approve or evaluate testing laboratories. In determining the acceptability of installations, procedures, equipment, or materials, the authority having jurisdiction may base acceptance on compliance with NFPA or other appropriate standards. In the absence of such standards, said authority may require evidence of proper installation, procedure, or use. The authority having jurisdiction may also refer to the listings or labeling practices of an organization that is concerned with product evaluations and is thus in a position to determine compliance with appropriate standards for the current production of listed items.

A.3.2.2 Authority Having Jurisdiction (AHJ). The phrase "authority having jurisdiction," or its acronym AHJ, is used in NFPA documents in a broad manner, since jurisdictions and approval agencies vary, as do their responsibilities. Where public safety is primary, the authority having jurisdiction may be a federal, state, local, or other regional department or individual such as a fire chief; fire marshal; chief of a fire prevention bureau, labor department, or health department; building official; electrical inspector; or others having statutory authority. For insurance purposes, an insurance inspection department, rating bureau, or other insurance company representative may be the authority having jurisdiction. In many circumstances, the property owner or his or her designated agent assumes the role of the authority having jurisdiction; at government installations, the commanding officer or departmental official may be the authority having jurisdiction.

A.3.2.4 Listed. The means for identifying listed equipment may vary for each organization concerned with product evaluation; some organizations do not recognize equipment as listed unless it is also labeled. The authority having jurisdiction should utilize the system employed by the listing organization to identify a listed product.

A.3.3.9.1 Investigator's Special Tools. Examples include heavy equipment, hydrocarbon detectors, ignitable liquid detection canine teams, microscopes, flash point testers, and so forth.

A.3.3.9.2 Standard Equipment and Tools. Investigator's standard equipment and tools include a high-resolution camera, flash, and film or media; a flashlight; a shovel; a broom; hand tools; a tape measure or other measuring device; safety clothing and equipment; and evidence collection equipment and supplies. Examples of safety clothing and equipment are found in 10.1.2 of NFPA 921.

A.4.1.1 Job Performance Requirements (JPRs) are organized according to duties. Duties describe major job functions and result from a job task analysis. JPRs, in total, define the tasks that investigators must be able to perform to be qualified; however, it is not logical, nor the committee's intent, that each and every JPR be performed during each investigation. Rather, that the investigator correctly applies selected JPRs as related to the investigation demands or the individual responsibilities.

A.4.1.2 For additional information concerning safety requirements or training, see applicable local, state, or federal occupational safety and health regulations, *Safety at Scenes of Fire and Related Incidents*, and IAAI Fire Investigator Safety Checklist. Chapter 10 of NFPA 921 also provides the investigator with guidance.

A.4.1.3 The basic methodology for fire investigation involves collecting data, then developing and testing hypotheses *(see Chapter 2 of NFPA 921)*. The methodology recommended is the scientific method. Key steps in the scientific method are as follows:

(1) Recognizing the need
(2) Defining the problem
(3) Collecting data
(4) Analyzing the data
(5) Developing the hypothesis
(6) Testing the hypothesis

Developing hypotheses is an ongoing process of data collection and evaluation that happens throughout the investigation. Hypotheses are generally developed and tested for evaluating fire spread and growth, evaluating the nature of fire patterns, and determining origin, cause, and responsibility.

Testing of hypotheses can be either experimental or cognitive. Ultimately, the hypotheses and conclusions reached are only as dependable as the data used or available. Each investigator must apply a level of confidence in that opinion.

A.4.1.4 For additional information concerning safety requirements or training, see applicable local, state, or federal occupational safety and health regulations and *Safety at Scenes of Fire and Related Incidents*.

A.4.1.5 Fire investigators are encouraged to interact with other professionals or organizations in their respective communities. The interaction is important for the effective transfer of information, which can be general, such as what is related in training seminars or journals, or specific to one particular incident.

A.4.1.6 It is understood that fire investigators with arrest powers, fire investigators without arrest powers, and private sector fire investigators can utilize this standard. The following is a list of those legal and regulatory requirements that are critical within the fire investigation field. It is the responsibility of the AHJ to select those issues that are pertinent to its respective agency or organization. Those selected issues should then serve as the measurement criteria or training guideline for the AHJ.

Due process issues (stated in task terms) are as follows: Conduct search and seizure, conduct arrests, conduct interviews, maintain chain of custody, utilize criminal and civil statutes applicable to the situation, and interpret and utilize contract law and insurance law. Show due process of civil rights laws, privacy laws, the fair credit reporting act, laws of trespass and invasion of privacy, laws of libel and slander, laws of punitive damages and attorney-client privilege, rules of evidence including spoliation, and other laws applicable to the AHJ.

A.4.2.8 Examples of tampered systems are fire doors propped open, sprinkler systems shut down, and detection systems disabled. Examples of system failures include construction features such as compartmentation or fire doors that do not confine a fire, sprinkler systems that do not control a fire, smoke control systems that do not function correctly, HVAC systems that do not perform adequately, and alarm systems that fail to provide prompt notification. It is always important to consider the design and intention of the system. Investigators should keep in mind the possibility that systems may not have failed to function, but rather, may have been overcome by the fire development.

A.4.3.2 The use of a high-resolution camera is highly recommended. The use of various video camera systems to supplement visual documentation can be utilized and is encouraged.

A.4.4.2 Fire investigators should determine and identify in advance what authority and specific need each may have to seize and hold item(s) considered to be evidence. Where such authority or need is lacking, items should not be seized.

Annex B Explanation of the Standard and Concepts of JPRs

This annex is not a part of the requirements of this NFPA document but is included for informational purposes only.

B.1 Explanation of the Standard and Concepts of Job Performance Requirements (JPRs). The primary benefit of establishing national professional qualification standards is to provide both public and private sectors with a framework of the job requirements for the fire service. Other benefits include enhancement of the profession, individual as well as organizational growth and development, and standardization of practices.

NFPA professional qualification standards identify the minimum JPRs for specific fire service positions. The standards can be used for training design and evaluation, certification, measuring and critiquing on-the-job performance, defining hiring practices, and setting organizational policies, procedures, and goals. (Other applications are encouraged.)

Professional qualification standards for a specific job are organized by major areas of responsibility defined as duties. For example, the fire fighter's duties might include fire suppression, rescue, and water supply; the public fire educator's duties might include education, planning and development, and administration. Duties are major functional areas of responsibility within a job.

The professional qualifications standards are written as JPRs. JPRs describe the performance required for a specific job. JPRs are grouped according to the duties of a job. The complete list of JPRs for each duty defines what an individual must be able to do in order to successfully perform that duty. Together, the duties and their JPRs define the job parameters; that is, the standard as a whole is a job description.

B.2 Breaking Down the Components of a JPR. The JPR is the assembly of three critical components. *(See Table B.2.)* These components are as follows:

(1) Task that is to be performed
(2) Tools, equipment, or materials that must be provided to successfully complete the task
(3) Evaluation parameters and/or performance outcomes

Table B.2 Example of a JPR

(1) Task	(1) Ventilate a pitched roof
(2) Tools, equipment, or materials	(2) Given an ax, a pike pole, an extension ladder, and a roof ladder
(3) Evaluation parameters and performance outcomes	(3) So that a 4-ft × 4-ft hole is created, all ventilation barriers are removed, ladders are properly positioned for ventilation, ventilation holes are correctly placed, and smoke, heat, and combustion by-products are released from the structure

B.2.1 The Task to Be Performed. The first component is a concise, brief statement of what the person is supposed to do.

B.2.2 Tools, Equipment, or Materials that Must be Provided to Successfully Complete the Task. This component ensures that all individuals completing the task are given the same minimal tools, equipment, or materials when being evaluated. By listing these items, the performer and evaluator know what must be provided in order to complete the task.

B.2.3 Evaluation Parameters and/or Performance Outcomes. This component defines how well one must perform each task — for both the performer and the evaluator. The JPR guides performance toward successful completion by identifying evaluation parameters and/or performance outcomes. This portion of the JPR promotes consistency in evaluation by reducing the variables used to gauge performance.

In addition to these three components, the JPR contains requisite knowledge and skills. Just as the term *requisite* suggests, these are the necessary knowledge and skills one must have to be able to perform the task. Requisite knowledge and skills are the foundation for task performance.

Once the components and requisites are put together, the JPR might read as follows.

B.2.3.1 Example 1. The Fire Fighter I shall ventilate a pitched roof, given an ax, a pike pole, an extension ladder, and a roof ladder, so that a 4-ft × 4-ft hole is created, all ventilation barriers are removed, ladders are properly positioned for ventilation, and ventilation holes are correctly placed.

(A) Requisite Knowledge. Pitched roof construction, safety considerations with roof ventilation, the dangers associated with improper ventilation, knowledge of ventilation tools, the effects of ventilation on fire growth, smoke movement in structures, signs of backdraft, and the knowledge of vertical and forced ventilation.

(B) Requisite Skills. The ability to remove roof covering; properly initiate roof cuts; use the pike pole to clear ventilation barriers; use an ax properly for sounding, cutting, and stripping; position ladders; and climb and position self on ladder.

B.2.3.2 Example 2. The Fire Investigator shall interpret burn patterns, given standard equipment and tools and some structural/content remains, so that each individual pattern is evaluated with respect to the burning characteristics of the material involved.

(A) Requisite Knowledge. Knowledge of fire development and the interrelationship of heat release rate, form, and ignitibility of materials.

(B) Requisite Skills. The ability to interpret the effects of burning characteristics on different types of materials.

B.3 Examples of Potential Uses.

B.3.1 Certification. JPRs can be used to establish the evaluation criteria for certification at a specific job level. When used for certification, evaluation must be based on the successful completion of JPRs.

First, the evaluator would verify the attainment of requisite knowledge and skills prior to JPR evaluation. This might be through documentation review or testing.

Next, the candidate would then be evaluated on completing the JPRs. The candidate would perform the task and be evaluated based on the evaluation parameters and/or performance outcomes. This performance-based evaluation can be either practical (for psychomotor skills such as "ventilate a roof") or written (for cognitive skills such as "interpret burn patterns").

Note that psychomotor skills are those physical skills that can be demonstrated or observed. Cognitive skills (or mental skills) cannot be observed but are evaluated on how one completes the task (process-oriented) or on the task outcome (product-oriented).

Using Example 1, a practical performance-based evaluation would measure the ability to "ventilate a pitched roof." The candidate passes this particular evaluation if the standard was met — that is, a 4-ft × 4-ft hole was created, all ventilation barriers were removed, ladders were properly positioned for ventilation, ventilation holes were correctly placed, and smoke, heat, and combustion by-products were released from the structure.

For Example 2, when evaluating the task "interpret burn patterns," the candidate could be given a written assessment in the form of a scenario, photographs, and drawings and then be asked to respond to specific written questions related to the JPR's evaluation parameters.

Remember, when evaluating performance, you must give the candidate the tools, equipment, or materials (e.g., an ax, a pike pole, an extension ladder, and a roof ladder) listed in the JPRs before he or she can be properly evaluated.

B.3.2 Curriculum Development/Training Design and Evaluation. The statements contained in this document that refer to job performance were designed and written as JPRs. While a resemblance to instructional objectives might be present, these statements should not be used in a teaching situation until after they have been modified for instructional use.

JPRs state the behaviors required to perform specific skill(s) on the job as opposed to a learning situation. These statements should be converted into instructional objectives with behaviors, conditions, and standards that can be measured within the teaching/learning environment. A JPR that requires a fire fighter to "ventilate a pitched roof" should be converted into a measurable instructional objective for use when teaching the skill. *[See Figure B.3.2(a).]*

Using Example 1, a terminal instructional objective might read as follows.

The learner will ventilate a pitched roof, given a simulated roof, an ax, a pike pole, an extension ladder, and a roof ladder, so that 100 percent accuracy is attained on a skills checklist. (At a minimum, the skills checklist should include each of the measurement criteria from the JPR.)

Figure B.3.2(b) is a sample checklist for use in evaluating this objective.

While the differences between job performance requirements and instructional objectives are subtle in appearance, the purpose of each statement differs greatly. JPRs state what is necessary to perform the job in the "real world." Instructional objectives, however, are used to identify what students must do at the end of a training session and are stated in behavioral terms that are measurable in the training environment.

By converting JPRs into instructional objectives, instructors will be able to clarify performance expectations and avoid confusion related to using statements designed for purposes other than teaching. Additionally, instructors will be able to add local/state/regional elements of performance into the standards as intended by the developers.

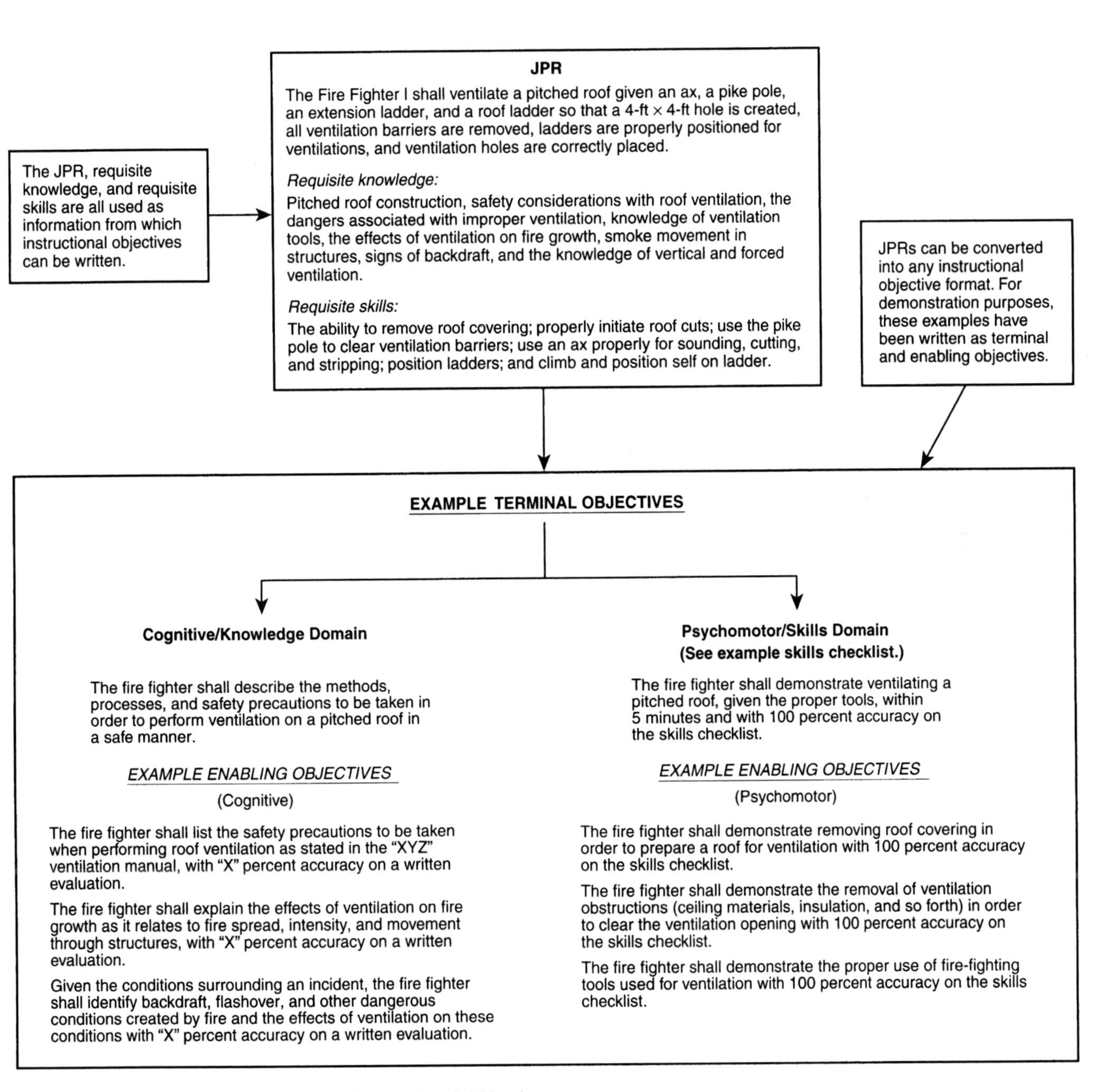

FIGURE B.3.2(a) Converting JPRs into Instructional Objectives.

OBJECTIVE: The fire fighter shall demonstrate ventilating a pitched roof, given the proper tools, within 5 minutes and with 100 percent accuracy on the skills checklist.

YES	NO	
❑	❑	1. 4-ft × 4-ft hole was created.
❑	❑	2. All ventilation barriers were removed.
❑	❑	3. Ladders were properly positioned.
❑	❑	4. Ventilation holes were correctly placed (directly over fire, at highest point, and so forth).
❑	❑	5. The task was completed within 5 minutes. (Time to complete task:________)

FIGURE B.3.2(b) Skills Checklist.

Requisite skills and knowledge should be converted into enabling objectives. The enabling objectives help to define the course content. The course content should include the requisite knowledge and skills. Using Figure B.3.2(b) as an example, the enabling objectives are pitched roof construction, safety considerations with roof ventilation, removal of roof covering, properly initiated roof cuts, and so forth. These enabling objectives ensure that the course content supports the terminal objective.

Note that it is assumed that the reader is familiar with curriculum development or training design and evaluation.

B.4 Other Uses. While the professional qualifications standards are principally used to guide the development of training and certification programs, there are a number of other potential uses for these documents. Because the documents are written in JPR terms, they lend themselves well to any area of the profession where a level of performance or expertise must be determined.

These areas might include the following:

(1) *Employee Evaluation/Performance Critiquing.* The JPRs can be used as a guide by both the supervisor and the employee during an evaluation. The JPRs for a specific job define tasks that are essential to perform on the job as well as the evaluation criteria to measure when those tasks are completed.
(2) *Establishing Hiring Criteria.* The professional qualifications standards can be used in a number of ways to further the establishment of hiring criteria. The AHJ could simply require certification at a specific job level (e.g., Fire Fighter I). The JPRs could also be used as the basis for pre-employment screening by establishing essential minimal tasks and the related evaluation criteria. An added benefit is that individuals interested in employment can work toward the minimal hiring criteria at local colleges.
(3) *Employee Development.* The professional qualifications standards can be useful to both the employee and the employer in developing a plan for an individual's growth within an organization. The JPRs and the associated requisite knowledge and skills can be used as a guide to determine additional training and education required for the employee to master the job or profession.
(4) *Succession Planning.* Succession planning or career pathing addresses the efficient placement of people into jobs in response to current needs and anticipated future needs. A career development path can be established for targeted individuals to prepare them for growth within an organization. The JPRs and requisite knowledge and skills can then be used to develop an educational path to aid in the individual's advancement within the organization or profession.
(5) *Establishing Organizational Policies, Procedures, and Goals.* The JPRs can be incorporated into organizational policies, procedures, and goals where employee performance is addressed.

B.5 Bibliography. See Section C.2.

Annex C Informational References

C.1 Referenced Publications. The following documents or portions thereof are referenced within this standard for informational purposes only and are thus not part of the requirements of this document unless also listed in Chapter 2.

C.1.1 NFPA Publications. National Fire Protection Association, 1 Batterymarch Park, P.O. Box 9101, Quincy, MA 02269-9101.

NFPA 921, *Guide for Fire and Explosion Investigations,* 2001 edition.

Fire Protection Handbook, 19th edition.

C.1.2 Other Publications.

C.1.2.1 IAAI Publication. International Association of Arson Investigators, 300 S. Broadway, Suite 100, St. Louis, MO 63102.

Fire Investigator Safety Checklist, 1997.

C.1.2.2 Other Publication. Munday, J. W. *Safety at Scenes of Fire and Related Incidents.* London: Fire Protection Association, 1994.

C.2 Informational References. The following documents or portions thereof are listed here as informational resources only. They are not a part of the requirements of this document. These references apply to Annex B material.

Boyatzis, R. E., *The Competent Manager: A Model for Effective Performance.* New York: John Wiley & Sons, 1982.

Castle, D. K., "Management Design: A Competency Approach to Create Exemplar Performers." *Performance and Instruction* 28:1989; 42–48.

Cetron, M., and O'Toole, T., *Encounters with the Future: A Forecast into the 21st Century.* New York: McGraw Hill, 1983.

Elkin, G., "Competency-Based Human Resource Development: Making Sense of the Ideas." *Industrial & Commercial Training* 22:1990; 20–25.

Furnham, A., "The Question of Competency." *Personnel Management* 22:1990; 37.

Gilley, J. W., and Eggland, S. A., *Principles of Human Resource Development.* Reading, MA: Addison-Wesley, 1989.

Hooton, J., *Job Performance = Tasks + Competency × Future Forces.* Unpublished manuscript, Vanderbilt University, Peabody College, Nashville, TN, 1990.

McLagan, P. A., "Models for HRD Practice." *Training & Development Journal.* Reprinted, 1989.

McLagan, P. A., and Suhadolnik, D., *The Research Report.* Alexandria, VA: American Society for Training and Development, 1989.

Nadler, L., "HRD on the Spaceship Earth." *Training and Development Journal,* October 1983; 19–22.

Nadler, L., *The Handbook of Human Resource Development.* New York: Wiley-Interscience, 1984.

Naisbitt, J., *Megatrends,* Chicago: Nightingale-Conant, 1984.

Spellman, B. P., "Future Competencies of the Educational Public Relations Specialist" (Doctoral dissertation, University of Houston, 1987). *Dissertation Abstracts International*, 49:1987; 02A.

Springer, J., *Job Performance Standards and Measures.* A series of research presentations and discussions for the ASTD Second Annual Invitational Research Seminar, Savannah, GA (November 5–8, 1979). Madison, WI: American Society for Training and Development, 1980.

Tracey, W. R., *Designing Training and Development Systems.* New York: AMACOM, 1984.

C.3 References for Extracts. (Reserved)

Index

Sample Forms

his appendix provides the reader with sample forms that are taken from NFPA 921, *Guide for Fire and Explosion Investigations.* These forms can be used help fire investigators in collecting data. These sample forms can be found in ragraph A.15.3.2 of NFPA 921.

FIRE INCIDENT FIELD NOTES

Agency: ______________________ File No: ______________

TYPE OF OCCUPANCY

Location/ Address						
Property Description	Structure	Residential	Commercial	Vehicle	Wildland	Other
Other Relevant Info						

WEATHER CONDITIONS

Indicate Relevant Weather Information					
	Visibility	Rel. humidity	GPS	Elevation	Lightning
	Temperature	Wind direction	Wind speed	Precipitation	

OWNER

Name			DOB
d/b/a (if applicable)			
Address			
Telephone	Home	Business	Cellular

OCCUPANT

Name			DOB
d/b/a (if applicable)			
Permanent Address			
Temporary Address			
Telephone	Home	Business	Cellular

DISCOVERED BY

Incident Discovered by	Name		DOB
Address			
Telephone	Home	Business	Cellular

 NFPA 921 (p. 1 of 2)

FIGURE C.1 Sample Form for Collecting Fire Incident Field Data
[Source: NFPA 921, 2004 edition, Figure A.15.3.2(a)]

FIRE INCIDENT FIELD NOTES (Continued)

File No: ______________________

REPORTED BY

Incident Reported by	Name		DOB
Address			
Telephone	Home	Business	Cellular

INVESTIGATION INITIATION

Request Date and Time	Date of request	Time of request
Investigation Requested by	Agency name	Contact person/Telephone no.
Request Received by	Agency name	Contact person/Telephone no.

SCENE INFORMATION

Arrival Information	Date		Time		Comments	
Scene Secured	Yes	No	Securing agency		Manner of security	
Authority to Enter	Contemporaneous to exigency		Consent		Warrant	
			Written	Verbal	Admin.	Crim. / Other
Departure Information	Date		Time		Comments	

OTHER AGENCIES INVOLVED

	Dept. or Agency Name	Incident No.	Contact Person/Phone
Primary Fire Department			
Secondary Fire Department(s)			
Law Enforcement			
Private Investigators			

ADDITIONAL REMARKS

____

____

____

____

____

 NFPA 921 (p. 2 of 2)

FIGURE C.1 *Continued*

CASUALTY FIELD NOTES

Agency: ______ Incident date: ______ Case number: ______

DESCRIPTION

Name: ______ DOB: ______ Sex/Race: ______
Address: ______ Phone: ______
Other identifiers: ______
Description of clothing and jewelry: ______
Occupation: ______ Place of employment: ______
Marital status: ______
Victim's doctor: ______ Victim's dentist: ______
Smoker: ❑ Yes ❑ No ❑ Unknown

CASUALTY TREATMENT

Treated at scene: ❑ Yes ❑ No By: ______
Transported to: ______ Remarks: ______

SEVERITY OF INJURY

❑ Minor ❑ Moderate ❑ Severe ❑ Fatal
Describe injury: ______

NEXT OF KIN

Name: ______ Address: ______ Phone: ______
Relationship: ______ Notified on ___ /___ /_____ By: ______

FATALITY INFORMATION

Where was victim initially found: ______
Who located victim: ______
Body position when initially found: ______
Victim's appearance: ______
Body removed by: ______ To: ______
Photographed in place: ❑ Yes ❑ No Significant blood present under/near victim: ❑ Yes ❑ No

MEDICAL EXAMINER/CORONER

Agency: ______
Date of examination: ___ /___ /_____ Location: ______
Autopsy requested: ❑ Yes ❑ No Autopsy completed: ❑ Yes ❑ No Copy attached: ❑ Yes ❑ No
Full body x-rays: ❑ Yes ❑ No Other x-rays: ______
Identification made from: ❑ Physical appearance ❑ Dental records ❑ Fingerprints ❑ Prior injury comparison
❑ Other: ______
Condition of trachea: ______
Evidence of prefire injury: ❑ Yes ❑ No Type/location: ______
Blood samples taken: ❑ Yes ❑ No Other specimens collected: ______
CO level: ______ Blood alcohol: ______ Other: ______
Cause of death: ______

COMPLETE BODY DIAGRAM ON REVERSE

 NFPA 921 (p. 1 of 2)

FIGURE C.2 Sample Form for Collecting Casualty Field Data
[Source: NFPA 921, 2004 edition, Figure A.15.3.2(b)]

CASUALTY FIELD NOTES (Continued)

REMARKS

____

____

____

____

____

____

BODY DIAGRAM

Indicate parts of body injured: ❑ None ❑ Blisters (red marker) ❑ Burns (black marker)

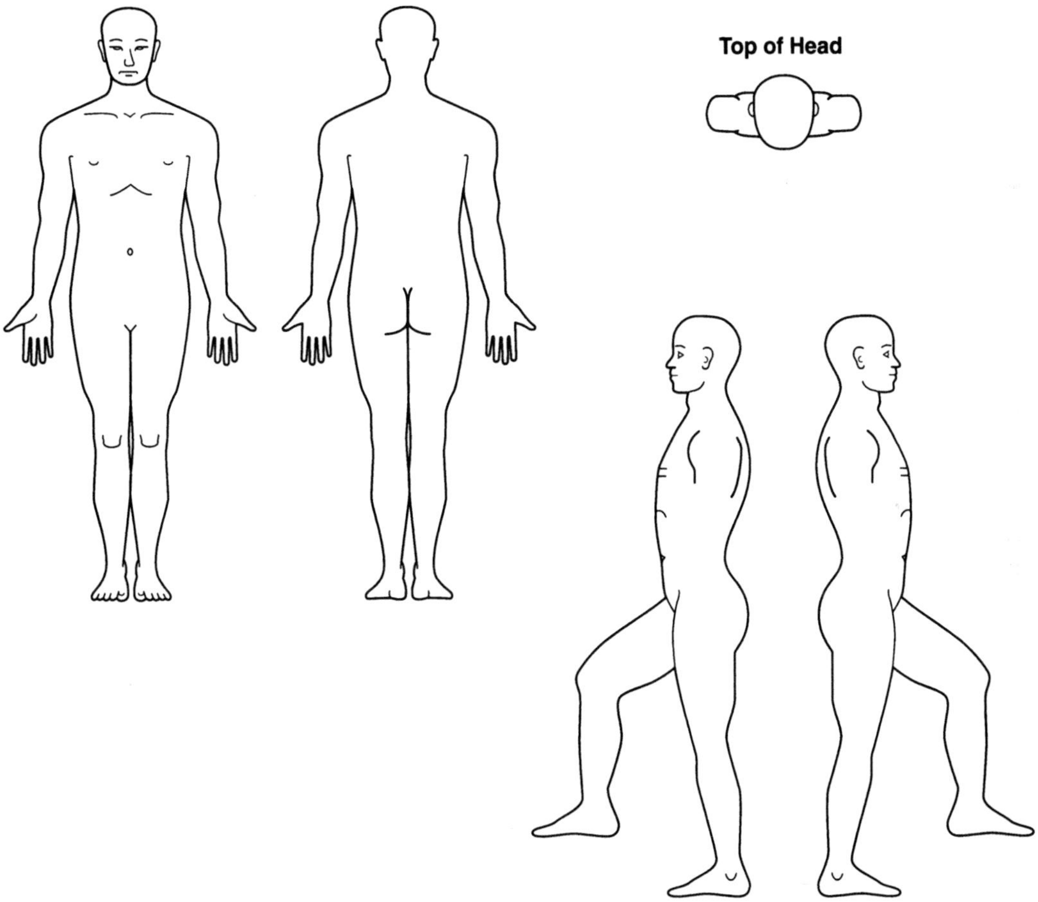

Top of Head

Fire Investigation Data Sheet/Attachment: ______________________ Initials: ________
Body Diagram

 NFPA 921 (p. 2 of 2)

FIGURE C.2 ***Continued***

WILDFIRE NOTES

Agency: ______________________ File number: ______________

PROPERTY DESCRIPTION

Fire damage: ❑ Less than acre ______ No. acres	Other properties involved:
Security: ❑ Open ❑ Fenced ❑ Locked gate	Comments:

FIRE SPREAD FACTORS

Type fire: ❑ Ground ❑ Crown	Factors: ❑ Wind ❑ Terrain	Comments:

AREA OF ORIGIN

PEOPLE IN AREA

At time of fire: ❑ Yes ❑ No ❑ Undetermined	Comments:

IGNITION SEQUENCE

Heat of ignition:

Material ignited:

Ignition factor:

If equipment involved: Make: Model: Serial no.:

Comments:

 NFPA 921 (p. 1 of 1)

FIGURE C.3 Sample Form for Collecting Wildfire Data
[Source: NFPA 921, 2004 edition, Figure A.15.3.2(c)]

EVIDENCE FORM

Case #: ____________________

Date of incidents: ____/____/____ Storage location: ______________________________

Item No.	Description	Location		
______	____________________	____________	Destroyed	Released
______	____________________	____________	Destroyed	Released
______	____________________	____________	Destroyed	Released
______	____________________	____________	Destroyed	Released
______	____________________	____________	Destroyed	Released
______	____________________	____________	Destroyed	Released
______	____________________	____________	Destroyed	Released
______	____________________	____________	Destroyed	Released
______	____________________	____________	Destroyed	Released
______	____________________	____________	Destroyed	Released

How was evidence received? Date received: ____/____/____ Date stored: ____/____/____

❑ Removed from scene by investigator.

❑ Received by investigator from: ______________________________
Name, Company, or Dept.

Received via: ❑ UPS ❑ FedEx ❑ Airborne ❑ US Mail ❑ In person ❑ Freight ____________
Name of Company

❑ Other: ______________________________
Describe

______________________________ Received by

______________________________ Case Investigator

LOCATION EVIDENCE REMOVED

______________________________ Owner

______________________________ State Zip Phone

______________________________ Company

______________________________ Address 2

______________________________ Address 1

______________________________ City

______________________________ City

______________________________ State Zip Phone

NFPA 921 (p. 1 of 2)

FIGURE C.4 Sample Form for Collecting Evidence
[Source: NFPA 921, 2004 edition, Figure A.15.3.2(d)]

EVIDENCE FORM (Continued)

INTERNAL EXAMINATION

Investigator	Date Pulled	Date Examined	Date Returned

EVIDENCE DESTRUCTION

Authorized by Date

Investigator's Authorization Date

Destroyed by Date

EVIDENCE RELEASE

Signature of Person Receiving Evidence

Person Receiving Evidence (Please Print) Date

Company Name

Address

City State Zip Code

Authorized by Date

Investigator's Authorization Date

Released VIA

REMARKS

EXAMINATION BY OTHERS

Name Date of Examination

Company

Address

City State Zip Phone

Authorized by

Investigator's Authorization Date

Name Date of Examination

Company

Address

City State Zip Phone

Authorized by

Investigator's Authorization Date

Name Date of Examination

Company

Address

City State Zip Phone

Authorized by

Investigator's Authorization Date

 NFPA 921 (p. 2 of 2)

FIGURE C.4 ***Continued***

VEHICLE INSPECTION FIELD NOTES

Job # ______ File # ______ Date of Occurrence ______
Insured ______ Date of Assignment ______
Address (City, State) ______ Date of Receipt ______
Loss Location ______ Date of Inspection ______
______ Insp Location ______
Stolen? ❑ Yes ❑ No Recovered By ______ at ______ on ______
Police Report ______ Fire Report ______
# of Keys ______ Alarm System? ❑ Yes ❑ No Alarm Type ______
Hidden Keys? ❑ Yes ❑ No Location ______

VEHICLE

Make ______ Model ______ Year ______
VIN ______ Odometer ______

EXTERIOR

Tires	Tire Type	Wheel Type	Tire Tread Depth	Lugs	Missing
LF					
LR					
RR					
RF					
SP					

Doors	Glass Y/N	Window UP/DOWN	(un)Locked	Open/Closed	Prior Damage
LF					
LR					
RR					
RF					

Body Panels	Construction	Condition	Prior Damage
F Bumper			
Grill			
LF Fender			
LR Quarter			
R Bumper			
RR Quarter			
RF Fender			
Hood			
Roof			
Trunk			

UNDER HOOD	Intact	Missing	Parts Missing	Condition
Engine				
Battery				
Belts & Hoses				
Wiring				
Accessories				

FLUIDS	Level	Condition	Sample Taken
Oil			
Transmission			
Radiator			
Pwr Steer			
Brake			
Clutch			

ATS 851B, 8/97 NFPA 921 (p. 1 of 2)

FIGURE C.5 Sample Form for Vehicle Inspection
[Source: NFPA 921, 2004 edition, Figure A.15.3.2(e); Applied Technical Services, Inc.]

VEHICLE INSPECTION FIELD NOTES (Continued)

Job # ______________

INTERIOR	Intact	Missing	Parts Missing	Condition
Dash Pod				
Glove Box				
Strg Column				
Ignition				
Front Seat				
Rear Seat				
Rear Deck				
			Make/Model	
Stereo				
Speakers				
Accessories				
FLOOR			**Sample Taken**	
LF				
LR				
RR				
RL				

PERSONAL EFFECTS IN THE INTERIOR

TRUNK OR CARGO AREA

AFTERMARKET ITEMS NOT PREVIOUSLY DESCRIBED

ATS 851B, 8/97

NFPA 921 (p. 2 of 2)

FIGURE C.5 *Continued*

PHOTOGRAPH LOG

Roll #: ______________________ Exposures: ______________________

Case #: Date:

Camera make/type: Film type: Film speed: ASA:

Number	Description	Location
1)		
2)		
3)		
4)		
5)		
6)		
7)		
8)		
9)		
10)		
11)		
12)		
13)		
14)		
15)		
16)		
17)		
18)		
19)		
20)		
21)		
22)		
23)		
24)		
25)		
26)		
27)		
28)		
29)		
30)		
31)		
32)		
33)		
34)		
35)		
36)		

Photos taken by: ______________________ Initials: __________

 NFPA 921 (p. 1 of 1)

FIGURE C.6 Sample Form for Photograph Log
[Source: NFPA 921, 2004 edition, Figure A.15.3.2(f)]

ELECTRICAL PANEL DOCUMENTATION

Fire location:	Date:	Case #:

Panel location:	Main size:	Fuses: ☐ / Circuit breakers: ☐

LEFT BANK

#	Rating Amps	Labeled Circuit	Status
1	—		—
3	—		—
5	—		—
7	—		—
9	—		—
11	—		—
13	—		—
15	—		—
17	—		—
19	—		—
21	—		—
23	—		—
25	—		—
27	—		—
29	—		—

Notes:

RIGHT BANK

#	Rating Amps	Labeled Circuit	Status
2	—		—
4	—		—
6	—		—
8	—		—
10	—		—
12	—		—
14	—		—
16	—		—
18	—		—
20	—		—
22	—		—
24	—		—
26	—		—
28	—		—
30	—		—

Notes:

Documented by: ______

NFPA 921 (p. 1 of 1)

FIGURE C.7 Sample Form for Electrical Panel Data
[Source: NFPA 921, 2004 edition, Figure A.15.3.2(g)]

STRUCTURE FIRE

Agency: ______________________ Case number: ______________________

TYPE OF OCCUPANCY

Residential	Single family	Multifamily	Commercial	Governmental
Church	School	Other:		

Estimated age:	Height (stories):	Length:	Width:

PROPERTY STATUS

Occupied at time of fire? ❑ Yes ❑ No	Unoccupied at time of fire? ❑ Yes ❑ No	Vacant at time of fire? ❑ Yes ❑ No
Name of person last in structure prior to fire:	Time and date in structure:	Exited via which door/egress:

Remarks:

BUILDING CONSTRUCTION

Foundation Type	Basement		Crawl space		Slab		Other:
Material	Masonry		Concrete		Stone		Other:
Exterior Covering	Wood	Brick/Stone	Vinyl	Asphalt	Metal	Concrete	Other:
Roof	Asphalt	Wood		Tile		Metal	Other:
Type of Construction	Wood frame	Balloon	Heavy timber	Ordinary	Fire resistive	Non-combustible	Other:

ALARM/PROTECTION/SECURITY

Sprinklers ❑ Yes ❑ No	Standpipes ❑ Yes ❑ No	Security camera(s) ❑ Yes ❑ No
Smoke Detectors ❑ Yes ❑ No	Hardwired ❑ Yes ❑ No	Battery ❑ Yes ❑ No
Were batteries in place? ❑ Yes ❑ No	Location(s):	

Hidden keys ❑ Yes ❑ No where:	Security bars: Windows? ❑ Yes ❑ No Doors? ❑ Yes ❑ No

Remarks:

NFPA 921 (p. 1 of 2)

FIGURE C.8 Sample Form for Structure Fire Data
[Source: NFPA 921, 2004 edition, Figure A.15.3.2(h)]

STRUCTURE FIRE (Continued)

CONDITIONS DOORS/WINDOWS

Doors	Locked	Unlocked but closed	Open	
	Forced entry? ❑ Yes ❑ No	Who forced if known?		
Windows	Secure	Unlocked but closed	Open	Broken
	Broken by first responders? ❑ Yes ❑ No	Remarks:		

FIRE DEPARTMENT OBSERVATIONS

Name of first on scene:	Department:
General observations:	
Obstacles to extinguishment?	First-In Report attached? ❑ Yes ❑ No

UTILITIES

Electric	On Off None	Overhead Underground	
	Company:	Contact:	Telephone:
Gas/Fuel	On Off None	Natural LP Oil	
	Company:	Contact:	Telephone:
Water	Company:	Contact:	Telephone:
Telephone	Company:	Contact:	Telephone:
Other	Company:	Contact:	Telephone:

COMMENTS:

NFPA 921 (p. 2 of 2)

FIGURE C.8 ***Continued***

COMPARTMENT FIRE MODELING

Room Number ________ Use ______________________________

Size (use diagrams if possible) Wall/floor/ceiling

Construction

Length ______________________ ______________________

Width ______________________ ______________________

Height ______________________ ______________________

Lining Materials (that represent over 10% of room lining)
(Include thickness, density, and other material characteristics if known)

Wall Material	Percentage of Walls or Area Involved

Ceiling Material	

Floor or Floor Covering Material	

Doors, Windows, and Other Openings [Enter all heights as distance above floor. If door sill is at floor, enter zero (0).]

Openings	to Top	to Sill	Width	Changes During Fire (How?)[1]

[1] For example: "Window broke at 10:33" or "Door was closed until opened by escaping occupant, then left open — Exit Time 10:30."

 NFPA 921 (p. 1 of 2)

FIGURE C.9 Sample Form for Compartment Fire Modeling Fire Data
[Source: NFPA 921, 2004 edition, Figure A.15.3.2(i)]

COMPARTMENT FIRE MODELING (Continued)

Heating, Ventilation, and Air Conditioning (HVAC). Include air flows from HVAC systems. Give rates and positions of supply and return or exhaust in this room. Also sizes and types of ducts/diffusers.

____

____

____

Tightness of Walls, Closed Windows, Door Fits, etc. (Unless fit is very loose, classify as tight, average, or loose. If fit is very loose, try to get size, number, and location of cracks, etc.)

Doors ________________________________

Windows ________________________________

Inside Walls ________________________________

Exterior Walls ________________________________

Fire History (List all significant events involving progress of the fire.)

	Time (hard or soft)	Event
e.g.	1:10 am	sofa involved, flames 3 feet high
	1:17 am	room flashover
	1:19 am	large fire plume into hallway
	1:23 am	smoke out of third floor window

Initial Fuel Item(s) Description

	Description	Size	Material
e.g.	sofa	full	polyurethane, with cotton upholstery

Suspected Ignitor (List ignitor if known with qualification on confidence.)

Ignitor: e.g. cigarette

Confidence: probable

 NFPA 921 (p. 2 of 2)

FIGURE C.9 ***Continued***

Answer Key

CHAPTER 1

1. **C and D.** A form is provided at the back of NFPA 921. This form can be copied and mailed to NFPA, or an individual can submit an online proposal at the NFPA website (www.nfpa.org).
2. **B.** Every 3 years. NFPA 921 is on a 3-year cycle. The latest edition was issued in 2004.
3. **A.** A guide for fire and explosion investigations. NFPA 921 was developed to establish recommendations and guidelines for fire investigations, as stated in Chapter 1. It was established to assist the investigator, not to mandate the investigator's protocol. A guide can contain mandatory statements, however, although it is not suitable to enforce as law.
4. **B.** False. NFPA 921, Section 1.2, "Purpose," clearly states that not every portion of this document is applicable to every situation. The investigator must determine the appropriate recommended procedures, depending on his or her role in the investigation.
5. **A, B, and C** are the most correct answers according to NFPA 921, Section 1.2. Although the fire chief might be impressed with a well-conducted investigation, that is not the essential reason for seeking the proper determination of a fire's origin and cause.

CHAPTER 2

1. (1) Recognize the need/identify the problem. (2) Define the problem. (3) Collect data. (4) Analyze the data. (5) Develop a hypothesis. (6) Test the hypothesis.
2. **A.** True.
3. **B.** False. The principle of deductive reasoning is used, in which the investigator compares the hypothesis to all known facts.
4. **B.** False. The evidence should be collected and preserved for further testing and evaluation or for use in a courtroom presentation.
5. Answers will vary: (1) Uncovers new factual data for analysis. (2) Provides organization. (3) Allows for an objective study of the scene.

CHAPTER 3

1. **D.** At the high temperatures associated with post-flashover conditions, flaming combustion will continue even with very low concentrations of oxygen.
2. **B.** Thermal inertia is the product of a material's thermal conductivity, density, and heat capacity. Therefore in this list, concrete has the highest thermal inertia.
3. **C.** The rigid polystyrene panel has the lowest thermal inertia among the items listed; therefore its surface would heat more quickly than the other materials.
4. **A.** Remember that there is a difference between flash point and ignition temperature. Although kerosene has a higher flash point than gasoline, its ignition temperature is lower.
5. The ventilation factor is the area of the compartment opening times the square root of the height of the compartment opening. It is used in the equations that calculate the minimum HRR needed to cause a compartment to flashover. There is substantial research on this concept, and numerous reference materials provide further information. It is important to note that NFPA 921 is merely a starting point and not the definitive text; that is why Chapter 3 is titled "Basic Fire Science."
6. The fire investigator should determine the basis for witness observations of fire behavior. Speed of fire development is a conclusion that should be supported by facts. Many times, the witness does not discover the fire in its incipient stage; more often, it is discovered during or after flashover when the flames are exiting the compartment. Fire growth during those later stages can seem fast. The investigator must examine the area of origin and determine the fuel load through direct observation of the remains. Interviews of people who are familiar with the fuel load and its location can aid in the determination of the fuel load. Based on the developed facts, known fuel HRR, and the use of models, the investigator can then determine whether the fire development exceeded what would be expected from the known fuel load.
7. **C.** Because Paragraph 5.5.4.2 of NFPA 921 states that the onset of flashover occurs when the radiant energy flux is 20 kW/m^2, you might wonder how the answer can be 2 W/cm^2. Remember to look to the units used: W/cm^2 equals kW/m^2, as there are 1000 cm^2 in 1 m^2 and 1000 W in 1 kW. This is important to notice in reading different references; make sure to compare numbers in the same units.
8. **E.** None of the above. The flame temperatures of most hydrocarbon and cellulose fuels are essentially the same. There are variations in temperature, but they are not sig-

nificant. The important difference between fuels is their HRR. When you are provided with a set of possible answers, it is important to realize that the right answer might not be included. Have confidence in your knowledge and be prepared to find mistakes in written material.

9. **B.** Answers A and C are irrelevant. Answer D has nothing to do with corners, as a corner does not affect the volume. The higher ratio of surface area to mass affects the amount of energy received and therefore lowers the time to ignition when surfaces are subjected to the same energy flux.
10. **D.** Be able to define the terms that describe basic fire science. Answer A is a unit of measurement per unit of time, and answer C is a unit of measurement per unit of area. An energy flux is typically given as kW/m^2 or W/cm^2 (1 watt = 1 joule/second).
11. **C.** First flame height is directly related to HRR but not to ceiling height. Ceiling height would affect the plume and the development of a ceiling jet. Ceiling height affects flashover because flashover is caused by the development of the hot gas layer. A high ceiling takes longer to reach the critical condition necessary for flashover.
12. **D.** Entrainment of air into the plume changes the plume by cooling it.
13. Reread the pertinent sections of Chapter 3 of NFPA 921 and other reference material until you understand flashover. Discuss flashover as a transition to a fully developed compartment fire, upper gas layer temperature of 1112°F (600°C), radiant energy flux from the gas layer to other fuels, fuel-controlled versus ventilation-controlled fire, HRR, untenable conditions in the compartment, extensions of flames from vents, movement of smoke out of the compartment, possible breakage of windows, and other factors.
14. Reread the pertinent sections of NFPA 921 Chapter 3 and other reference material until you understand HRR. Discuss peak HRR, total HRR, time to peak HRR, flame height relationship to HRR, critical HRR for flashover, HRR relation to ventilation, fuel package HRR, and HRR related to flashover.

CHAPTER 4

1. Because patterns can provide information of the fire origin, spread, and location of fuel packages.
2. **A.** True. The initial burn patterns at the area of origin can be altered during the course of the fire. The initial patterns can be affected by factors such as duration of burning, drop down, post-flashover ventilation patterns, suppression actions, and so forth.
3. **D.** This concept is discussed in NFPA 921, Chapter 4.
4. Airflow over coals or embers can raise temperatures, but the patterns can be misleading because ventilation could cause a fire to spread from slow-burning fuels to rapid-burning fuels, confusing the issue of point of origin.
5. (1) Movement patterns: Flame and heat movement patterns are produced by the growth and movement of fire and the products of combustion away from an initial heat source. (2) Intensity patterns: Flame and heat intensity patterns are produced by the response of materials to the effects of various intensities of heat exposure.
6. **B.** False.
7. Charring refers to the carbonaceous material that has been burned and has a blackened appearance. Spalling refers to the breakdown in surface tensile strength of concrete, masonry, or brick, caused by exposure to high temperatures and rates of heating that result in mechanical forces within the material. Areas might appear lighter in color than adjacent areas. (Spalling and spalling breakdown are covered in Chapter 4 of NFPA 921.)
8. **A.** True.
9. The key variables that affect the validity of depth of calcinations analysis are as follows:
   - Single versus multiple heat or fuel sources
   - Comparisons made only from the same material
   - Finish of the gypsum wallboard should be considered
   - Measurements should be made in a consistent fashion
   - Gypsum wallboard can be damaged during suppression, overhaul, and postfire by hose streams and standing water
10. **B.** False.

CHAPTER 5

1. **B.** Limiting a fire to its area of origin. *Compartmentation* refers to the design features of a building that limit a potential fire to the area or room of origin. These features include but are not limited to fire doors and fire walls.
2. Live load: The weight of temporary loads that need to be designed into the weight-carrying capacity of the structures, such as furniture, furnishings, equipment, machinery, snow, and rainwater.
3. Dead load: The weight of materials that are part of a building, such as the structural components, roof coverings, and mechanical equipment.
4. **A.** Nominal lumber: 2 by 4-in. lumber.
5. **B.** Dimensional lumber: Glued pieces of lumber.
6. **B.** Residential or lightweight commercial construction.
7. **C.** Does not have any concealed spaces. Almost all construction has concealed spaces, which should be a concern for suppression as well as investigation personnel. Concealed areas of concern in platform frame construction are the soffit areas.
8. **D.** Does not need to be protected by gypsum board.
9. **C.** Bearing walls are required to have a minimum 1-hour rating. The correct answer is a 2-hour rating.
10. **B.** False. The effects of weather can greatly diminish a laminated wood beam's ability to carry a load.
11. **C.** Failure of one truss does not affect adjacent trusses. The load supported by the failed truss is transferred to the adjacent truss.

2. **D.** Floor assemblies are tested for fire spread from below and not from above.
3. **C.** A fire-rated door must be installed in a fire-rated wall, to maintain the fire rating of the fire wall.
4. **D.** Utility and other penetrations must be firestopped to maintain the rating of the fire wall.
5. **C.** Concrete and masonry are weak, not strong, under tension loads.

CHAPTER 6

1. **C.** The flow is defined as current, but the measurement of the current is defined as amperage. Amperage is determined by dividing the voltage by the resistance of the circuit.
2. **C.** Electricity flows within circuits and conductors. In the United States, the majority of electricity is alternating current (ac) in which the electrons flow back and forth 60 times per second in the circuit.
3. **D.** Using Ohm's Law to find current, you divide the voltage by the resistance, or 120/10 = 12 amps.
4. **B.** To determine watts, you multiply the voltage by the current use, or 120 × 8.3 = 1000 watts.
5. **B.** The current flowing out is the same as the return current.
6. **D.** Each of the hot conductors carries 120 V when measured from the hot conductor to ground. The difference in the voltage between the two hot conductors is 240 V.
7. **C.** At some point in the system, the neutral wires will be tied to earth or ground. In most residential systems, this occurs at the electrical utility transformer, at the electrical service disconnect, or by connecting to the building's water pipe system. The ground can consist of a rod driven in the earth or, as in many residential homes, the copper water pipe that extends from the building's water system underground.
8. **B.** The *National Electrical Code®* (*NEC®*) provides a safety factor in sizing the conductors in residential homes. The NEC requires that in residential construction, 14 AWG conductors be protected by no higher than a 15-A circuit.
9. **C.**
10. **C.** Arcing through char is usually very localized, with beads or local melting on the conductor. Fire melts larger lengths of the conductor and is not localized. Fire can later destroy evidence of the arcing.
11. **A.** Tests indicate that 14-gauge wires carrying 20 amps may have a slightly elevated temperature that could cause slow deterioration of the insulation, but would not cause it to melt or fall off without additional circumstances occurring to the conductor.
12. **A.** Poor connections can cause increased heating at the contact and the development of oxides at the connection. The oxide has a higher resistance to electric flow and can develop heat hot enough to glow.
13. **A.** The movement of water droplets, dust, and other material in the atmosphere builds up electrostatic charges in the clouds. Lightning bolts average 24,000 A with the potential to range up to 15 million V.
14. **B.** Electrical activity is generally localized and if a long length of conductor is melted, this indicates that the damage was done by fire.
15. **B.** A conductor-to-conductor fault is considered low resistance with high current fault, which will open the circuit protection in a short period of time. This fault does not provide enough heat to ignite nearby ordinary combustibles.
16. **C.** The arcs in ordinary residential construction are a very high temperature but are not competent ignition sources for many fuels because they are so brief and localized. (See "Heat Release Rate in Chemistry.")

CHAPTER 7

1. The ranges of the properties of natural gas and commercial propane are as follows:

   Natural Gas
   - *Relative weight to air:* Lighter.
   - *Vapor density:* 0.59 to 0.72.
   - *Lower explosive limit (LEL):* 3.9 percent to 4.5 percent.
   - *Upper explosive limit (UEL):* 14.5 percent to 15 percent.
   - *Ignition temperature:* 900°F to 1170°F (483°C to 632°C).

   Commercial Propane
   - *Relative weight to air:* Heavier.
   - *Vapor density:* 1.5 to 2.0.
   - *Lower explosive limit (LEL):* 2.15 percent.
   - *Upper explosive limit (UEL):* 9.6 percent.
   - *Ignition temperature:* 920°F to 1120°F (493°C to 604°C).
2. **A.**
3. The American Society of Mechanical Engineers (ASME) and the U.S. Department of Transportation (DOT).
4. The common requirements of fuel gas appliances are as follows:
   - *Installation:* Basic requirements are similar.
   - *Approvals:* Approved acceptable to the authority having jurisdiction.
   - *Type of gas:* Equipment must be used with the specific type of gas for which it was designed.
   - *Areas of flammable vapors:* Appliances may not be installed in residential garage locations where flammable vapors are likely to be present.
   - *Gas appliance pressure regulators:* Necessary when the building gas supply pressure is higher than that at which the gas utilization equipment is designed to operate or varies beyond the design pressure limits of the equipment.
   - *Accessibility for service:* Located for accessible maintenance, service, and emergency shutoff.
   - *Clearance to combustible materials:* Appliances and vents should be installed with sufficient clearance from combustible materials.

- *Electrical connections:* Appliances should be electrically safe and comply with NFPA 70.
- *Venting and air supply:* Venting to the exterior is required for most appliances.
- *Appliance controls:* Common controls such as temperature, ignition and shutoff devices, gas appliance pressure regulators, and gas flow control accessories.

5. Repeat the test with a small gas burner open and ignited to show whether the meter is working properly.
6. Long-existing underground leaks, which have been permeating the soil and dissipating into the air, can be identified by the presence of dead grass or other vegetation over the area of the leak.

CHAPTER 8

1. (A) Age, (B) physical disabilities, (C) intoxication, (D) incapacitating or limiting injuries, and (E) medical conditions.
2. (A) Age, (B) level of rest, (C) alcohol use, (D) drug use, and (E) inhalation of smoke and toxic gases.
3. A. Person will tend to exit by the method used to enter.
   B. People will have an orderly response to the threat.
   C. The people will have conflicting responses, be less orderly.
   D. Women report a fire more quickly than men.
   E. Men attempt to put out a fire or delay reporting it.
   F. Confusion and heightened anxiety may result.
   G. Person may have the belief that he or she is less safe in a fire.
   H. Person will respond more readily and quickly.
   I. Person will have a more delayed response than to a verbal alarm.
   J. Person may have increased response time, but the suppression may also impede escape.
   K. Person may have a false sense of security and may not act to escape as quickly.
   L. A person's perception of the threat may be minimized.
   M. Person may have decreased strength and mental acuity.
4. (A) An alert word, (B) a statement of the danger, (C) a statement of how to avoid the danger, and (D) explanations of the consequences of the danger.
5. *Child firesetters (ages 2 to 6)*
   Motive: curiosity. *Location:* hidden and out of sight of their guardian.

   Juvenile firesetters (ages 7 to 13)
   Motive: broken family environment or physical or emotional trauma. *Location:* educational setting.

   Adolescent firesetters (ages 14 to 16)
   Motive: stress, anxiety, or anger. *Location:* schools, churches, vacant buildings, fields, and vacant lots.
6. (A) Presence of flames, (B) presence of smoke, (C) effects of toxic gases and oxygen depletion.
7. Larger groups have a tendency not to respond in appropriate time frames because individuals do not want to be first to disrupt the group.

CHAPTER 9

1. **B.** Documentary evidence.
2. **B.** False. Questions regarding access to a fire scene are protected under the Fourth Amendment to the U.S. Constitution.
3. **C.** In most states, the arson statutes refer to structures and property. The intentional act of setting fire to a person would fall under the assault statutes. The investigator is advised, however, to become familiar with the arson statutes that pertain to his or her jurisdiction.
4. A. *Exigent circumstances:* Legal authority to enter property in an emergency, such as to control and extinguish a hostile fire
   B. *Consent:* Person in lawful control of the property grants permission
   C. *Administrative search warrant:* Obtained from a court of competent jurisdiction authorizing entry to fulfill an administrative responsibility—for example, to conduct a fire scene search for the limited purpose of origin and cause only
   D. *Criminal search warrant:* Obtained on the traditional showing of probable cause to believe that a crime has occurred and that evidence of the crime will be found on the premises to be searched
5. **B.** True.
6. **A.** True.
7. **B.** False. For the admission into evidence of fire debris samples, it is absolutely necessary to prove a documented chain of custody to maintain the integrity of the evidence.
8. **B.** Consent can be taken back at any time without reason by the lawful property owner.
9. **C** (as soon as practical). The Supreme Court did not specifically define a "reasonable amount of time." If in doubt, the investigator should use another means of entry such as an administrative search warrant or consent from the person who has control over the premises.
10. **B.** False. Fire investigation usually requires removing some evidence from the scene. In and of itself, this removal is not necessarily considered spoliation.

CHAPTER 10

1. **A.** True.
2. Biological, physical, chemical, and radiological hazards.
3. **C.** Incident Scene Commander.
4. The status of all utilities—electricity, gas, and water.
5. **A.**

CHAPTER 11

1. **D.** Interviews with individuals who have a specific interest in the outcome of the fire investigation should be approached with an attitude of distrust.
2. **C.** Attorney-client communications are considered privileged, and are protected from disclosure in a court of law.

3. **A.** The U.S. Fire Administration is a federal agency that maintains an extensive database related to fire incidents through the National Fire Incident Reporting System.
4. **B.** The Society of Fire Protection Engineers (SFPE) is involved with engineering protection from fire hazards. The NFIRS, IAAI, and FBI all maintain records of arson incidents.
5. **C.** Interviews are verbal sources. The NOAA maintains weather data, television studios have video footage, and lightning detection systems have data on lightning strikes.

CHAPTER 12

1. Conflict of interest.
2. **B.** False.
3. Safety gear as well as tools and equipment appropriate to the fire scene.
4. Have a preinvestigation team meeting to address questions of jurisdictional boundaries and assign specific responsibilities to team members. Conditions of the scene and safety considerations should also be addressed.
5. Answers may vary. Any five from the following list would be correct:
   - Materials engineer or scientist
   - Mechanical engineer
   - Electrical engineer
   - Chemical engineer or scientist
   - Fire science and engineering
   - Fire protection engineer
   - Fire engineering technician
   - Industry expert
   - Attorney
   - Insurance agent/adjuster
   - Canine teams

CHAPTER 13

1. Answers will vary and can include any five of the following: photographs, videotape, diagrams, maps, overlays, tape recordings, notes.
2. True.
3. **B.** Transfer the images from the memory stick/card to a nonalterable medium, such as a CD-ROM.
4. **B.** 100–400.
5. A. Sequential photographs.
6. Answers will vary and can include any five of the following: activities during the fire and the fire itself, crowds watching the fire, fire suppression activities, exterior photographs, structural photographs, interior photographs, utility and appliance photographs, evidence photos, victim photographs, photos taken from witness viewpoints, aerial photographs.
7. **D.** Interior photographs (because of lighting concerns).

CHAPTER 14

1. **A.** True.
2. **B.** False.
3. **B.** False.
4. **A.** True.
5. **A.** True.
6. **A.** True.
7. **A.** True.
8. **A.** True.
9. **B.** False.
10. **A.** True.
11. **B.** False.
12. Any three of the following will be acceptable:
    - Name of the investigator collecting the evidence
    - Date and time collected
    - Identification name or number
    - Case number
    - Brief description of the evidence
    - Where the item was discovered
13. **B.** Chain of custody.
14. **B.** False.
15. **A.** True.

CHAPTER 15

1. (1) The physical marks left by the fire; (2) the observations reported by persons who witnessed the fire or who were aware of the conditions at the time of the fire; (3) the physics and chemistry of fire initiation, development, and growth as they relate to the analysis of known or hypothesized fire conditions capable of producing the conditions that exist; and (4) the location where electrical arcing has caused damage and the electrical circuit involved.
2. There may be irrefutable evidence that determines the area of origin, such as an article of physical evidence or information provided by a dependable eyewitness to the initiation of the fire.
3. Most often, the investigator will need to identify and document fire patterns and review all available data before forming a hypothesis when determining the area of origin. Whenever the hypothesis contradicts any portion of the data, it is important for the investigator to determine whether the scenario or the evidence is erroneous.
4. The area of origin is most often determined by examining the fire patterns, starting with the area of the least damage and moving to the areas of greatest damage. Only after the area of origin has been determined, based on the patterns produced by the movement of heat, flames, and smoke, combined with a determination of the duration of burning at one or more points, can the specific location of the point or points of origin be identified. During the majority of investigations, following the fire patterns from the area(s) of least damage to the area(s) of the most severe damage will lead the investigator to one or more points where the fire originated.
5. A. *Notes:* Type, location, description, and measurements of the patterns; the material on which the patterns are displayed; and the investigator's analysis of the direction and intensity of the patterns.

B. *Photography:* Patterns should be photographed in several different ways to show their shape, size, relationship to other patterns, and the location within the fire scene. Include changes in the viewing angle of the camera and different lighting techniques to highlight the texture of the pattern.

C. *Vector diagrams:* Construct a diagram of the scene including walls, doors and doorways, windows, and any pertinent furnishings or contents. Use arrows to show the direction of heat or flame spread based on fire patterns present. The arrows can be labeled to show temperature, duration of heating, heat flux, or intensity.

D. *Depth-of-char survey grid diagrams:* Construct a depth-of-char diagram showing char measurements on graph paper with a convenient scale.

6. The preliminary scene assessment includes the identification of the area(s) of least damage compared to the area(s) of most damage. The interior and exterior of a structure should always be documented. One purpose of the initial examination is to determine the scope of the investigation, and it should include considerations such as equipment and staffing needs. The investigator may develop areas of interest and a preliminary scenario as to how the fire spread through the structure. This scenario allows the investigator to organize and plan the work to be done.
7. **A.** True.
8. The purpose of fire scene reconstruction is to recreate the state that existed prior to the fire. It allows the investigator to observe patterns and identify prefire conditions such as fuel load, surfaces, and fire, smoke, and heat movement—and may also uncover data indicating areas that warrant further study and that assist in forming a preliminary hypothesis. Scene reconstruction is accomplished by returning the contents and other items within the structure to their original prefire positions after debris has been removed. Any remains or contents that were uncovered during debris removal should be noted in their original location and returned to their prefire positions.
9. Contradictions should be recognized and resolved by using the scientific method. The investigator should reexamine the data to determine whether other reasonable hypotheses can exist or whether there may be other explanations that still support the original hypothesis. The investigator must evaluate all the data and identify and resolve all contradictions to validate the determined area of origin.
10. **B.** False. Even though it is usually not possible to determine an origin and cause determination and even though an analysis is more difficult to accomplish, studying a total burn site will still produce valuable information about the fire and what happened to the contents.

CHAPTER 16

1. **B.** False.
2. **A.** True.
3. Ignition source, first fuel ignited, oxidizer present, and ignition sequence.

CHAPTER 17

1. **C.** Suspicious.
2. **A.** True.

CHAPTER 18

1. **A.** True.
2. **A.** Benchmark events are particularly valuable as a foundation for the timeline or may have significant relation to the cause, spread, detection, or extinguishment of a fire.
3. **C.** To effectively evaluate and document the sequence of events precipitating the fire, the actual fire incident, and postfire activity, it is quite possible that two or more timelines will be required. A macro evaluation of events incorporates activities that occurred prior to the fire or event, whereas a micro evaluation is specific in time and focuses on some discrete segment of the total timeline.
4. **B.** Fault trees are developed by using deductive reasoning. Deductive reasoning is used by the fire investigator to compare a hypothesis to all know facts.
5. **B.** False. (See NFPA 921, Chapter 20, for more information.)
6. **C.** (See NFPA 921, Chapter 20, Paragraph 20.4.8, for more information.)
7. **D.** (See NFPA 921, Chapter 20, Paragraph 20.4.8.3(b), for more information.)
8. **A.** True. (See NFPA 921, Chapter 20, Paragraph 20.6.2, for more information.)
9. **A.** True. (See NFPA 921, Chapter 20, Paragraph 20.6, for more information.)

CHAPTER 19

1. **A.** True.
2. **B.** False. Although an explosion is almost always accompanied by noise, noise itself is not an essential element. The generation and violent escape of gases are the essential elements of an explosion.
3. **A.** True.
4. **B.** False. The ignition source of an explosion may be blasting caps or other pyrotechnic devices. The wires and component parts of the device will sometimes survive the explosion.
5. **A.** True.
6. **A.** True. The two distinct phases include the positive pressure phase and the negative pressure phase.
7. **A.** True. There are two major types of explosions with which investigators are routinely involved: mechanical and chemical.
8. **B.** False. The seat of an explosion is defined as the crater or area of greatest damage, located at the point or epicenter of an explosion. Grain dust explosions most

often occur in confined areas of relatively wide dispersal, such as grain elevators. The large areas of origin preclude the production of pronounced seats of explosion.

9. **B.** False. Explosions that occur in mixtures at or near the lower explosive limit (LEL) or upper explosive limit (UEL) of a gas or vapor produce less violent explosions than those near the optimum concentration.
10. **B.** False. It was once widely thought that if the walls of a particular structure were blown out at floor level, the fuel gas was heavier than air, and conversely, if the walls were blown out at ceiling level, the fuel was lighter than air. However, the level of the explosion damage within a conventional room is a function of the construction strength of the wall headers and bottom plates, the least resistant giving way first.
11. **C and D.** Deflagrations are propagating reactions in which the energy transfer is accomplished through ordinary process such as heat and mass transfer. The rate of such reactions is controlled by the rate of the particular transfer phenomena but at a velocity that is less than the speed of sound (subsonic). Deflagrations usually results in structural damage that is uniform and omnidirectional and thus a widespread evidence of burning, scorching, and blistering.
12. **A and C.** Mechanical, physical, and chemical are the three major types of explosions. However, there are several subtypes within each classification.
13. **F.** All of the above.
14. **A, B, C, D, and E.** For a given mass of dust material, the total surface area increases as the particle size decreases (the finer the dust, the more violent the explosion). An increase in the moisture content of the dust particle increases the energy required for ignition and the ignition temperature of a dust suspension. The magnitude of turbulence within the suspended dust–air mixture greatly increases the rate of combustion and thereby the rates of pressure rise. The concentration of the dust in air has a great effect on the ignitibility and violence of the blast pressure wave. Reaction rates are controlled more by the surface-area-to-mass-ratio than by a maximum concentration. Ignition temperatures for most materials range from 600°F to 1100°F (320°C to 590°C).
15. **C and D.** The terms *high-order damage* and *low-order damage* have been used to characterize explosion damage. These terms are recommended to reduce confusion with similar terms used to describe the energy release from explosions.
16. **H.** All of the above.
17. **B and C.** A boiling liquid expanding vapor explosion will produce a seated explosion if the confining vessel (such as a barrel or small tank) is of a small size and if the rate of pressure release at failure is rapid enough.
18. **B, C, and D.** The distance to which missiles can be propelled outward from an explosion depends greatly on their initial direction. Other factors include their weight and aerodynamic characteristics.
19. **C.** All of the above.
20. **B, C, and D.**

CHAPTER 20

1. **B.**
2. **A.** True.
3. **D.**
4. **A.**
5. What an investigator may consider "excessive," "unnatural," or "abnormal" can actually occur in an accidental fire, depending on the geometry of the space, the fuel characteristics, and the ventilation of the compartment.
6. Multiple fires can have the effect of overtaxing the fire suppression system beyond its design capabilities.
7. A. Geographic area, clusters.
   B. Temporal frequency.
   C. Materials and methods.
8. **C.**
9. **B.**
10. Motive refers to the reason that an individual or group may do something; intent refers to the purposefulness or deliberateness of the person's actions.
11. **D.**
12. **B.** False.

CHAPTER 21

1. **A.** True. (See NFPA 921, Chapter 23, Paragraph 23.3.1.)
2. **B.** False. (See NFPA 921, Chapter 23, Paragraph 23.3.2.)
3. **A.** True. (See NFPA 921, Chapter 23, Paragraph 23.3.4.)
4. **A.** True. (See NFPA 921, Chapter 23, Paragraph 23.3.5.1.)
5. **B.** False. (See NFPA 921, Chapter 23, Paragraph 23.3.6.2.)
6. **B.** False. (See NFPA 921, Chapter 23, Paragraph 23.4.1.)
7. **A.** True. (See NFPA 921, Chapter 23, Paragraph 23.5.1.)
8. **A.** True. (See NFPA 921, Chapter 23, Paragraph 23.4.6.)
9. **B.** False. (See NFPA 921, Chapter 23, Paragraph 23.5.2.)
10. **A.** True. (See NFPA 921, Chapter 23, Paragraph 23.3.6.1.)
11. **A.** True.
12. **A.** True. (See NFPA 921, Chapter 23, Section 23.5.)
13. **B.** False. (See NFPA 921, Chapter 23, Section 23.5.)
14. **A.** True.
15.
    - Remains identification
    - Victim identification
    - Cause of death
    - Manner of death
    - Victim activity
    - Postmortem changes
    - (See NFPA 921, Chapter 23, Section 23.4 through Paragraph 23.4.6.)

CHAPTER 22

1. **C.** If the appliance is determined to be in the area of origin, it is important to document the condition of

the appliance. The investigators are cautioned not to disassemble the appliance until all parties have been notified. (Disassembly should be conducted only by someone who has the proper expertise.)

2. **C.** It may be necessary for the investigator to obtain information about the appliance so that other parties can be notified and for documentation. It is oftentimes necessary to move the appliance to obtain these data and should be done with minimal disturbance. The appliance must be properly documented before being removed.
3. **B.** False. (See NFPA 921, Chapter 24, Paragraph 24.5.1.5.)
4. **C.** Failure of an appliance may generate energy and leave patterns on the appliance. However, burn patterns on appliances may be the result of other material burning near the appliance or other factors not related to the cause.
5. **A.** The investigator needs to determine whether there is enough energy provided by the appliance to ignite the available fuels. If the investigator cannot determine the ignition scenario, additional examination might be needed, or the cause should be classified as undetermined.
6. **C.** Appliances manufactured after 1962 are required to have a ground, except some double-insulated appliances. They may only have two conductors.
7. **C.** These cords are often termed *polarized* plugs. The neutral blade can only be plugged into the neutral side of the receptacle. In older homes, these cords can be used with an adapter that provides a ground.
8. **B.** The prongs may show no signs of smoke staining, indicating that they were protected inside the receptacle, compared to prongs that were unplugged and show signs of smoke staining.
9. **C.** Most heat-producing appliances contain thermal protective devices that are often discovered following the fire. Examination of the appliance may provide information on whether it opened, had been bypassed, or remained closed (with current capable of passing through it).
10. **B.** The most common is bimetallic thermostats, in which two dissimilar metals are bonded together. One of the metals expands as the temperature increases, causing the switch to open.
11. **D.** TCOs are single-use devices that are at a higher temperature rating than the thermostat of the appliance. When the thermal pellet inside the TCO melts, the device needs to be replaced before the appliance is used again.
12. **C.** Fluorescent ballasts manufactured after 1978 are equipped with a thermal protection device. Some transformers are filled with oil or potting compound, yet many others are not. Under normal operating conditions, transformers can operate for a long period of time. However, with long-term use and when ventilation is restricted, windings can short out to each other, causing additional heating.
13. **A.** Newer motors often are equipped with some type of thermal protection devices, many of which reset after they have cooled. Electrical appliances often provide overcurrent circuit protection in the circuits leading to the motor and monitor the current flow to the motor.
14. **D.** These heaters are low wattage and operate at regular intervals so that the internal surfaces do not sweat (antisweat).
15. **A.** Thermostats that can automatically reset have been installed in ballasts manufactured since 1978. These ballasts are usually identified by a "P" on the label or stamped on the case. Some of the newer ballasts are electronic.

CHAPTER 23

1. **B.** False. The more fuels that become involved, the more intense the fire.
2. **A.** True. Atomized mists caused from a variety of hydrocarbon fuels under pressure are easily ignited in this form, as any open flame or heat source is usually sufficient to ignite this form of fuels.
3. **D.** (See NFPA 921, Chapter 25, Paragraph 25.5.3.2.)
4. **B.** (See NFPA 921, Chapter 25, Paragraph 25.4.4.)
5. **B.** (See NFPA 921, Chapter 25, Paragraph 25.5.1.2.)
6. **B.** (See NFPA 921, Chapter 25, Paragraphs 25.5.1 through 25.5.5.)
7. **B.** (See NFPA 921, Chapter 25, Paragraph 25.5.3.)
8. **B.**
9. **B.** False. (See NFPA 921, Chapter 25, Section 25.10.)
10. **A.** True. (See NFPA 921, Chapter 25, Section 25.12.)

CHAPTER 24

1. It is important to know the fire weather conditions for wildfire investigations because of the effect on fuels. Fine fuel, such as grasses, are very reactive to relative humidity. Temperature, relative humidity, wind direction and speed, precipitation, and lightning downstrike information are all important factors.
2. **B.** False. Flashover is the transition from a developing fire to a fully developed fire in a compartment. Rapid ignition of a wildfire fuel complex, such as continuous brush or tree crowns, is not flashover. There are terms that are unique to compartment fires or wildfires, and they should be used accordingly.
3. **C.** Fire intensity is affected by moisture content, oil content of the fuel, and wind direction.
4. **D.** A V-shaped pattern in a compartment fire is caused by the interaction of the fire plume with a vertical surface. A wildfire V-shaped pattern is caused by the fire spread on a horizontal ground surface.
5. **B.** False

CHAPTER 25

1. Major investigations are described as incidents in which there are several interested parties or multiple agencies with interests in conducting an investigation.

These may involve investigations of fatal fires, fires in high-rise buildings, and other fires involving major structural damage and large monetary losses. Major incidents are not always large in size or magnitude but tend to be complex.

2. **C.** A protocol regarding evidence collection, processing, and storage should be agreed on by all interested parties prior to conducting the investigation. This will often require that one person assume the responsibilities and perform the duties of the evidence technician and maintain the security of the evidence.
3. **B.** Joint investigations allow all interested parties an equal opportunity to establish facts and examine their own hypotheses.
4. **A.** Joint investigations conducted by representatives of both the public and private sectors can occur wherein public officials can maintain the security of the scene. Various concerns can be addressed and alleviated through planning and communication. Under some circumstances or in certain jurisdictions, this may require public officials to conduct an investigation while other interested parties observe. Regardless of the circumstances, every investigator's goal should be to conduct an investigation that seeks the facts and that allows his or her observations to be viewed by all interested parties. Allowing all interested parties to participate in an investigation can provide resources and expertise that are of benefit to public officials.
5. **D.** A consensus among the parties that identifies the various concerns should be developed prior to conducting an investigation. An outline addressing the issues related to conducting a scene examination should be part of this agreement. The agreement should not provide the details of the scene examination but should provide an understanding as to how the examination will be conducted. This agreement is often called a "Memorandum of Understanding."
6. Any three of the following:
   - Control and access to the site.
   - How and which information discovered will be shared.
   - Joint custody and examination of evidence. Interested parties should be notified prior to destructive examination.
   - Acquisition and processing of needed nonproprietary information.
   - Release of information to the public.
   - Protocol for scene examination and debris removal.
   - Development of a flow chart to provide guidance for the general scope of the investigation.
7. **B.** It is often possible to evaluate portions of a building and gather necessary data so that they can then be released for occupant access. Safety issues must be addressed before allowing occupants access to any space. Each investigator should collect all relevant data from a specific tenant space before allowing occupants access to the scene.
8. **C.** During major investigations, interested parties normally bring in their own investigation teams. The team members often include investigators, forensic technicians, legal counsel, and other specialized personnel. Each interested party should have one spokesperson at the team leader committee meeting or the organization meeting. Other representatives are allowed to attend the meeting but they should voice their concerns through their representatives. This allows for an orderly meeting that will provide positive results.
9. **C.** Safety is of the utmost importance in all investigations and is the responsibility of each person on the scene. Some scenes require one or more safety officers to monitor scene conditions that are subject to change caused by weather conditions or structure alterations.
10. Any three of the following would be correct:
    - Fire chief
    - Fire prevention personnel
    - Suppression personnel
    - Police officers
    - Passersby
    - Neighbors
    - Property owners
    - Employees
    - Tenants

Glossary

Absolute Temperature A temperature measured in Kelvins (K) or Rankins (R). Absolute zero is the lowest possible temperature, with 0 K being equal to –273°C, and 0 R equal to –460°F; 273 K corresponds to 0°C, and 460 R corresponds to 0°F.

Accelerant An agent, often an ignitible liquid, used to initiate a fire or increase the rate of growth or spread of fire.

Accident (1) An unplanned event that interrupts an activity and sometimes causes injury or damage. (2) A chance occurrence arising from unknown causes; an unexpected happening due to carelessness, ignorance, and the like.

Air Drop The aerial application of suppressant either on the fire or on unburned fuels.

Alloy A solid or liquid mixture of two or more metals, or of one or more metals with certain nonmetallic elements, caused by fusing the components. The properties of alloys are often greatly different from those of the component metals. (Source: *The Condensed Chemical Dictionary*)

Ambient Someone's or something's surroundings, especially as they pertain to the local environment, for example, ambient air and ambient temperature.

Ampacity The current, in amperes (A), that a conductor can carry continuously under the conditions of use without exceeding its temperature rating.

Ampere The unit of electric current that is equivalent to a flow of 1 coulomb per second (C/s). One coulomb is defined as 6.24×10^{18} electrons.

Approved Acceptable to the authority having jurisdiction.

Arc A high-temperature luminous electric discharge across a gap.

Arcing Through Char Arcing associated with a matrix of charred material (e.g., charred conductor insulation) that acts as a semiconductive medium.

Area of Origin The room or area where a fire began. *See also* Point of Origin.

Arrow Pattern A fire pattern displayed on the cross section of a burned wooden structural member.

Arson The crime of maliciously and intentionally, or recklessly, starting a fire or causing an explosion.

Autoignition Initiation of combustion by heat but without a spark or flame.

Autoignition Temperature The lowest temperature at which a combustible material ignites in air without a spark or flame.

Backdraft An explosion resulting from the sudden introduction of air (e.g., oxygen) into a confined space containing oxygen-deficient, superheated products of incomplete combustion.

Bead A rounded globule of resolidified metal at the end of the remains of an electrical conductor that was caused by arcing. It is characterized by a sharp line of demarcation between the melted and unmelted conductor surfaces.

Blast Pressure Front The expanding leading edge of an explosion reaction that separates a major difference in pressure between normal ambient pressure ahead of the front and potentially damaging high pressure at and behind the front.

BLEVE Boiling liquid expanding vapor explosion.

British Thermal Unit (Btu) The quantity of heat required to raise the temperature of 1 pound (lb) of water 1°F at the pressure of 1 atm and a temperature of 60°F. A Btu is equal to 1055 joules, 1.055 kilojoules, and 252.15 calories.

Burning Rate *See* Heat Release Rate.

Burning Velocity The velocity at which a flame reaction front moves into the unburned mixture as it chemically transforms the fuel and oxidant into combustion products. It is an inherent characteristic of a combustible and is a fixed value. It is only a fraction of the flame speed.

Calorie (cal) The amount of heat necessary to raise 1 g of water 1°C at 15°C. A calorie is 4.184 joules, and there are 252.15 calories in a British thermal unit (Btu).

Cause The circumstances, conditions, or agencies that brought about or resulted in the fire or explosion incident, damage to property resulting from the fire or explosion incident, or bodily injury or loss of life resulting from the fire or explosion incident.

Ceiling Layer A buoyant layer of hot gases and smoke produced by a fire in a compartment.

Char Carbonaceous material that has been burned and has a blackened appearance.

Char Blisters Convex segments of carbonized material separated by cracks or crevasses that form on the surface of char, forming on materials such as wood as the result of pyrolysis or burning.

Clean Burn A fire pattern on surfaces where soot has been burned away.

Code A standard that is an extensive compilation of provisions covering broad subject matter or that is suitable for adoption into law independently of other codes and standards.

Combustible Capable of burning, generally in air under normal conditions of ambient temperature and pressure, unless otherwise specified. Combustion can occur in cases where an oxidizer other than the oxygen in air is present (e.g., chlorine, fluorine, or chemicals containing oxygen in their structure).

Combustible Gas Indicator An instrument that samples air and indicates whether there are combustible vapors present. Some units may indicate the percentage of the lower explosive limit of the air–gas mixture.

Combustible Liquid A liquid having a flash point at or above 100°F (37.8°C). *See also* Flammable Liquid.

Combustion Products Heat, gases, solid particulates, and liquid aerosols produced by burning.

Compartmentation Design features of a building that limit fire growth to the room or building section of origin. These features include but are not limited to fire walls and fire doors.

Conduction Heat transfer to another body or within a body by direct contact.

Controlled Burn *See* Prescribed Fire.

Convection Heat transfer by circulation within a medium such as a gas or a liquid.

Convection Column *See* Plume.

Current A flow of electric charge.

Deductive Reasoning Reasoning from theories to account for specific experimental results.

Deflagration A combustion reaction in which the velocity of the reaction front through the unreacted fuel medium is less than the speed of sound.

Detection (1) Sensing the existence of a fire, especially by a detector from one or more products of the fire, such as smoke, heat, ionized particles, infrared radiation, and the like. (2) The act or process of discovering and locating a fire.

Detonation A reaction in which the velocity of the reaction front through the unreacted fuel medium is equal to or greater than the speed of sound.

Drop Down The spread of fire by the dropping or falling of burning materials. Synonymous with *Fall Down*.

Effective Fire Temperatures Identifiable temperatures reached in fires that reflect physical effects that can be defined by specific temperature ranges.

Electric Spark A small, incandescent particle created by some arcs.

Empirical (1) Originating in or based on observation or experience. (2) Relying on experience or observation alone often without due regard for system and theory (empirical data). (3) Capable of being verified or disproved by observation or experiment (empirical laws). (Source: *Merriam Webster's Collegiate Dictionary*)

Entrainment The process of air or gases being drawn into a fire, plume, or jet.

Evidence Any species of proof or probative matter, legal presented at the trial of an issue, by the act of the pa ties and through the medium of witnesses, records, doc ments, concrete objects, etc., for the purpose of inducin belief in the minds of the court or jury as to their conte tion. (Source: *Black's Law Dictionary*, 6th edition)

Explosion The sudden conversion of potential energ (chemical or mechanical) into kinetic energy with the pro duction and release of gases under pressure or the releas of gas under pressure. These high-pressure gases then d mechanical work such as moving, changing, or shatterin nearby materials. *See also* Backdraft; High-Order Explo sion; Low-Order Explosion; Secondary Explosion.

Explosion Dynamics Analysis The process of using forc vectors to trace backward from the least to the most dam aged areas following the general path of the explosio force vectors.

Explosion Seat The crater or area of greatest damage locate at the point of initiation (epicenter) of an explosion.

Explosive Any chemical compound, mixture, or device th functions by explosion. *See also* Explosive Material; Hig Explosive; Low Explosive.

Explosive Material Any material that can act as fuel for a explosion.

Exposed Surface The side of a structural assembly or objec that is directly exposed to the fire.

Extinguish To cause to cease burning.

Failure Distortion, breakage, deterioration, or other fau in an item, component, system, assembly, or structure th results in unsatisfactory performance of the function fc which it was designed.

Failure Analysis A logical, systematic examination of a item, component, assembly, or structure and its place an function within a system, conducted in order to identif and analyze the probability, causes, and consequences o potential and real failures.

Fall Down *See* Drop Down.

Finish Rating The time in minutes, determined under spe cific laboratory conditions, at which the stud or joist i contact with the exposed protective membrane in a pro tected combustible assembly reaches an average tempera ture rise of 250°F (121°C) or an individual temperatur rise of 325°F (163°C) as measured behind the protectiv membrane nearest the fire on the plane of the wood.

Fire A rapid oxidation process with the evolution of ligh and heat in varying intensities.

Fire Analysis The process of determining the origin, caus development, and responsibility as well as the failure ana ysis of a fire or explosion.

Firebreak A natural or human-made barrier to stop th spread of a fire.

Fire Cause The circumstances, conditions, or agencies tha bring together a fuel, ignition source, and oxidizer (suc as air or oxygen) resulting in a fire or a combustion explo sion. *See also* Cause.

Fire Dynamics The detailed study of how chemistry, fire science, and the engineering disciplines of fluid mechanics and heat transfer interact to influence fire behavior.

Fire Head The portion of a fire that is moving most rapidly, subject to influences of slope and other topographic features.

Fire Heel The opposite side of the fire from the head that is less intense and easier to control.

Fire Investigation The process of determining the origin, cause, and development of a fire or explosion.

Fire Patterns The physical manifestation of the effect of fire on materials.

Fire Propagation *See* Fire Spread.

Fire Scene Reconstruction The process of recreating the physical scene during fire scene analysis through the removal of debris and the replacement of contents or structural elements in their prefire positions.

Fire Science The body of knowledge concerning the study of fire and related subjects (such as combustion, flame, products of combustion, heat release, heat transfer, fire and explosion chemistry, fire and explosion dynamics, thermodynamics, kinetics, fluid mechanics, fire safety) and their interaction with people, structures, and the environment.

Fire Spread The movement of fire from one place to another.

Firing Out Deliberate burning of unburned fuel between the fire and a firebreak.

Flame Speed The local velocity of a freely propagating flame relative to a fixed point.

Flash Fire A fire that spreads rapidly through a diffuse fuel, such as dust, gas, or the vapors of an ignitible liquid, without the production of damaging pressure.

Flame The luminous portion of burning gases or vapors. *See also* Premixed Flame.

Flame Front The leading edge of burning gases of a combustion reaction.

Flameover The condition where unburned fuel (pyrolysate) from the originating fire has accumulated in the ceiling layer to a sufficient concentration (i.e., at or above the lower flammable limit) that it ignites and burns. It can occur without ignition and prior to the ignition of other fuels separate from the origin.

Flammable Capable of burning with a flame.

Flammable Limits The upper or lower concentration limits at a specified temperature and pressure of a flammable gas or a vapor of an ignitible liquid and air, expressed as a percentage of fuel by volume that can be ignited.

Flammable Liquid A liquid having a flash point below 100°F (37.8°C) (tag closed cup) and having a vapor pressure not exceeding 40 psi (2068 mm Hg) at 100°F (37.8°C). *See also* Combustible Liquid.

Flammable Range Concentration range of a flammable gas or a vapor of a flammable liquid in air that can be ignited.

Flashover A transition phase in the development of a contained fire in which surfaces exposed to thermal radiation reach ignition temperature more or less simultaneously and fire spreads rapidly throughout the space.

Flash Point of a Liquid The lowest temperature of a liquid, as determined by specific laboratory tests, at which the liquid gives off vapors at a sufficient rate to support a momentary flame across its surface.

Forensic Legal; pertaining to courts of law.

Fuel A material that yields heat through combustion. *See also* Target Fuel.

Fuel-Controlled Fire A fire in which the heat release rate and growth rate are controlled by the characteristics of the fuel, such as quantity and geometry, and in which adequate air for combustion is available.

Fuel Gas Natural gas, manufactured gas, LP-Gas, and similar gases commonly used for commercial or residential purposes such as heating, cooling, or cooking.

Fuel Load The total quantity of combustible contents of a building, space, or fire area, including interior finish and trim, expressed in heat units or the equivalent weight in wood.

Gas The physical state of a substance that has no shape or volume of its own and will expand to take the shape and volume of the container or enclosure it occupies.

Glowing Combustion Luminous burning of solid material without a visible flame.

Ground Fault A current that flows outside the normal circuit path, such as (a) through the equipment grounding conductor, (b) through conductive material other than the electrical system ground (metal, water, plumbing pipes, etc.), (c) through a person, or (d) through a combination of these ground return paths.

Guide A document that is advisory or informative in nature and that contains only nonmandatory provisions. A guide may contain mandatory statements such as when a guide can be used, but the document as a whole is not suitable for adoption into law.

Hazard Any arrangement of materials and heat sources that presents the potential for harm, such as personal injury or ignition of combustibles.

Heat A form of energy characterized by vibration of molecules and capable of initiating and supporting chemical changes and changes of state.

Heat Flux The measure of the rate of heat transfer to a surface, expressed in kilowatts/m^2, kilojoules/$m^2 \cdot s$, or Btu/$ft^2 \cdot s$.

Heat of Ignition The heat energy that brings about ignition.

Heat Release Rate (HRR) The rate at which heat energy is generated by burning.

High Explosive A material that is capable of sustaining a reaction front that moves through the unreacted material at a speed equal to or greater than that of sound in that medium [typically 3300 ft/s (1000 m/s)]; a material capable of sustaining a detonation. *See also* Detonation.

High-Order Explosion A rapid pressure rise or high-force explosion characterized by a shattering effect on the confining structure or container and long missile distances.

Ignitible Liquid Any liquid or the liquid phase of any material that is capable of fueling a fire, including a flammable liquid, combustible liquid, or any other material that can be liquefied and burned.

Ignition The process of initiating self-sustained combustion. *See also* Autoignition; Self-Ignition; Spontaneous Ignition.

Ignition Energy The quantity of heat energy that should be absorbed by a substance to ignite and burn.

Ignition Temperature Minimum temperature a substance should attain in order to ignite under specific test conditions.

Ignition Time The time between the application of an ignition source to a material and the onset of self-sustained combustion.

Inductive Reasoning Reasoning from specific observations and experiments to more general hypotheses and theories. (Source: *Funk & Wagnall's Multimedia Encyclopedia*)

Isochar A line on a diagram connecting points of equal char depth.

Joule (J) The preferred SI unit of heat, energy, or work; there are 4.184 J in 1 calorie (cal), and 1055 J in 1 British thermal unit (Btu). A watt (W) is a joule/second (J/s). *See also* British Thermal Unit; Calorie.

Kilowatt A measurement of energy release rate.

Kindling Temperature *See* Ignition Temperature.

Liquid *See* Combustible Liquid; Flammable Liquid; Ignitible Liquid.

Low Explosive An explosive that has a reaction velocity of less than 3300 ft/s (1000 m/s).

Low-Order Explosion A slow rate of pressure rise or low-force explosion characterized by a pushing or dislodging effect on the confining structure or container and short missile distances.

Material First Ignited The fuel that is first set on fire by the heat of ignition. To be meaningful, both a type of material and a form of material should be identified.

Noncombustible Material A material that, in the form in which it is used and under the condition anticipated, will not ignite, burn, support combustion, or release flammable vapors when subjected to fire or heat. Also called *incombustible material* (not preferred).

Nonflammable (1) Not readily capable of burning with a flame. (2) Not liable to ignite and burn when exposed to flame. Its antonym is *flammable.*

Ohm (Ω) The unit of electrical resistance (R) that measures the resistance between two points of a conductor when a constant difference of potential of 1 volt (V) between these two points produces in this conductor a current of 1 ampere (A).

Origin *See* Point of Origin; Area of Origin.

Overcurrent (1) Any current in excess of the rated current of equipment or the ampacity of a conductor. It may result from an overload, short circuit, or ground fault. (2) A momentary excess current flow, such as a refrigerator motor starting up.

Overload Operation of equipment in excess of normal full-load rating or of a conductor in excess of rated ampacity, which, when it persists for a sufficient length of time would cause damage or dangerous overheating. A fault such as a short circuit or ground fault, is not an overload.

Oxygen Deficiency Insufficiency of oxygen to support combustion. *See also* Ventilation-Controlled Fire.

Physical Evidence Any physical or tangible item that tends to prove or disprove a particular fact or issue.

Piloted Ignition Temperature *See* Ignition Temperature.

Plastic Any of a wide range of natural or synthetic organic materials of high molecular weight that can be formed by pressure, heat, extrusion, and other methods into desired shapes.

Plume The column of hot gases, flames, and smoke rising above a fire. Also called *Convection Column, Thermal Updraft,* or *Thermal Column.*

Point of Origin The exact physical location where a heat source and a fuel come in contact with each other and a fire begins.

Premixed Flame A flame for which the fuel and oxidizer are mixed prior to combustion, as in a laboratory Bunsen burner or a gas cooking range. Propagation of the flame is governed by the interaction between flow rate, transport processes, and chemical reaction.

Prescribed Fire A fire resulting from intentional ignition by a person or a naturally caused fire that is allowed to continue to burn according to approved plans to achieve resource management objectives.

Preservation Application or use of measures to prevent damage, change or alteration, or deterioration.

Products of Combustion *See* Combustion Products.

Proximate Cause The cause that directly produces the effect without the intervention of any other cause.

Pyrolysis The chemical decomposition of a compound into one or more other substances by heat alone; pyrolysis often precedes combustion.

Radiant Heat Heat energy carried by electromagnetic waves longer than light waves and shorter than radio waves. Radiant heat (electromagnetic radiation) increases the sensible temperature of any substance capable of absorbing the radiation, especially solid and opaque objects.

Radiation Heat transfer by way of electromagnetic energy.

Rate of Heat Release *See* Heat Release Rate.

Recommended Practice A document that is similar in content and structure to a code or standard but that contains only nonmandatory provisions using the word *should* to indicate recommendations in the body of the text.

Rekindle A return to flaming combustion after apparent but incomplete extinguishment.

Relative Humidity The amount of moisture in a given volume of air, compared to how much the air could hold at that same temperature.

Responsibility The accountability of a person or other entity for the event or sequence of events that caused the fire or explosion, spread of the fire, bodily injuries, loss of life, or property damage.

Risk (1) The degree of peril; the possible harm that might occur. (2) The statistical probability or quantitative estimate of the frequency or severity of injury or loss.

Rollover *See* Flameover.

Scientific Method (1) The systematic pursuit of knowledge involving the recognition and formulation of a problem, the collection of data through observation and experiment, and the formulation and testing of a hypothesis. (2) Principles and procedures for the systematic pursuit of knowledge involving the recognition and formulation of a problem, the collection of data through observation and experiment, and the formulation and testing of hypotheses. (Source: *Merriam-Webster's Collegiate Dictionary*)

Seat of Explosion A craterlike indentation created at the point of origin of an explosion.

Seated Explosion An explosion with a highly localized point of origin, such as a crater.

Secondary Explosion Any subsequent explosion resulting from an initial explosion.

Self-Ignition Temperature The minimum temperature at which the self-heating properties of a material lead to ignition.

Self-Heating The result of exothermic reactions, occurring spontaneously in some materials under certain conditions, whereby heat is liberated at a rate sufficient to raise the temperature of the material.

Self-Ignition Ignition resulting from self-heating. Its synonym is *Spontaneous Ignition*.

Short Circuit An abnormal connection of low resistance between normal circuit conductors where the resistance is normally much greater. A short circuit is an overcurrent situation, but it is not an overload.

Smoke An airborne particulate product of incomplete combustion suspended in gases, vapors, or solid and liquid aerosols.

Smoke Condensate The condensed residue of suspended vapors and liquid products of incomplete combustion.

Smoke Explosion *See* Backdraft.

Smoldering Combustion without flame, usually with incandescence and smoke.

Soot Black particles of carbon produced in a flame.

Spalling Chipping or pitting of concrete or masonry surfaces.

Spark A small, incandescent particle. *See also* Electric Spark.

Spoliation The loss, destruction, or material alteration of an object or document that is evidence or potential evidence in a legal proceeding by one who has the responsibility for its preservation.

Spontaneous Ignition Initiation of combustion of a material by an internal chemical or biological reaction that has produced sufficient heat to ignite the material.

Spontaneous Heating Process whereby a material increases in temperature without drawing heat from its surroundings.

Standard A document, the main text of which contains only mandatory provisions using the word *shall* to indicate requirements and that is in a form generally suitable for mandatory reference by another standard or code or for adoption into law. Nonmandatory provisions shall be located in an appendix, footnote, or fine print note and are not to be considered a part of the requirements of a standard.

Static Electricity The electrical charging of materials through physical contact and separation, and the various effects that result from the positive and negative electrical charges formed by this process.

Suppression The sum of all the work done to extinguish a fire from the time of its discovery.

Target Fuel A fuel that is subject to ignition by thermal radiation such as from a flame or a hot gas layer.

Temperature The intensity of sensible heat of a body as measured by a thermometer or similar instrument. *See also* Absolute Temperature; Autoignition Temperature; Effective Fire Temperatures; Ignition Temperature; Self-Ignition Temperature.

Thermal Column *See* Plume.

Thermal Expansion The proportional increase in length, volume, or superficial area of a body with rise in temperature.

Thermal Inertia The properties of a material that characterize its rate of surface temperature rise when exposed to heat; related to the product of the material's thermal conductivity (k), its density (ρ), and its heat capacity (c).

Thermal Updraft *See* Plume.

Thermoplastic Plastic materials that soften and melt under exposure to heat and can reach a flowable state.

Thermoset Plastics Plastic materials that are hardened into a permanent shape in the manufacturing process and are not commonly subject to softening when heated. Thermoset plastics typically form char in a fire.

Timeline Graphic representation of the events in the fire incident displayed in chronological order.

Transitional Velocity The sum of the velocity of the flame front caused by the volume expansion of the combustion products due to the increase in temperature and an increase in the number of moles and any flow velocity due to motion of the gas mixture prior to ignition.

Units of Measure Metric units of measurement in this standard are in accordance with the modernized metric system known as the International System of Units (SI). The unit *liter* (L) is outside of but recognized by SI and is commonly used in international fire protection. The common conversion units are as follows:

1 in. = 2.54 cm

1 ft = 0.3048 m

1 ft^2 = 0.09290 m^2

1 ft^3 = 7.481 gal

1 ft^3 = 0.02832 m^3

1 U.S. gal = 3.785 L

1 lb = 0.4536 kg

1 oz (weight) = 28.35 g

1 ft/s = 0.3048 m/s

1 lb/ft^3 = 16.02 kg/m^3

1 gpm = 0.06308 L/s

1 atm = pressure exerted by 760 mm of mercury of standard density at 0°C, 14.7 $lb/in.^2$ (101.3 kPa)

1 Btu/s = 1.055 kW

1 Btu = 1055 J

1 kW = 0.949 Btu/s

1 in. w.c. = 248.8 Pa = 0.036 psi

1 psi = 27.7 in. water column

Upper Layer *See* Ceiling Layer.

Vapor The gas phase of a substance, particularly of those that are normally liquids or solids at ordinary temperatures. *See also* Gas.

Vapor Density The ratio of the average molecular weight of a given volume of gas or vapor to the average molecular weight of an equal volume of air at the same temperature and pressure.

Vector An arrow used in a fire scene drawing to show the direction of heat, smoke, or flame flow.

Vent An opening for the passage of, or dissipation of, fluids, such as gases, fumes, smoke, and the like.

Ventilation (1) Circulation of air in any space by natural wind or convection or by fans blowing air into or exhausting air out of a building. (2) A fire-fighting operation of removing smoke and heat from the structure by opening windows and doors or making holes in the roof.

Ventilation-Controlled Fire A fire in which the heat release rate or growth is controlled by the amount of air available to the fire.

Venting The removal of combustion products as well as process fumes (e.g., flue gases) to the outer air.

Volt (V) (1) The unit of electrical pressure (electromotive force) represented by the symbol E. (2) The difference in potential required to make a current of 1 ampere (A) flow through a resistance of 1 ohm (Ω).

Watt (W) The unit of power, or rate of work. It is equal to 1 joule per second (J/s), or the rate of work represented by a current of 1 ampere (A) under the potential of 1 volt (V).

Bibliography

STM E-119, *Standard Test Methods for Fire Tests of Building Construction and Materials.* West Conshohocken, Penn.: ASTM International, 2003.

ote, Arthur and Percy Bugbee, *Principles of Fire Protection.* Quincy, Mass.: National Fire Protection Association, 1988.

ote, Arthur E., *Fire Protection Handbook,* 19th edition. Quincy, Mass.: National Fire Protection Association, 2003.

ote, Arthur E., *Fundamentals of Fire Protection.* Quincy, Mass.: National Fire Protection Association, 2004.

ouglas, John E. et al., *Crime Classification Manual.* San Francisco, Calif.: Jossey-Bass Publishers, 1997.

araday, M., *Chemical History of a Candle.* New York: E.P. Dutton & Co., 1920.

unk & Wagnalls Multimedia Encyclopedia. http://www.funkandwagnalls.com.

arner, Bryan A. and Henry L. Black, *Black's Law Dictionary,* 6th edition. St. Paul, Minn.: West Group, 1990.

ewis, Richard J., Sr., *Hawley's Condensed Chemical Dictionary,* 14th edition. New York: Wiley-Interscience, 2002.

Iish, Frederick C., editor-in-chief, *Merriam-Webster's Collegiate Dictionary,* 11th edition. Springfield, Mass.: Merriam-Webster, Inc., 2003.

elson, Harold E., *Fire Growth Analysis of the Fire of March 20, 1990, Pulaski Building, 20 Massachusetts Avenue, N.W., Washington DC.* Gaithersburg, MD.: National Institute of Standards and Technology, 1994. Available at http://nist.gov/bfrlpubs/fire94/PDF/f94057.pdf.

FPA Publications. NFPA, 1 Batterymarch Park, P.O. Box 9101, Quincy, MA 02269-9101.

NFPA 1, *Uniform Fire Code*™, 2003 edition.

NFPA 54, *National Fuel Gas Code,* 2002 edition.

NFPA 58, *Liquefied Petroleum Gas Code,* 2004 edition.

NFPA 68, *Guide for Venting of Deflagrations,* 2002 edition.

NFPA 70, *National Electrical Code*®, 2002 edition.

NFPA *101*®, *Life Safety Code*®, 2003 edition.

NFPA 211, *Standard for Chimneys, Fireplaces, Vents, and Solid, Fuel-Burning Appliances,* 2003 edition.

NFPA 471, *Recommended Practice for Responding to Hazardous Materials Incidents,* 2002 edition.

NFPA 472, *Standard for Professional Competence of Responders to Hazardous Materials Incidents,* 2002 edition.

NFPA 473, *Standard for Competencies for EMS Personnel Responding to Hazardous Materials Incidents,* 2002 edition.

NFPA 1033, *Standard for Professional Qualifications for Fire Investigator,* 2003 edition.

NFPA 1500, *Standard on Fire Department Occupational Safety and Health Program,* 2002 edition.

NFPA 1561, *Standard on Emergency Services Incident Management System,* 2005 edition.

NFPA 1670, *Standard on Operations and Training for Technical Search and Rescue Incidents,* 2004 edition.

NFPA 1977, *Standard on Protective Clothing and Equipment for Wildland Fire Fighting,* 1998 edition.

A Pocket Guide to Accelerant Evidence Investigation Collection, 2nd edition. International Association of Arson Investigators, Massachusetts chapter, 1992.

Society of Fire Protection Engineers, *The SFPE Handbook of Fire Protection Engineering,* 3rd edition. Quincy, Mass.: NFPA and Society of Fire Protection Engineers, 2002.

Summers, Wilford I. and Terrell Croft, *The American Electrician's Handbook,* 14th edition. New York: McGraw-Hill/TAB Electronics, 2002.

U.S. Department of Labor. Occupational Safety and Health Administration, 200 Constitution Avenue, NW, Washington, DC 20210.

29 CFR 1910.120, *HAZWOPER (Hazardous Waste Operations and Emergency Response).*

29 CFR 1910.146, *Permit-Required Confined Spaces.*

29 CFR 1910.147, *The Control of Hazardous Energy (Lockout/Tagout Standard).*

Index

Printed in the United Kingdom by
Lightning Source UK Ltd., Milton Keynes
137565UK00001B/7-28/P

9 780763 744021